Self-Organizing Control of Stochastic Systems

CONTROL AND SYSTEMS THEORY

A Series of Monographs and Textbooks

Editor

JERRY M. MENDEL

University of Southern California
Los Angeles, California

Associate Editors

Karl J. Åström
Lund Institute of Technology
Lund, Sweden

Michael Athans
Massachusetts Institute of Technology
Cambridge, Massachusetts

David G. Luenberger
Stanford University
Stanford, California

OTHER VOLUMES IN PREPARATION

Self-Organizing Control of Stochastic Systems

by GEORGE N. SARIDIS

Professor of Electrical Engineering
Purdue University
West Lafayette, Indiana

MARCEL DEKKER, INC.   New York and Basel

To

ANNA

Library of Congress Cataloging in Publication Data

Saridis, George N   1931–
    Self–organizing control of stochastic systems.

    (Control and systems theory ; v. 4)
    Includes bibliographical references and index.
    1. Control theory. 2. Self–organizing systems.
3. Stochastic processes. I. Title.
QA402.3.S265   629.8'312   75-40645
ISBN 0-8247-6413-7

MARCEL DEKKER, INC.

270 Madison Avenue, New York, New York 10016

Current printing (last digit):
10 9 8 7 6 5 4 3 2 1

PRINTED IN THE UNITED STATES OF AMERICA

PREFACE

Control systems is a discipline of engineering developed in
the last thirty years into an analytic and synthetic science. Its
growth has always been proportional to the demands of modern technological problems and the development of new technologies. In the
last few years, these demands have shifted into the treatment of
processes with completely or partially unknown dynamics which require control systems with higher-level decision-making capabilities.

In the last twenty years, the control literature has been expanded by a large variety of publications treating the problems of
systems with uncertain dynamics. Such research efforts, however, have
appeared in diverse and disjoint forms under different names, and
have been characterized as "controversial" by many respectable researchers. The purpose of this book is to unify the area of control
systems with uncertain dynamics by properly defining and consolidating
the existing methodologies under their common features and goals. This
ambitious undertaking would lead to the extension of the existing
hierarchy of control systems and establish the analytic relations
with the rest of the control methodologies. The success or failure
of this task can only be judged by the reader.

This book was not intended to be a self-contained treatise on
control systems and their associated mathematics. Actually, it was
originally conceived as a research monograph. However, with growing experience in the field as a guide, it has been shaped as an
advanced textbook and a research reference book that can be used in
a graduate-level course or in the research institute and the industrial laboratory. The extensive bibliography should also serve as
a source of reference in the area. The reader is required to have
a mathematical knowledge of linear systems, state variables, transform theory, probability theory and elementary stochastic processes,
optimal deterministic and stochastic control, and finite state

systems both in the continuous and discrete time space. Implementation of the examples was made mostly on a digital or a hybrid
computer, and therefore programming of such machines would be useful
to the reader who would like to familiarize himself with the methodologies. Because of the nature and length of the processes involved,
no problems are given in the text for exercise. The interested
reader will gain sufficient experience by resimulating the existing
examples with sufficient intellectual stimulation.

The book is divided into three parts: The first part covers
the definitions of the area and a survey of a number of deterministic
Self-Organizing Control Systems which have not been treated in this
book because of space limitation. The second part provides the
reader with a basic treatment on estimation theory, stochastic and
dual optimal control, and identification methods. The reader
familiar with this material can skip directly to the third part
where Parameter Adaptive and Performance Adaptive Self-Organizing
methods are presented, compared, and applied to various problems.
The relation to Self-Organizing Controls with Learning and Intelligent systems is also discussed. The terminology and notation used
have been the most commonly adopted in the modern optimal control
systems literature. The material of the book has been developed
from the lectures in the course "Self-Organizing Control Systems"
at Purdue University during the last ten years.

It is at this part of the book that the author customarily
presents his acknowledgments to the people that helped him spiritu-
ally and materially to write this book. Being no exception, I would
like to present my gratitude first to Prof. K. S. Fu, for the inspira-
tion, motivation, and the exposure to his numerous contributions in
the area through personal communication, which made this book possible. I would also like to thank Profs. J. M. Mendel and J. Meditch
for their moral support during the endless years of the development
of the material. I would like to thank my colleagues, Dr. Z. J.
Nikolic, Dr. G. Stein, Dr. H. Gilbert, Dr. P. Badavas, Dr. P. Fensel,
Dr. R. Kitahara, Dr. R. Lobbia, Dr. R. Hofstadter, Dr. D. Ricker,

and Dr. H. Stephanou and Mr. T. K. Dao for their contribution in the development of parts of the material presented, and the students of my EE 687 and EE 689 classes at Purdue University for running most of the simulations and performing the comparison studies. Special thanks to Messrs. H. Stephanou, T. K. Dao, A. Desrochers, and G. Lee for their comments and correction of the manuscript, and Mmes. Terry Brown, Wanitta and Wanda Booth, Molly Harrington, and Becky Fagan for typing the manuscript. Finally I would like to thank Purdue University and The National Science Foundation for providing the support (NSF Grant GK-36808) and the facilities to conceive, develop, and produce this book.

George N. Saridis
West Lafayette, Indiana

CONTENTS

CHAPTER 2

TABLES

PART I

FOUNDATIONS OF SELF-ORGANIZING CONTROL SYSTEMS

Chapter 1

MOTIVATION AND DEFINITIONS

1.1 INTRODUCTION

The study of control systems as a means to produce desirable
responses from a process, and their various applications, can be
traced back to the records of ancient civilizations. However, World
War II was a milestone for the field since the whole area of scat-
tered investigations consolidated into a discipline by the pioneering work of men like Bode, Nichols, Popov, Nyquist and others. The
major theme of the new discipline was the concept of feedback with
its philosophical consequences. The emphasis at the time was on
servomechanisms and other control hardware designed to drive simple
electromechanical systems. Since then, the field of control systems
has expanded, along with developments of our technology, to include
sophisticated theoretical studies and complicated hardware applications. One of the most spectacular accomplishments of the field was
its contribution to the feat of putting a man on the moon. The
major credit for the growth and the success of the field, especially
in the later years, should be given to the digital computer and its
fantastic capabilities.

In more recent years, the interest of the modern system engineer
has considerably shifted from purely hardware engineering problems
to a wider spectrum covering urban, societal, and economic systems,
traffic and transportation systems, biomedical systems, etc. The
contribution of systems theory will, it is to be hoped, produce
generic solutions to problems that have been created in the modern
world as a result of the explosion of technology. However, several
adjustments are necessary to the existing theoretical results in

3

order to deal with the problems in the new areas. These adjustments
pertain to our lack of knowledge of concrete models of these systems,
a knowledge we had with the ones described by the laws of physics.
To anticipate solutions for these problems, a discipline should be
defined to treat systems with uncertain models.

Self-Organizing Control (S.O.C.) is the branch of system theory
developed especially to treat systems with completely or partially
unknown dynamics operating in a deterministic or a stochastic environment. Such a discipline exhibits the additional feature of explicit
or implicit identification and in many cases <u>on-line</u> modeling of the
unknown system under consideration. This information then is used
to adjust the controller for an approximately optimal control of the
system. This discipline, being the subject of research by the author
and other scientists, has only recently matured enough to be pre-
sentable in a unified form. This book actually represents the first
attempt in U.S. literature to bring together the various methods
available that treat unknown systems from the self-organizing point
of view.

Self-organizing control may also be considered as the first
step toward the controls of the future, that is, the *Intelligent
Controls*. Such controls would replace the human mind in making
decisions, planning control strategies, and learning new functions
by training and performing other intelligent functions whenever the
environment does not allow or does not justify the presence of a
human operator.

The purpose of this chapter is to explain the need of Self-
Organizing Controls by a discussion of their potential applications
and establish their proper foundations by providing the correct
definitions and consolidating the field within which self-organizing
control is not only absolutely necessary but also irreplaceable as a
control system. This discussion will provide the introduction to the
discipline and lead to the systematic presentation of a collection
of various methods that qualify for the discipline under the definitions provided. By doing so, a new home is provided for a large
number of techniques which have appeared in the literature under a

variety of names like adaptive, optimalizing, self-optimizing, etc.,
systems that were sometimes opposing each other due to conflicts in
their philosophy. Because it was necessary to create a unifying
new philosophy on control systems to encompass the appropriate
methods, a large number of techniques were left out. As a result,
many people may disagree with some of the ideas presented in this
book before reading it. I would personally recommend being patient
and forming final opinions only after reading to the end of the
last chapter.

1.2 SYSTEMS THEORY, CONTROLS, AND THEIR DEVELOPMENT

In order to place this subject in the proper scientific perspective, its relation to the generic fields of systems engineering
and control systems must be identified. Therefore, an account is
presented here which will, it is hoped, establish the continuity of
thinking and the causality of the proposed new discipline of Self-
Organizing controls with respect ot the other systems-engineering
methodologies.

1.2.1 Systems Engineering

Systems-theoretic approaches have always been an integral part
of the methodologies applied by the practicing engineer. This has
been the case mainly because in contrast to other sciences, the
essence of his work is not to discover the functional relationships
of physical entities, a process that utilizes analytic thinking, but
to design systems composed of many components that must be put together to function as a whole [1.38].

The systems-theoretic point of view takes always a macroscopic
look at nature and characteristically utilizes only those features
that are necessary for the design and the physically correct or legal
performance of the system. System-theoretic thinking is demonstrated
by the mechanical engineer who uses averaging thermodynamic laws
instead of molecular kinetics when he designs an engine, or the
electrical engineer who applies Ohm's law instead of electron dynamics
when he designs an electric circuit.

The technological explosion of the late 1950s and early 1960s, and the invention of the digital computer found the engineer in the middle of creative activities which necessitated the expansion and formalization of the system-theoretic approach. What appeared to be a large-scale system that encompassed all aspects of engineering, required the development of a concrete abstract theory of systems applicable to each one of the engineering disciplines which con- sidered the solutions of the problems regardless of the differences of their originating physical laws. The science of mathematics was there to provide the tools and a name was given to the new disci- pline: *Systems Engineering*.

In order to understand the potential of system theory and ex- plore its application, one should define the pertinent concepts from the system engineer's point of view. Such an approach is necessary since these definitions are considered to be functional and are intended for solving existing problems, in contrast with other cases. Such concepts have been investigated by many scientists [1.22], [1.48], [1.51], [1.52].

The best way to introduce system theory is by presenting its underlying concepts, e.g., those of a system and of the state of a system. The dictionary defines a system as "the collection of ob- jects united by some form of interdependence." On the other hand, the state of a system at any given time is "the information necessary to determine the behavior of the system from that time on." However, these definitions are not functional because they are not mathemati- cally concrete enough to permit the development of a mathematical model. For instance, according to these definitions, everything is a system and a state is too vague to handle.

A more precise mathematical definition of such terms is given by Zadeh [1.51].

Definition 1.1

"A system is designed as a collection of ordered pairs of time functions, representing inputs and outputs, and defining *abstract objects*. Such pairs must satisfy the condition that if one of them belongs to the system every segment of it belongs to the system.

This definition characterizes every abstract object as well as

every system by its input-output relation and not by its physical
identity specific to the respective process. This way, one may treat
various systems as a whole by their causal relationships only, thus
dealing with a large variety of problems in a unified way. In order
to be able to mathematically structure a system the following
definitions are necessary:

Definition 1.2

> "The present *state* of a system may be thought of as the minimal
> amount of past information required to completely describe the
> future behavior (e.g., outputs) of the system when the inputs
> to the system for all present and future time are known."

Definition 1.3

> "A state variable is a mathematical variable dependent on time
> which completely describes the state of the system."

Using these definitions one can describe a system as a set of
integrodifferential equations which may describe the state or output
of the system at a future time when the present state and past,
present, and future inputs of the system are given. Such systems
are called *deterministic*. In many cases, however, the states of a
system can be assigned to a set of values with a certain probability
while an uncertain environment may introduce random inputs to the
system. In such cases, the future behavior of the system can only
be described with a certain probability and the system is called
stochastic. If a system is uncertainly described while it operates
in a stochastic environment, and if it may reduce the uncertainty
of its description as the process evolves, the system is called
learning. When a system exhibits highly intelligent functions like
advanced decision making, planning, etc., it is called *intelligent*.
Further discussion about the latter systems will be presented in the
sequel.

1.2.2 The System-Theoretic Methodologies

Most problems of current interest are dealing with systems that
contain a large number of state variables. These problems appear in
energy and communication systems, traditionally treated by engineering, as well as socioeconomic, urban, ecological, societal, trans-

portation, health-care, administrative, and managerial systems which
belong to disciplines unrelated to engineering. Systems engineering
seem to offer solutions to such problems after the success of the
systems-theoretic approach in the space program.

The system-theoretic approach models each system with an
appropriate mathematical model, based on definition 1.1 which is
independent of the system's physical identity. The mathematical
model is then analyzed and predictions of its behavior are made.
Based on these predictions, appropriate control can be designed and
tested for the model. Finally, those controls are implemented for
the real problem by construction based on the mathematical model.
In a more formal way, this procedure may be summarized by five steps:
1.   Modeling of the process.
2.   Mathematical analysis.
3.   Mathematical synthesis.
4.   Testing of the controls.
5.   Real-system design.

Only the first and the last steps depend on the specific
physical process to be considered. The other three can be performed by mathematical methods and with the assistance of a computer.
This gives a high degree of flexibility to the method. In more
details the steps are described as follows:

Modeling is the process of obtaining a mathematical model for
the physical systems under consideration. It requires a strong
interaction and teamwork between the expert in the area and the
systems engineer. Special attention is required for the systems
with a certain degree of uncertainty. It usually requires two
steps:
1.   Structural identification, which involves the interpretation of
      the underlying physical laws.
2.   Parameter identification, which involves the acquisition of
      the values of the system's functional parameters.

Mathematical analysis is the process of studying the past,
present, and future performance of the system of a particular structure, independent of its physical identity. Response-analysis

procedures are suitable for deterministic systems, and state-estimation methods are usually applicable to stochastic systems. Learning
methods are preferred for systems with a large degree of uncertainty.
Since processes of current interest involve, more or less, a large
number of state variables, various analytic methods have been
developed to study the systems involved. A tentative list of these
methods follows:

1.   Hierarchical methods.

2.   Decomposition methods.

3.   Flow-graph and network methods.

4.   Learning theoretic methods, etc.

Mathematical synthesis pertains to the problem of designing
controls to force the systems to perform in a desired way that, in
most cases, is the main goal of an engineering investigation. The
mathematical model which has been previously analyzed is forced to
behave in a desired way by the incorporation of an external controller, either in an open-loop or feedback form. Various methods of
designing mathematical controls for systems of current interest are
listed below:

1.   Optimization methods of mathematical programming.

2.   Operations research approaches of mathematical programming.

3.   Stability studies.

4.   Stochastic and dual control.

5.   Self-organizing, learning, and intelligent controls.

Testing represents a very important component of systems-
theoretic methodologies. It usually provides a vehicle to establish
the validity of the analysis and synthesis procedures previously
discussed. Large computer systems have enormously simplified the
task of testing. Some testing techniques are listed below:

1.   Small-scale model testing.

2.   Analog computer simulation.

3.   Digital computer simulation.

4.   Hybrid computer simulation.

The real-system design procedure involves the conversion of the
design information from the model to the original real life problem

under consideration. It involves teamwork between the expert in the area and the systems engineer and it represents the final step of the system-theoretic approach.

1.2.3 Automatic Control Systems

Automatic control systems have always been an integral part of system engineering and their first examples can be traced back to the ancient days of civilization. The modern developments of automatic controls were initiated during and after World War II and should be credited to scientists and engineers such as Wiener, Lyapunov, Bode, Nichols, Popov, Nyquist, Pontryagin, Kalman, Fel'dbaum and others in [1.7], [1.8], [1.17], [1.32], [1.34], [1.35]. Their work has consolidated the discipline of controls in one of the most exciting areas in modern engineering with contributions ranging from landing a man on the moon to outer-space exploration, and to the controls of biomedical and bionic devices to help incapacitated people. Automatic controls have been treated at different times by a variety of mathematical methods and they have thus acquired various names and characterizations according to the preference of the designer. In order to bring them under the proper perspective suitable for this book, the following definition is given:

Definition 1.4

   "Control of a process implies driving the process to effectively attain a prespecified goal."

   The words "effectively" and "goal" are used in a broad sense and may mean satisfying specifications or may imply optimality of a performance criterion in the case of optimal control.

   The first attempt to solve the control problem as defined above was to apply a direct external steering to the system involved that had been calculated in advance in order to attain the prespecified goal. This crude approach is known as *open-loop control*. The next step which revolutionized the field was achieved in electronic amplifiers but had been used in steam regulators before, where the *actual output of the system is compared with the desired output*

or steering function and their difference is used to drive the sys-
tem [1.17].   This is known as *feedback control* and implies an <u>on-</u>
<u>line</u> correction of the steering to accomplish the desired goal with-
out any external supervision. This new method created the era of
automation which is responsible for the technological accomplishments
of our days. More sophisticated approaches were then required to
implement this new philosophy and drive systems that were previously
considered unmanageable. By including all the design specifications
in a *performance index* (in the form of a cost functional), *optimal*
control was used to solve the control problem by applying variational
methods to minimize the performance index[1.3], [1.5], [1.20], [1.33],
[1.36], [1.47]. Such an approach found wide acceptance because of its mathematical elegance and the use of the computers to implement the resulting algorithms. However, the technological expansion brought about by
the solution of such problems demanded the investigation of systems
with more complicated structure than previously considered. Such is
the class of systems in which unpredictable signals or parameter
variations with known statistical description must be included in the
control process. This required the formulation of the *stochastic*
control problem defined below, the *optimal solution* of which was one
of the most powerful approaches used to tackle the problem [1.1], [1.2],
[1.6], [1.23], [1.36], [1.43], this will be reviewed in Chap. 4.
Definition 1.5
> "A stochastic control problem is one in which inherent uncer-
> tainties of the process under investigation are statistically
> *irreducible* from measurable quantities as the process evolves."

There is a large class of control systems that suffer from
degradation of their performance caused by environmental changes
outside the range of compensation of a conventional feedback controller. Other control systems suffer from inadequate performance
because of poor or inaccurate modeling of the system's behavior. In
both cases, the problem is created by uncertainties in the system's
description or in the inaccuracies in the signals involved, the
statistical description of which is not known in advance. Many
systems have been designed to compensate automatically for such
deterioration of performance in a way similar to modification that

the designer would recommend when he obtained additional information about the behavior of the system. The investigation of these control systems is the purpose of this book and will be introduced in the next section.

1.3 DECISION-MAKING UNDER REDUCIBLE UNCERTAINTIES

In the simple account of automatic control systems, the emphasis was given to the problem to be solved and not to the mathematical procedure used for the analysis or synthesis of the controller, the latter being a matter of personal preference when mathematical limitations are not present [1.9]. However, the account of automatic controls was concluded with the definition of the stochastic optimal control problem as it applies to processes operating in random environments that have been probabilistically or statistically analyzed. Such noisy interactions are typically represented by a wind profile in an aircraft design or the channel noise in a telemetry system [1.5]. The randomness of these quantities has been established by long experimentation and has been modeled by probability density functions with known coefficients which the control designer must account for in his study. The randomness of these processes is mainly attributed to limitations of describing the environment, and their modeling cannot be further improved once the appropriate probability density is computed.

In the more sophisticated systems of the latest technology as well as the other broad areas of systems research, a new type of uncertainty is frequently encountered. These uncertainties are due to the temporary ignorance of the designer at the time of the design, which is reducible as the actual process evolves. Such *reducible uncertainties* are obviously [1.11], [1.45] a product of the generalization of the feedback philosophy. This philosophy extends the design procedure into the real time of the evolution of the process, and accounts for unknown a-priori information due to partial ignorance. For example the coefficients of gravity or atmospheric variables during the automatic landing procedure of a space vehicle on an unknown planet, the attitude control of a booster stage of a space vehicle, the flight control of a supersonic

aircraft at different altitudes, the orbit stabilization of an orbiting satellite, and, more recently, the lower-dimensional modeling of
a high-dimension large-scale system for simplified study of its
behavior are all cases of reducible uncertainties. This extension
of the design is usually obtained by providing the controller with
self-organizing [1.29] capabilities to reduce the uncertainties related to the model of the plant or the function of the improvement
of performance of the system. Thus, the need of a class of control
systems with *self-organizing* capabilities that are beyond the area
of the automatic control systems discussed in the previous section,
has been established. It is important for the reader to realize
that the separation of the areas arose from the engineering-applica-
tion point of view and not from the creation of a new mathematical
discipline, since mathematics very similar to the ones used in
previous automatic control problems will be used in this new area.

1.3.1 Modeling and System Identification

    The main feature of self-organization, in view of the reducible
uncertainty of the plant, is the on-line improvement of the knowledge
of the model of the system. As discussed in Sec. 1.2.2, the process
of modeling comprises a *structural system identification* when the
structure of the plant and the underlying physical laws are unknown,
or *parameter identification* if the uncertainties of the plant can be
grouped in a parameter vector.

    On-line structural identification is an extremely difficult
function to perform when the plant is completely unknown. It
involves the identification of the nonlinearities involved, their
location, the order and couplings of various parts of the plant, and
there is no exact mathematical scheme that can perform all these
functions. A semiheuristic method has been proposed by Saridis and
Hofstadter [1.40] to classify nonlinearities according to their membership class, like saturation class, etc., and their relative
location in the plant. This method may be performed on-line but it
is, in general, time-consuming and therefore not suitable for this
type of identification. The method could be used off-line, however,

with greater success. Another solution to the structural identification is to assume off-line a series representation of the plant
with adjustable parameters. This approach requires a large number
of terms to yield a reasonable model, which also has the disadvantage
of not resembling the physical plant. In general, the area of
structural identification has not been satisfactorily explored as yet
to be successfully used in self-organizing control systems.

On-line parameter identification has been considerably re-
searched in the last few years and a large number of algorithms exist
in the literature. A collection of such algorithms [1.18], [1.24],
[1.27], [1.37], [1.42] will be presented in Chap. 6, where the subject
is quantitatively reviewed.

System identification may not appear explicitly in the self-
organization of a control system. The reason is that only the information regarding the improvement of the performance of the system
may be judged necessary. In such cases, it can be said that identification is performed implicitly, since the accumulation of information about the plant has been used directly by the controller without any intermediate modeling. However, this statement may be used
to unify most of the so-called "adaptive" and "learning" control
algorithms that have appeared under different forms in the literature [1.6], [1.9], [1.12], [1.31].

1.3.2 "Adaptive" and "Learning" Control Systems

The first attempt to treat problems with a built-in self-organization scheme was made by Fel'dbaum in his celebrated work [1.11]
under the name of *dual control*. There, active information about the
process was anticipated by the controller for the first time.
However, since transient optimality was required, the resulting
algorithms were too hard to implement. This will be made clear in
a brief presentation in Chap. 5.

Other attempts based on various heuristic or semiheuristic
concepts are found in the literature under the name of *adaptive
control systems*. The problem with these systems is that they are
diversified, inhomogeneous and sometimes redundant, and have always
been characterized as controversial. The reasons will be made clear
in the sequel.

The self-organizing characteristic required for controls that
drive systems with reducible uncertainties may be viewed as a
learning capability of the system. The property of improving the
model or identifying the system during the evolution of the process
is definitely the new quality of the controller, vaguely resembling
the qualities of a human controller. However, this opens up a new
area for control systems, where learning is the major property of
the computer [1.12]. With the aid of very fast large digital computers, many researchers have already designed *learning control
systems* that can be trained to make advanced decisions in an
unknown environment or even *intelligent control systems*, like robots
with anthropomorphic functions [1.10], [1.13], [1.14], [1.33], [1.46].
Therefore, this is an area of the future and has already attracted
the attention of the pioneer researcher.

Self-organization, with its active accumulation of information
intended to improve the performance of the system on-line, in other
words as the process evolves, can be considered as the first step
toward advanced learning and intelligent control systems. In order
to make a more quantitative presentation of the new discipline a
rigorous definition is necessary so that only the methodologies that
qualify may be studied in detail. This will be done after discussing
the controversies concerning the earlier name "adaptive controls,"
hoping to use this experience to avoid their shortcomings.

1.3.3   The Controversy over the Term "Adaptive Control"

Several definitions have been proposed for the area "beyond
stochastic control systems." The most successful terms, such as
"adaptive control" [1.21], [1.17], [1.24], [1.31], [1.43] and "learning
control" [1.13], [1.46], were borrowed from psychology and physiology to
indicate a "behavioristic" approach to the problem. Others, such as
"self-optimizing" [1.47], "optimalizing" [1.50], "self-organizing"
[1.29], etc., are based mainly on technical considerations and did
not receive a very wide acceptance. Since these terms have been
used for a large variety of superficially uncorrelated problems, they
have been the subject of controversy and misunderstanding. However,
the major disadvantage is that none of these definitions is general

enough to include all the work done in this area and make room for
expansion.

It is known that many of the above mentioned "adaptive systems"
may be redesigned as purely feedback-control systems without any
degradation in their performance by using ingenious filters inside
or outside the feedback loop or by expanding the compensation range
of the feedback controller. Such arguments are extensively presented
by Gibson [1.17].

On the other hand, the optimality requirement of some other
definitions impose heavy limitations on their implementation, due to
the complexity of the resulting algorithms. Furthermore, the uncertainties of some other systems may be reduced by an off-line
investigation of the system's behavior in the appropriate environment. The many controversies regarding which system is adaptive or
not are due mainly to the inadequacy of one of the first definitions
of such a system.

A few of the most popular definitions are quoted here without
comment to emphasize the difference of opinions on the subject of
definition of the new discipline.

Definition 1.6

Adaptive Systems: "An adaptive system is any system which has
been designed with an adaptive viewpoint." [Truxal in [1.31]].

Definition 1.7

Adaptive Control: "An adaptive control system must provide
continuous information about the present state of the plant
that is to identify the process; it must compare present system
performance to the desired or optimum performance and make a
decision to adapt the system so as to tend toward optimum
performance; and finally it must initiate a proper modification
so as to drive the system towards the optimum. These three
functions are inherent in an adaptive system." [Gibson [1.72]].

Definition 1.8

Adaptive Control Problem: "The problem of determining the
control law to meet specifications given only partial a-priori
information on plant dynamics" (Controversial.) [GAC Newsletter
[1.21]].

Definition 1.9

Adaptive Controller: "(1) (Controversial.) Controller which
is designed to solve the adaptive control problem. (2) Controller which includes means to measure functions of one or more

variables of the control system and which uses these measurements to adjust the system characteristics in a manner to
yield a system response within specifications." [GAC
Newsletter [1.21]].

Definition 1.10

Learning Control: "The design of controllers which are capable
of estimating the unknown information during the system's
operation and an optimal control will be determined on the
basis of the estimated information, such that the performance
of the system will be gradually improved." [Fu [1.13]].

Definition 1.11

Learning Systems: "A system that learns the unknown information about a process and uses it as an experience for future
decisions or controls, so the performance of the system will be
gradually improved." [Fu [1.13]].

Definition 1.12

Adaptive Control Processes: "We consider processes in which we
can learn about the nature of the unknown elements as the process proceeds. The multistage aspects of the process must be
made to compensate for the initial lack of information. We
call the mathematical theory developed to treat control processes of this type the 'theory of adaptive control processes'."
[Bellman [1.4]].

Definition 1.13

Dual Control: "Optimal systems with incomplete information
about the plant and with active accumulation of information
during the control process." [Feld'baum [1.11]].

Definition 1.14

Adaptation: "Adaptation is considered to be a process of modi-
fying the parameters of the structure of the system and the
control actions. The current information is used to obtain a
definite (usually optimal) state of the system when the operating conditions are uncertain and time-varying." [Tsypkin [1.45]].

Definition 1.15

Learning: "Under the term learning in a system, we shall consider a process of forcing the system to have a particular
response to a specific input signal (action) by repeating the
input signals and then correcting the system externally."
[Tsypkin [1.45]].

Definition 1.16

Self-learning: "Self-learning is learning without external
corrections, i.e., without punishments or rewards. Any additional information regarding the correctness or incorrectness
of the system's reaction is not given." [Tsypkin [1.45]].

Definition 1.17

    Adaptive and Learning Control: "The possibility of controlling
the plants under incomplete information is based on the application of adaptation and learning in automatic systems which
reduces initial uncertainty by using information obtained
during the process of control." [Tsypkin [1.45]].

The definitions given in the next section have been developed
to mitigate the disadvantages and controversies created by the
previous definitions, using the unifying property of <u>on-line</u> reduction of uncertainties through explicit or implicit identification
as the common denominator.

1.4 NEW DEFINITIONS

In 1973, a subcommittee on standards and definitions of the
Adaptive Learning and Pattern Recognition area of the Institute of
Electrical and Electronics Engineers (IEEE) Control Systems Society
published a report in the Society's Newsletter on new definitions of
the area called "Self-Organizing Processes and Learning Systems."
The new definitions were formulated by the subcommittee and reviewed
by the readership in the area before their publication, when they
were proposed for final adoption by the Society. The name *self-
organizing controls,* used by Mesasovic [1.29], was used to replace
Adaptive control in previous definitions to avoid controversial
interpretation. However, the word "adaptive" has been adopted as an
adjective to characterize functions similar to its original
"behavioral" meaning [1.49]. The area of *Learning systems* was redefined with minor modifications, with a new more precise definition
of *Learning control systems* to be compatible with definitions of
self-organizing controls. This was done in view of the considerable
overlap between the two control areas while, in general, the term
"learning systems" is more inclusive containing advanced decision-
making, pattern recognition, and other functions resembling human
behavior [1.13], [1.14].

1.4.1 Self-Organizing Control Systems

The area of *self-organizing control* (S.O.C.) *systems* is hereby
defined to describe systems with characteristic features "beyond
stochastic control systems." A common denominator of these features,
namely, the system's uncertainties and accumulation of information,
has been used in the definitions in order to provide room for all the
researchers in the area. A duality has been retained in the name of
the area in order to preserve both the "technical" as well as the
"behavioral" nature of the problems under consideration, and their
common area of application is derived to emphasize the unity of the
area. The name "self-organizing control process" has been used to
define the problems of technical origin, because of its technical
meaning of providing on-line structural as well as parametric adjustment to the system. The word "adaptive" if of behavioral origin and
has not been used as a name to avoid the existing arguments on its
interpretation. Instead, it has been used as an adjective to denote
the type of adaptation. The concept of identification and the memory
requirement which, for some researchers are the predominant features
of the area, are implicitly contained in the terms "accumulation of
information" and "learning."

The following definitions are listed to describe a self-organizing control process and controller.

Definition 1.18

> *Self-Organizing Control Process:* "A control process is called
> '*self-organizing*' if reduction of the a-priori uncertainties
> pertaining to the effective control of the process is accom-
> plished through information accrued from subsequent observa-
> tions of the accessible inputs and outputs as the control pro-
> cess evolves."

Definition 1.19

> *Self-Organizing Controller:* "A controller designed for a
> self-organizing control process will be called '*self-organizing*'
> if it accomplishes on-line reduction of the a-priori uncertain-
> ties pertaining to the effective control of the process as it
> evolves."

The word "optimal" has been replaced in the definitions by the
word "effective" or "improving" since in many problems strict

optimality is not achievable. No specific mention is made regarding
the deterministic or stochastic nature of the uncertainties so that
both cases may be treated equally.

It was previously mentioned that self-organization is achieved
either by reducing the uncertainties pertaining to the dynamics of
the plant, where an explicit identification scheme is present, or by
decreasing the uncertainties directly related to the improvement of
the performance of the system. In the latter case, the information
retrieved from the plant is used directly in the controller and the
appropriate performance evaluator. In order to distinguish between
the two categories of self-organizing controls, which represent two
distinct design procedures, the following definitions are given
where the word adaptive has been used to qualify the appropriate
function.

Definition 1.20

> *Parameter-Adaptive S.O.C. Process:* "A self-organizing control
> process will be called '*parameter-adaptive*' if it is possible
> to reduce the a-priori uncertainties of a parameter vector
> characterizing the process through subsequent observations of
> the inputs and outputs as the control process evolves."

Definition 1.21

> *Performance-Adaptive S.O.C. Process:* "A self-organizing
> control process will be called '*performance-adaptive*' if it is
> possible to reduce directly the uncertainties pertaining to the
> improvement of the performance of the process through subse-
> quent observations of the inputs and the outputs as the control
> process evolves."

Definition 1.22

> *Parameter-Adaptive Self-Organizing Controller:* "A self-
> organizing controller will be called '*parameter-adaptive*' if it
> accomplishes <u>on-line</u> reduction of the uncertainties of the
> parameter vector through subsequent observations of the inputs
> and the outputs as the control process evolves."

Definition 1.23

> *Performance-Adaptive Self-Organizing Controller:* "A self-
> organizing controller will be called '*performance-adaptive*' if
> it accomplishes reduction of the uncertainties related to the
> direct improvement of the performance of the process through
> subsequent observations of the inputs and outputs as the con-
> trol process evolves."

A detailed quantitative description of the most representative
stochastic parameter-adaptive S.O.C. algorithms are given in Chap.
7, while the equivalent performance-adaptive S.O.C. algorithms are
given in Chap. 8, where the idiosyncrasies of each category are
emphasized. Some of their deterministic counterparts are summarized
in Chap. 2.

1.4.2 Learning Control Systems

The name "learning systems" has been used to define problems
approached in a behavioral manner, encompassing the spectrum of
problems from simple control of systems with uncertainties in their
dynamics, to advanced decision-making pattern recognition, classification and organization, etc. [1.16], [1.13]. Therefore, they characterize a class more sophisticated than control systems. However,
a subclass of learning control systems may be technically interpreted as a performance-adaptive S.O.C. system. These are the
class of controllers designed to drive completely or partially
unknown systems using on-line "unsupervised learning." This overlap
provides the link that brings self-organizing and learning controls
together. On the other hand, by claiming that S.O.C.s possess some
crude learning properties one may consider self-organization as a
part of the broader family of learning systems which may treat processes of much higher degrees of sophistication than the ones
encountered in engineering problems.

The following definitions are presented to clarify the meaning
of learning systems and controls [1.41].

Definition 1.24

> *Learning Systems:* "A system is called *learning* if the infor-
> mation pertaining to the unknown features of a process or its
> environment is learned, and the obtained experience is used
> for future estimation, classification, decision, or control
> such that the performance of the system will be improved."

Definition 1.25

> *Learning Control Systems*: "A learning system is called *learning control system* if the learned information is used to control a process with unknown features. Learning control may be performed off-line, in which case the controller is train-able, or on-line, in which case it is synonymous to the per-formance-adaptive S.O.C. when it applies to the same kind of processes."

A further discussion of learning is given in Chap. 10, repre-senting the future trends in system design.

1.5 SOME THOUGHTS ON SELF-ORGANIZING CONTROLS, LEARNING, AND IN-TELLIGENT SYSTEMS

The set of definitions presented in the preceding sections have been designed to establish the area outside the "classical" auto-matic control systems and unify the algorithms that have been developed for this purpose. This area, called *Self-organizing control systems*, has been originally defined to treat all systems with completely or partially unknown plant descriptions. As such, S.O.C. referred mainly to the control of engineering systems for which on-line adaptation is required. The parameter-adaptive S.O.C. problem is the one where the uncertainties are described by a parameter vector which can be recovered by explicit parameter iden-tification as suggested by Gibson [1.17], implied by Widrow in his "open-loop adaptation" [1.47], and specified by Tsypkin as "adaptation" [1.45]. The performance-adaptive S.O.C. problem deals with the on-line control adjustments through performance measurements without explicit identification, and corresponds to Widrow's "close-loop adaptation" [1.47], and Tsypkin's "self-learning" [1.45], while it has not been included in Gibson's definition. Mathematical techniques under the general name of "Mathematical programming" have been developed by many researchers and will be discussed in the sequel.

Recognizing that S.O.C.s contain some kind of crude learning in their process, their association with the area of learning sys-tems was investigated. Learning systems have been defined in the

previous section to represent systems in which dynamic accumulation
of information takes place with or without supervision, e.g., off-line
or on-line [1.13], [1.28], [1.46] in order to improve the understanding of
a process, make advanced-level decisions, and in general imitate some
lower "behavioral" function of the human operator.   Such an area
covers research done on learning control systems, pattern recognition, automata theory, system classification and organization, etc.
On-line learning control systems satisfy the definitions of performance-adaptive S.O.C. when applied to engineering processes.   The
existing overlap established the relation between S.O.C.s and
learning systems, and on the other hand, indicated the next step in
the direction of the future of automatic control systems.   Advanced
decision-making which is part of learning systems is the most likely
candidate for the job.   Such systems, which may or may not include
the human operator as part of the controller, would perform decisions
of a higher degree of sophistication, like learning their functions
and changing forms in different environments, thus, crudely imitating
the functions of a human controller [1.13], [1.14], [1.46].

Further up in degree of sophistication there are reports in
the literature on intelligent machines and robots represented by an
area named "artificial intelligence" [1.10], [1.30].   Such systems, implemented by the fast, large, modern digital computer, can solve problems, identify objects, or plan a strategy for a complicated function of a system.   The controls involved in such systems have been
proposed by Fu [1.14] as the next step after learning control systems.
It is obvious that S.O.C.s fit very nicely in the picture and can
occupy the lowest level in a hierarchical intelligent control system
with the other two levels occupied by learning and intelligent controls.   This subject will be further discussed in Chap. 10 where a
hierarchically intelligent control system is proposed for bionic
uses.

In conclusion, the S.O.C. systems have been put in the right
perspective, at the threshold between "classical" automatic controls
and modern "intelligent" control systems from the theoretic point of

view. In the next section, they will be discussed in view of their
potential applications.

1.6 DISCUSSION ON APPLICATIONS

In all aspects of engineering, no new area is sufficiently
justified unless it is demonstrated that there is enough demand from
the applications point of view. Self-organizing controls are no
exception to this rule. However, it must be understood that one
should look for applications that are of the same level of sophistication as the S.O.C.s. This principle of not overdesigning a controller can be considered as a golden rule in engineering and applies
to its other disciplines as well. The major criticism that the
older "adaptive" control systems have drawn, is that either they were
representing trivial designs replaceable by a feedback loop, or that
they were overdesigned controls for simple first- or second-order
systems [1.9], [1.31]. It is true, that in classical hardware engineering
problems there are very few uncertainties about the plant that cannot be completely reduced before even the design process starts.
However, with the technological progress of the later years, the
growing importance of interactions among systems, and the interest
of the so-called "soft sciences" for the system-theoretic methods,
the processes to be controlled increased in sophistication, complexity, and size. Progress in modern space technology requires
unmanned exploration vehicles operating in unknown environments,
with unknown force fields, etc., far too deep in space for effective
remote control. Self-organizing control for steering, power output,
etc., is essential for such a vehicle. Industrial-process control
is another field where, due to the complexity of the overall system
and the uncertain description of the technoeconomic, scheduling, and
other processes, S.O.C. may find interesting applications.
Biomedical and bionic processes with the complex incomplete modeling
of biological systems, large-scale power systems with their huge
dimensionality are also potential applications of the new area.

Finally, the areas currently explored by the system theorists, representing an aggregate of diversified subsystems of huge dimensions like socioeconomic system, urban and environmental systems, transportation, ecological and pollution control systems, and others belonging to disciplines characterized by poor modeling, may use S.O.C. processes, if one models them by a low-order model with unknown parameters in a hierarchical design procedure [1.39].

It is quite clear from the preceding discussion that the time of enough demand for S.O.C. and learning systems has arrived. As a matter of fact, there have already been several attempts for their utilization which will be discussed in Chap. 10.

Therefore, it is justifiable to proceed with the quantitative presentation of the various S.O.C. algorithms. Because of the diversity of the deterministic algorithms that qualify as self-organizing, they are only superficially treated in Chap. 2. The most representative stochastic S.O.C. algorithms are mathematically treated in Part III of this book, after an introduction to the necessary background material is given in Part II for the sake of completeness of presentation.

1.7 REFERENCES

[1.1]   Aoki, M., *Optimization of Stochastic Systems*, Academic Press, New York, 1967.

[1.2]   Åström, K. J., *Introduction to Stochastic Control Theory*, Academic Press, New York, 1970.

[1.3]   Athans, M., and Falb, P. L., *Optimal Control: An Introduction to the Theory and its Applications*, McGraw-Hill, New York, 1966.

[1.4]   Bellman, R., *Adaptive Control Processes, A Guided Tour*, Princeton University Press, Princeton, N.J., 1961.

[1.5]   Bryson, A. E., and Ho, Y. C., *Applied Optimal Control*, Blaisdell, Boston, 1969.

[1.6]   Cooper, G. R., Gibson, J. E., Eveleigh, J. E., et al., "A Survey of the Philosophy and State of the Art of Adaptive Systems," Tech. Report PRF-2358, Purdue University, Lafayette, Ind., July 1960.

[1.7]   Deutz, R., *Estimation Theory*, Prentice-Hall, Englewood-Cliffs, N.J., 1965.

[1.8]   Eckman, D., *Automatic Controls*, Wiley, New York, 1958.

[1.9]   Eveleigh, V. W., *Adaptive Control and Optimization Techniques*, McGraw-Hill, New York, 1967.

[1.10] Feigenbaum, E. A., and Feldman, J. (eds.), *Computers and Thought*, McGraw-Hill, New York, 1963.

[1.11] Fel'dbaum, A., *Optimal Control Systems*, Academic Press, New York, 1965.

[1.12] Fu, K. S., "Learning Control Systems," in *Advances in Information Systems Science*,(J. T. Tou, ed.), Plenum Press, New York, 1969.

[1.13] Fu, K. S., "Learning Control Systems-Review and Outlook," *IEEE Trans. on Automatic Controls*, AC-15 (2), 210-221 (1970).

[1.14] Fu, K. S., "Learning Control Systems and Intelligent Control Systems - An Intersection of Artificial Intelligence and Auto. Control," *IEEE Trans. on Automatic Controls*, AC-16 (1), 70-72 (1971).

[1.15] Fu, K. S., *Sequential Methods in Pattern Recognition and Machine Learning*, Academic Press, New York, 1968.

[1.16] Fu, K. S., Gibson, J., Hill, J. D., et al., "Philosophy and State of the Art of Learning Control Systems," Tech. Report TR-EE-63-7, Purdue University, Lafayette, Ind., Nov. 1963.

[1.17] Gibson, J. E., *Non-linear Automatic Control*, Chap. 11, McGraw-Hill, New York, 1962.

[1.18] Graupe, D., *Identification of Systems*, Van Nostrand Reinhold, New York, N.Y., 1972.

[1.19] Horowitz, I. C., and Shaked, U., "Superiority of Transfer Function Over State-Variable Methods in Linear Time-Invariable Feedback System Design," *IEEE Trans. Automatic Control*, <u>AC-20</u> (1), 84-87 (1975).

[1.20] Isaacs, R., *Differential Games*, Wiley, New York, 1965.

[1.21] *IEEE, Automatic Control Group Newsletter*, July 1965, p. 7.

[1.22] Klir, G. (ed.), *Trends in General Systems Theory*, Wiley-Interscience, New York, 1972.

[1.23] Kushner, H., *Introduction to Stochastic Control*, Holt, Reinhart, and Winston, New York, 1971.

[1.24] Lee, R. C. K., *Optimal Identification, Estimation, and Control*, MIT Press, Cambridge, Mass., 1964.

[1.25] Leondes, C. T., and Mendel, J. M., "Artificial Intelligence Control," *Survey of Cybernetics*, (R. Rose, ed.), ILLIFE Press, London, 1969.

[1.26] Meditch, J., *Stochastic Optimal Linear Estimation and Control*, McGraw-Hill, New York, 1969.

[1.27] Mendel, J. M., *Discrete Techniques of Parameter Estimation*, Marcel Dekker, New York, 1973.

[1.28] Mendel, J. M., and Fu, K. S. (ed.), *Adaptive Learning and Pattern Recognition Systems*, Academic Press, New York, 1970.

[1.29] Mesarovic, M. D., "On Self-Organization Systems," *Self-Organizing Systems-1962*, (M. C. Yovits, G. T. Jacobi and G. D. Goldstein, eds.), Spartan Books, New York, 1962.

[1.30] Minsky, M., *Artificial Intelligence*, McGraw-Hill, New York, 1972.

[1.31] Mishkin, E., and Braun, Jr., L., (eds)., *Adaptive Control Systems*, McGraw-Hill, New York, 1961.

[1.32] Nichols, N. B., in *Theory of Servomechanisms*, (H. M. James, N. B. Nichols, and R. J. Phillips, eds.), M.I.T. Radiation Technology Lab. Publication, Cambridge, Mass., 1946.

[1.33] Nilsson, N. J., *Learning Machines*, McGraw-Hill, New York, 1965.

[1.34] Pontryagin, L. S., Boltyanskii, V. G., Gamkrelidge, R. V., and Mishchenko, E. F., *The Mathematical Theory of Optimal Processes*, Interscience Publishers, New York, 1962.

[1.35] Popov, E. P., *The Dynamics of Automatic Control Systems*, Addison-Wesley Publishing Co., Reading, Mass., 1962.

[1.36] Peschon, J. (ed.), *Disciplines and Techniques of Systems Control*, Blaisdell, Boston, 1965.

[1.37] Sage, A. P. and Melsa, J. L., *System Identification*, Academic Press, New York, 1971.

[1.38] Saridis, G. N., "The Development of Systems Engineering," Paper presented at the 2d Annual Meeting of General Systems Research Soc., University of Maryland, 1973.

[1.39] Saridis, G. N., "Learning System Theory for General System Studies," Proc. of 3d Annual Meeting of General Systems Research Society, University of Maryland, 1974.

[1.40] Saridis, G. N., and Hofstadter, R. F., "Pattern Recognition Approach to the Classification of Nonlinear Systems," *IEEE Trans. on Systems, Man and Cybernetics*, SMC-21 (4), 362-371, (1974).

[1.41] Saridis, G. N., Mendel, J. M., and Nikolic, Z. J., "Report on Definitions on S.O.C. Processes and Learning Systems," *IEEE S-CS Newsletter*, February 1973.

[1.42] Saridis, G. N., and Stein, G., "Stochastic Approximation Algorithms for Linear Discrete-Time System Identification," *IEEE Trans. Automatic Control*, AC-13, (5), 515-523, (1968).

[1.43] Sworder, D. D., *Optimal Adaptive Control Systems*, Academic Press, New York, 1966.

[1.44] Tou, J. T., *Modern Control Theory*, McGraw-Hill, New York, 1964.

[1.45] Tsypkin, Ya., *Adaptation and Learning in Automatic Control* (Z. J. Nikolic, transl.) Academic Press, New York, 1971.

[1.46] Tsypkin, Ya., *Foundation of the Theory of Learning Systems* (Z. J. Nikolic, transl.), Academic Press, New York, 1973.

[1.47] Widrow, B., "Pattern Recognition and Adaptive Control," *IEEE Trans. Appl. Industry*, 83 (74), 269, (1964).

[1.48] Wymore, W., *A Mathematical Theory of Systems Engineering Elements*, Wiley, New York, 1964.

[1.49] Wilde, D., *Optimum Seeking Methods*, Prentice-Hall, Englewood-Cliffs, N.J., 1964.

[1.50] Zaborszky, J., and Berger, R. L., "Self-Optimalizing Adaptive Control," *IEEE Trans. Appl. Industry*, 81 (63), 256 (1962).

[1.51] Zadeh, L. A., and Desoer, C., *Linear System Theory*, McGraw-Hill, New York, 1963.

[1.52] Zadeh, L. A., and Polak, E. (eds.), *System Theory*, McGraw-Hill, New York, 1969.

Chapter 2

SOME DETERMINISTIC SELF-ORGANIZING CONTROL ALGORITHMS

2.1 INTRODUCTION

The definitions presented in the preceding chapter have established the borderlines of the area of S.O.C.s and have provided a
home for many algorithms that have appeared in the literature. In
order to qualify they must meet the requirements imposed by the
definitions. Deterministic as well as stochastic systems qualify
as S.O.C. systems, but the emphasis of this book is primarily on
stochastic S.O.C.s, because it is more reasonable to assume that
where there are reducible uncertainties in the plant, there should
be irreducible uncertainties entering the system either as
disturbances or measurement errors.

This chapter presents the highlights of some of the many
deterministic S.O.C. systems that have been discussed in the
literature. The selection is far from being complete and it
should be stressed that it is not intended to be, because of the
number of the existing schemes, the diversity of their applications,
and the occasional lack of theoretical foundations. Therefore, a
small number of algorithms with solid theoretical foundation,
representative of rather general schools of thought, were selected
which may serve as examples of the area. The selection, therefore,
is completely subjective, and no slight was inteneded to the
authors whose work is not mentioned herein.

As expected, the first S.O.C. systems appeared under various
names and were dealing with systems operating in deterministic
environments. The first paper in 1951, containing the original
systematic formulation of the so-called "self-optimizing systems"

29

was by Draper and Li [2.10] and it was concerned with the cruise control of the engines of long-range aircraft. The problem was to
obtain maximum cruising range per pound of fuel by automatically
manipulating such parameters as throttle settings, fuel-air
mixtures, etc.

Other systems that followed dealt mostly with adaptive auto-
pilots like the General Electric Adaptive Autopilot [2.14], based on
modification of the loop gain by measuring the rate of error axis-
crossings, or the Honeywell Adaptive Autopilot [2.14], based on a
parameter (gain) perturbation that would result in a limit cycle
of the adaptive loop and therefore modify the adjustable gain to
keep the overall gain constant at every deviation from the limit-
cycle frequency. Such systems have been flight-tested and successfully used, but contribute very little to the formulation of the
self-organizing problem because they represent heuristically
developed ingenious devices limited to specific problems. Their
application is limited, on the other hand, by the lack of generalizations to higher-order systems. Reviews on such systems may be
found in Refs. [2.5], [2.14], and [2.37].

In most of the work in the literature of deterministic S.O.C.
systems, the parameter-identification aspect is mainly emphasized,
being the major new feature introduced by these systems. However,
identification must be performed <u>on-line</u>, in order for the
system to satisfy the definitions of self-organization. Therefore
the work of researchers such as Dreifke and Hougen [2.11] and others,
does not qualify as S.O.C. A typical Parameter-Adaptive S.O.C.
system is given in Fig. 2.1. Optimality is not required by those
systems, and the mathematical approach employed to represent the
plant, i.e., transfer function or state variable, is a matter of
preference of the researcher.

An implicit assumption is usually made that either the system
operates at steady state, or that the dynamics of the system change
very slowly compared to the dynamics of the adaptive scheme in such
a way that the identification is performed in a short time and its

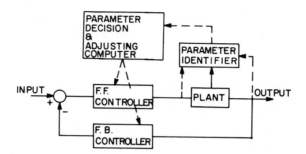

FIG. 2.1.   Block Diagram of a Deterministic Parameter-Adaptive
S.O.C. System.

results fed to the controller before any significant change in the
system.   This is a strict limitation, but separates the analysis of
the adaptive loop from that of the main system and thus facilitates
the study of the problem.

Adaptation by direct performance improvement has also been
used, and a class of deterministic systems, with the model reference
adaptive controls being the most representative, have qualified as
performance adaptive S.O.C.   A block diagram of a Performance
Adaptive S.O.C. system is given in Fig. 2.2, while the Model
Reference Adaptive Control is briefly discussed in the sequel.

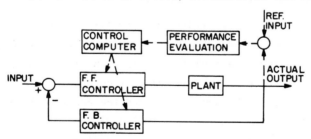

FIG. 2.2.   Block Diagram of a Deterministic Performance-
Adaptive S.O.C. System.

Other general types of deterministic S.O.C. discussed are the systems
using gradient methods for identification systems modeled by
polynomials, and systems modeled by functionals.

More recent work in the area by Horowitz, which is not
reported here and can be found in Ref. [2.23].

2.2 GRADIENT INDENTIFICATION S.O. CONTROLS

The S.O.C. belonging to this category have in common a parameter
identification algorithm based on a gradient search [2.60]. This
search can either be performed by a digital computer or implemented
by analog components, and may be used for linear and sometimes
nonlinear systems. These techniques will be outlined in the next
two sections. Since the original research stressed the identification aspects, only the utilization of this information has been
left to the "computer programmer" which programs the appropriate
controllers. However, in order to make the picture complete an
S.O.C., using a gradient search technique is designed to control a
system for which identification is performed <u>on-line</u> and the impulse
response is estimated. The combination of a "gradient identification"
and a gradient controller to form an integrated Parameter-Adaptive
S.O.C. system is left as an exercise for the reader.

2.2.1. Parameter Tracking for Identification

Margolis and Leondes [2.34] were the first to develop a method
for the design of an adaptive control system which determines the
values of the significant parameters in a physical process and uses
these values to program the controller according to specified control
laws. The process, the controller, and the adaptive loop are
depicted in a block diagram given in Fig. 2.3.

Assume that the physical process is described by a linear
nth-order differential equation with unknown coefficients given by

$$\sum_{i=0}^{n} f_i \frac{d^{(i)}}{dt^i} y(t) = u(t) \tag{2.1}$$

where $y(t)$ is the scalar output and $u(t)$ is the scalar input of the
plant, and f_i is a set of unknown coefficients.

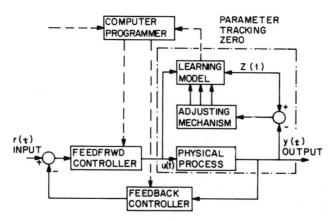

FIG.   2.3.   Block Diagram of the Parameter Tracking Adaptive
Control System of Margolis and Leondes.

A model called "learning model" may simulate the actual model
by updating the adjustable coefficients α_i of the differential
equation

$$\sum_{i=0}^{n} \alpha_i \frac{d^{(i)}}{dt^i} z(t) = u(t) \qquad (2.2)$$

where $z(t)$ is the output of the model.

The parameter tracking servo compares the outputs of both
processes and updates the parameters of the model according to their
difference.   Define the error between outputs by:

$$e(t) = z(t) - y(t) \qquad (2.3)$$

Then the method of steepest descent is used to minimize a convex
function of $e, f(e)$.   Such a function may be chosen to have a
minimum if and only if $\alpha_i = f_i$.   However, if the input signal is
zero $y = z = e = 0$, then the gradient of $f(e)$ may be zero for
$\alpha_i \neq f_i$.   Thus the servo will track only if there is a forcing
function.   The law governing the parameters of the model is

$$\frac{d\alpha_i}{dt} = -k \frac{\partial f(e)}{\partial \alpha_i} \quad \text{for } k > 0 \qquad (2.4)$$

The following first-order example illustrates the method

$$\frac{dy}{dt} + fy = u(t) \tag{2.5}$$

The learning model is defined by

$$\frac{dz}{dt} + \alpha z = u(t) \tag{2.6}$$

The error square e^2 has been chosen as a suitable $V(e)$ function. To implement the adjusting mechanism,

$$\frac{d\alpha}{dt} = -k \frac{\partial V(e)}{\partial \alpha} = -2ke \frac{\partial e}{\partial \alpha} = -2ke \frac{\partial z}{\partial \alpha} \tag{2.7}$$

Since $e = z - y$ is available for measurement, the only problem is generating the gradient $\frac{\partial e}{\partial \alpha}$. Two methods have been proposed to approximate $\frac{\partial V}{\partial \alpha}$. The first one uses a Taylor series expansion. If an auxiliary equation is used,

$$\frac{dz_1}{dt} + (\alpha + \Delta\alpha) \, z_1 = u(t) \tag{2.8}$$

Then

$$\frac{\partial z}{\partial \alpha} \cong \frac{z_1 - z}{\Delta\alpha} \tag{2.9}$$

Therefore, the adaptive scheme may be implemented by Eqs. (2.6), to (2.9). The second method may be implemented by taking $\frac{\partial}{\partial \alpha}$ of both sides of Eq. (2.6) and defining $w \triangleq \frac{\partial z}{\partial \alpha}$. Then

$$\frac{dw}{dt} + \alpha w = -z \tag{2.10}$$

Therefore, the adaptive scheme may be implemented by Eqs. (2.6), (2.7), and (2.10). It should be noticed, though, that the last equation is only an approximation, since α is not actually a constant.

The method has been further extended by Donalson and Leondes [2.9], and Donalson and Kishi [2.8], without any major changes in the approach. In every case, stability presented the most demanding problem of the system. The controller to drive the system in Fig. 2.3 is assumed to use the above information and meet a certain set of specifications. An improvement over this method is described in the next section.

2.2.2.   Analog Search for Identification

A S.O.C. system may be designed by using an analog parameter-identification scheme and a controller computer to implement the control adjustments as was done with the preceding method.   This system is depicted in Fig. 2.4.

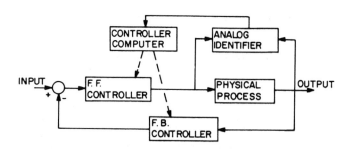

FIG.   2.4.   Block Diagram of the Analog Identification Method.

The analog identification scheme is based on the original work by Hoberock and Kohr [2.22] which treats the identification of the coefficients and nonlinearities of nonlinear differential equations. However, the method was developed as an _off-line_ scheme for systems with one or at most two nonlinearities.   The important features of the approach are the use of a set of linear filters to generate the derivatives of the inaccessible systems, and the definition of an error function in terms of a mismatch in the differential equation that guarantees unimodality of the minimization surface in the parameter space.

The method was modified by Saridis and has been used by Saridis and Ricker [2.52] for _on-line_ system identification in both deterministic and stochastic environments: Let the system dynamics be given by the following nonlinear differential equation:

$$\sum_{i=0}^{a} f_i(y,\dot{y},\cdots,y^{(n-1)}), \frac{d^{(i)}}{dt^i} y = u + \sum_{j=1}^{m} b_j(u,\dot{u},\cdots,u^{(m-1)})$$

$$\frac{d^{(j)}u}{dt^j} \qquad\qquad (2.11)$$

where u is the scalar input and y is the scalar output of the system,
the only variables available for measurement, and f_i, $i = 1, \cdots, n$ and
and b_j, $j = 1, \cdots, m$, $m < n$, are nonlinear functions of the outputs,
inputs, and their derivatives, respectively, and are *unknown* to the
designer. The identification problem consists of finding these f_i
and b_j on-line and feeding the information into the computer
controller to complete the loop.

In order to identify the nonlinear functions f_i and b_j, they
may be approximated by the following series expansions:

$$f_i = \sum_{j=1}^{M} c_{ij}\phi_j \quad i = 0, \cdots, n \qquad b_j = \sum_{k=1}^{N} k_{jk}\psi_k \quad j = 1, \cdots, m$$

(2.12)

where $\phi_j = \phi_j(y, \dot{y}, \ldots, y^{(n-1)})$ are known independent functions of the
output, and its derivatives, $\psi_k = \psi_k(u, \dot{u}, \cdots, u^{(m-1)})$ are known
independent functions of the input, and its derivatives, c_{ij} and
k_{jk} are constant unknown coefficients to be identified, and M and
N are suitable integers that guarantee a good approximation of the
true parameters. In many cases the ϕ_j and ψ_k may be chosen to be
polynomials of the states, i.e.,

$$\phi_1 = 1, \ \phi_2 = y, \ \phi_3 = y^2, \ \ldots, \ \phi_n = y^{n-1}$$

$$\phi_q = \dot{y}, \ \phi_{q+1} = \dot{y}^2, \qquad \ldots, \quad \ldots$$

(2.13)

$$\phi_r = y\dot{y}, \ \phi_{r+1} = y\ddot{y}, \qquad \ldots, \quad \ldots$$

$$\ldots, \qquad \qquad \ldots, \ \phi_N = y^{(n-1)N}$$

A decision on the form of ϕ_j and ψ_k and N and M would be made if
some information on the structure of the system is available, i.e.,
polynomials in an electromechanical system, etc.

Define the following error function:

$$I^2 = \left\{ \sum_{i=0}^{n} \sum_{j=1}^{M} \hat{c}_{ij}\phi_j \frac{d^{(i)}}{dt^i} y - \sum_{j=1}^{m} \sum_{k=1}^{N} \hat{k}_{jk}\psi_k \frac{d^{(j)}}{dt^i} u - u \right\}^2$$

(2.14)

where $\hat{c}_{ij}$ and $\hat{k}_{jk}$ are estimates of the parameters c_{ij} and k_{jk} that
minimize I^2, and y and u are the real output and input to the system.
It is obvious that I^2 is a quadratic function of $\hat{c}_{ij}$ and $\hat{k}_{jk}$ and is

therefore a unimodal function in the parameter space. It can then
be concluded that it has a unique minimum $I^2[c_{ij}, k_{jk}] =$
$\underset{c,k}{\text{Min}}\ I^2[\hat{c}_{ij}, \hat{k}_{jk}]$. Therefore, a gradient scheme can be implemented to
search for c_{ij} and k_{jk} that minimize I^2, provided that the dynamics
of the system are slower than the gradient scheme.

In order to obtain the derivatives of y and u on the analog
computer, which are not readily available and poorly generated by
differentiators, the following technique, employed by Hoberock and
Kohr is used to rewrite the equation of the system as

$$P[y] = Q[u] \qquad\qquad (2.15)$$

where P and Q are nonlinear differential operators defined as in
Eq. (2.11). If one could devise a linear filter $L(d/dt)$ applied to
the input and output, respectively, then one could generate new
variables y_c and u_c and their respective derivatives which are
related to the original variable by:

$$y_c = L[y] \quad u_c = L[u] \qquad\qquad (2.16)$$

This new filter L must commute with the unknown nonlinear operator
in order to satisfy Equ. (2.15), i.e.,

$$L[P[y]] = P[L[y]] = P[y_c]$$
$$L[Q[u]] = Q[L[u]] = Q[u_c] \qquad\qquad (2.17)$$

The only filter with such properties is a pure time delay, i.e.,

$$L = e^{-Ts} \qquad\qquad (2.18)$$

where s is the complex frequency variable. Such a filter may be
linearly approximated to a desired degree of accuracy by a transfer
function

$$H(s) = \frac{\omega_0^n}{s^n + a_{n-1}\omega_0 s^{n-1} + \cdots + a_1 \omega_0^{n-1} s + \omega_0^n} \qquad\qquad (2.19)$$

where ω_0 is a cut-off frequency, the coefficients of which may be
evaluated to minimize an integral time absolute error (ITAE)
criterion [2.15] or by Padé approximation [2.19]. Using L in
Eq. (2.18), Eq. (2.15) may be rewritten as

$$P[y_c] = Q[u_c] \qquad\qquad (2.20)$$

where P and Q are the same nonlinear operators as before, depending

on the same $a_i(y_c, \cdots, y_c^{(n-1)})$ and $b_j(u_c, \cdots, u_c^{(n-1)})$ which
are to be identified but with variables that are available for
measurement. A new error function is now defined

$$
I_c^2 = \left\{ \sum_{i=0}^{n} \sum_{j=1}^{N} \hat{c}_{ij} \phi_j(y_c, \cdots, y_c^{(n-1)}) \frac{d^{(i)}}{dt^i} y_c - \right.
$$
$$
\left. \sum_{j=1}^{m} \sum_{k=1}^{M} \hat{k}_{jk} \psi_k(u_c, \cdots, u_c^{(n-1)}) \frac{d^{(i)}}{dt^i} u_c - u_c \right\}^2 \qquad (2.21)
$$

which is also a unimodal function on the parameter space. A steepest
descent method may be used now to obtain c_{ij} and k_{jk} of the form

$$
\frac{d\hat{c}_{ij}}{dt} = -\gamma \frac{\partial z_c^2}{\partial \hat{c}_{ij}} = -2\gamma I_c \phi_j(y_c, \cdots, y_c^{(n-1)}) y^{(i)}
$$
$$
\qquad (2.22)
$$
$$
\frac{d\hat{k}_{jk}}{dt} = -\gamma \frac{\partial z_c^2}{\partial \hat{k}_{jk}} = +2\gamma I_c \psi_k(u_c, \cdots, u_c^{(m-1)}) u^{(j)}
$$

which is easily implementable. The analog diagram of the identification scheme is given in Fig. 2.5.

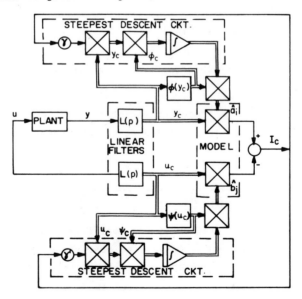

FIG. 2.5. The Analog Identification System Described by:

$$
\frac{dy}{dt} + 0.5(1 + y^2)\, y = u(t)
$$

The model used for the particular example was

$$f_1 \frac{dy}{dt} + f_0 y = bu(t)$$

with $u(t) = b \sin \alpha t$

$$a_0 = c_{00} + c_{01} y + c_{02} y^2$$

$$a_1 = c_{10} + c_{11} y + c_{12} y^2$$

For the L[P] filters, a third-order ITAE approximation was used for the pure time delay:

$$\frac{y_c(s)}{y(s)} = \frac{\omega_0^3}{s^3 + 1.75 \, \omega_0 s^2 + 2.45 \, \omega_0^2 s + \omega_0^3}$$

with $\omega_0 = 1$ $^{rad}/sec$. The method converged to the true parameters in less than 20 sec.

The main shortcoming of the method is that it requires a large number of terms in the series expansions in order to provide an accurate representation of the plant. A hybrid computer may improve the computation aspects of such a procedure.

2.2.3. Polynomial Expansion for Real-time Control

Meditch and Gibson [2.35] proposed a self-organizing controller that utilizes information about the impulse response of the unknown linear time-varying system, to improve its operation, with the purpose of optimizing a performance criterion defined over an interval $0 \leq t \leq T$ of the overall time. The subdivision of the process time into *control intervals* is absolutely necessary for the feasibility of the on-line optimization required, and does not guarantee an optimal performance over the duration of the whole process. A prediction of the input for the interval $0 \leq t \leq T$ is obtained using the information from the preceding T sec. and the control input to the plant is synthesized using this predicted signal. Thus the control part of the self-organizing system is performed on the input signal to the plant. The synthesis of the control signal is accomplished using a sum of orthonormal polynomials defined over $0 \leq t \leq T$ and computing the respective coefficients which are then

assumed constant over the control interval. The use of the concepts
in interval control and orthonormal series expansion reduces the
optimization problem from a two-point boundary value problem to a
relatively simple calculus problem, to be treated in realtime using
analog computer techniques. The usefulness of polynomial expansions
to model unknown functions, which has been explored in the last two
cases, is of major importance and will be explored further in the
next section. A block diagram of the overall system is given in
Fig. 2.6.

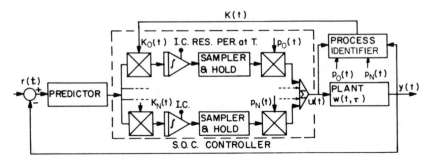

FIG. 2.6.   Real-Time Control Using Series Expansion.

The optimal controller is expressed as a finite sum of
Legendre polynomials which are defined on a finite interval
$0 \leq t \leq 1$:

$$u(t) = \sum_{k=0}^{N} u_k P_k(t), \quad t_i \leq t \leq t_i + T \tag{2.23}$$

The polynomials $P_k(t)$ are modified Legendre polynomials defined
over the interval $0 \leq t \leq T$:

$$P_k(t) = \sqrt{\frac{2k+1}{T}} P_k\left(\frac{2}{T}t - 1\right) \tag{2.24}$$

where $P_k(t)$ are the original Legendre polynomials defined over
$0 \leq t \leq 1$.

The per-interval performance index is defined as

$$I = \int_0^T \{w(t)[y^d(t) - y(t)]^2 + u^2(t)\} \, dt \tag{2.25}$$

where $y^d(t)$ is the desired output, $y(t)$ the actual output, and $w(t)$ is a weighting coefficient of the system error.

The actual output at the $(i + 1)$st interval can be expressed as follows, due to the linearity of the system:

$$y(t) = y_i(t) + \int_0^t u(\tau)c(t,\tau)\, d\tau \quad 0 \le t \le T \tag{2.26}$$

where $y_i(t)$ is the response due to initial conditions at $t = 0$, i.e., to the previous intervals, and $c(t,\tau)$ is the impulse response of the systems obtained from the identifier.

To minimize the performance index I and find the optimal u_k:

$$\frac{\partial I}{\partial u_k} = 2 \int_0^T \left\{ w(t)\, y_0(t) - y(t) \left[-\frac{\partial y(t)}{\partial u_k} \right] + u(t)\, \frac{\partial u(t)}{\partial u_k} \right\} dt = 0$$

$$k = 0, 1, \ldots, N \tag{2.27}$$

where

$$\frac{\partial u(t)}{\partial u_k} = P_k(t) \tag{2.28}$$

$$\frac{\partial y(t)}{\partial u_k} = \int_0^t \frac{\partial u(\tau)}{\partial u_k} c(t,\tau)\, d\tau = \int_0^t P_k(\tau)c(t,\tau)\, d\tau$$

On the other hand, due to the orthonormality condition:

$$\int_0^T u(t)\, \frac{\partial u(t)}{\partial u_k}\, dt = \int_0^T \left[\sum_{i=0}^N u_i P_i(t) \right] P_k(t)\, dt = u_k \tag{2.29}$$

$$\frac{\partial I}{\partial u_k} = 2 \int_0^T \left\{ w(t)[y^d(t) - y(t)] \left[-\int_0^T P_k(\tau)c(t,\tau)\, d\tau \right] \right\} dt$$

$$+ 2u_k = 0 \tag{2.30}$$

which yields the $N + 1$ necessary equations to evaluate u_k, $k = 0, 1, \ldots, N$.

The unknown quantities in (2.30) can be obtained as follows:

1. Use a predictor $p\,[\bullet]$ to evaluate the best estimate of $[y^d(t) - y(t)]$ from the past interval information.

2. Obtain the impulse response $c(t,\tau)$ by some identification scheme, as, for instance, the ones previously described. Define the function

$$K_k(t) = w(t) \int_0^t P_k(\tau) c(t,\tau) \ d\tau \qquad (2.31)$$

Then the coefficients of the controller are given by

$$u_k = \int_0^T K_k(t)[y^d(t) - y(t)] \ dt \qquad (2.32)$$

The three free parameters in the system may be defined as follows:

1. $w(t)$ to weight the error properly, i.e., $w(t) = At^n$.
2.   The control interval T, to ensure stability, and prediction and identification accuracy.
3.   The order of the expansion N to accurately represent the controller.

It was pointed out by Meditch and Gibson [2.35] that decreasing the length of the time interval T improves the quality of the control.

It should be pointed out again that the solution obtained is not overall time optimal and it requires a large number of terms $P_k(t)$ to model the control accurately. However, this method as well as the two discussed previously, propose general trends in S.O.C. such as the use of polynomials for modeling of unknown systems, the per-interval control for on-line optimization, and the use of gradient search algorithm for on-line identification and control.

2.3 POLYNOMIAL REPRESENTATION FOR SYSTEM IDENTIFICATION AND MODELING

The importance of the use of polynomial expansions to model the unknown system differential equations or impulse response, has been recognized in the previous sections and will be further explored in this section. Such expansions have been used by engineers for a long time, i.e., Fourier series for signal representation, etc. System representation by polynomial functions should be credited to N. Wiener [2.59]. This representation is still one of the most in- genious modeling procedures and will be given in the sequel. Since

polynomial series expansion will be used throughout the book a brief
definition of their properties is appropriate. The idea originates
from a theorem, due to Weierstrass, stating that under proper conditions closed-form unknown functions $f(t)$ may be represented by an
infinite sum of polynomial functions $f_i(t)$, defined over the same
time interval $t_a \leq t \leq t_b$ over which $f(t)$ is defined.

$$f(t) = \sum_{i=0}^{\infty} c_i f_i(t) \tag{2.33}$$

The functions $f_i(t)$ that satisfy the conditions of orthogonality
are called *orthogonal functions* and have very interesting properties
[2.37].

Definition 2.1

Two functions $f_i(t)$ and $f_j(t)$ are *orthogonal* with respect to a
weighting function $w(t)$, defined over $t_a \leq t \leq t_b$, if and only
if (iff):

$$\int_{t_a}^{t_b} w(t) f_i(t) f_j(t) \, dt = a\delta_{ij} \tag{2.34}$$

where δ_{ij} is the Kroennecker delta, $\delta_{ij} = 1$, $i = j$, $\delta_{ij} = 0$,
$i \neq j$, and a is some nonzero number.

Definition 2.2

If $a = 1$, the functions $f_i(t)$ and $f_j(t)$ are called "orthonormal."
Orthogonal functions are important because of the *finality of
their coefficients* c_i and the simplicity of their evaluation.
By the finality of the coefficients it is meant that their
evaluation is independent of each other and the length of the
series used to approximate $f(t)$. Such approximations assume
the form

$$\tilde{f}(t) = \sum_{i=0}^{n} c_i f_i(t) \tag{2.35}$$

where the length n is specified to yield a reasonable small
weighted integral square error:

$$\text{ISE} = \int_{t_a}^{t_b} w(t) [f(t) - \tilde{f}(t)]^2 \, dt \tag{2.36}$$

The coefficients c_i may be computed to minimize (2.35) for a
fixed n:

$$c_i = \int_{t_a}^{t_b} f(t) f_i(t) w(t) \, dt \qquad i = 0, \ 1, \cdots, n \qquad\qquad (2.37)$$

The simplicity of the form of (2.37) is due to the assumption that $f_i(t)$ are orthonormal functions of t.

For different intervals $t_a \leq t \leq t_b$ and weighting functions w(t) one may obtain different orthonormal polynomial expansions. Table 2.1 lists the most popular ones used in engineering practice, for more details see [2.37]. Although orthonormal functions appear complex, and although they are not necessary for the modeling of a function f(t), their properties and simplicity of implementation by actual hardware in the frequency domain makes them very attractive for modeling and unknown system impulse response or other function of time, if t represents time in f(t), or even for nonlinear differential equations in the state space if t is thought to be a vector of state variables and its derivatives. An example of such a modeling procedure for the unknown impulse response of a linear system, which emphasizes the hardware feasibility of a polynomial representation, is given next.

2.3.1 The Automatic Spectrum Analyzer

The automatic spectrum analyzer is a realization of an on-line identifier of the coefficients c_k, k = 1,...,n, of the approximation of a linear system's response y(t) over a finite interval $0 \leq t \leq T$.

$$y(t) = \sum_{k=1}^{n} c_k \, \phi_k(t) \qquad\qquad (2.38)$$

The method has been proposed by Truxal, Mishkin, and Braun and is presented in detail in [2.37]. A bank of linear systems with impulse responses $g_k(t)$, k = 1,...,n, is designed as shown in Fig. 2.7, and is called the "*spectrum analyzer.*" The $g_k(t)$ may be identified to represent transformations of orthonormal functions which would correspond to (2.38).

Assuming that $g_k(t)$ represent linear time-invariant systems:

Table 2.1. Orthonormal Polynomials and Their Coefficients.

Number	Type of Expansion	Interval	Weighting Factor, $w(t)$	nth Polynomial term	nth Expansion Coefficient
1	Fourier	$t_1 \le t \le t_1+T$	$w(t) = 1$	$\phi_n(t) = e^{j\frac{2\pi nt}{T}}$	$A_n = \frac{1}{T}\int_{t_1}^{t_1+T} f(\tau)e^{-j\frac{2\pi n\tau}{T}}dt$
2	Laguerre	$0 \le t < \infty$	$w(t) = e^{-t}$	$L_n(t) = \frac{1}{n!}e^t\frac{d^n}{dt^n}(t^n e^{-t})$	$A_n = \int_0^\infty f(\tau)L_n(\tau)e^{-t}dz$
3	Legendre	$-1 \le t \le 1$	$w(t) = 1$	$P_n(t) = \frac{\sqrt{2n+1}}{2^n\sqrt{2}n!}\frac{d^n}{dt^n}(t^2-1)^n$	$A_n = \int_{-1}^1 f(\tau)P_n(\tau)d\tau$
4	Tchebychev	$-1 \le t \le 1$	$w(t) = \frac{1}{\sqrt{1-t^2}}$	$T_n(t) = \frac{\sqrt{2}}{\pi}\cos(n\cos^{-1});\ n\ge1$	$A_n = \int_{-1}^{+1} f(\tau)T_n(\tau)\frac{1}{\sqrt{1-\tau^2}}d\tau$
5	Hermite	$-\infty < t < \infty$	$w(t) = e^{-t^2}$	$H_n(t) = \frac{(-1)^n}{\sqrt{2^n}n!}\frac{e^{t^2}}{\sqrt{\pi}}\frac{d^n}{dt^n}e^{-t^2}$	$A_n = \int_{-\infty}^\infty f(\tau)H_n(\tau)e^{-\tau^2}d\tau$
6	Jacobi	$0 \le t \le 1$	$w(t) = 1$	$J_n(t) = \frac{\sqrt{1+2n}}{n!}\frac{d^n}{dt^n}[t^n(1-t)^n]$	$A_n = \int_0^1 f(\tau)J_n(\tau)d\tau$

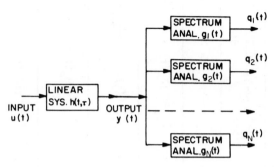

FIG. 2.7.   The Automatic Spectrum Analyzer as a System
            Response Identifier.

$$q_k(t) = \int_0^\infty y(\tau) g_k(t - \tau) \, d\tau \quad y(t) = 0, \ t < 0$$

and at t = T

$$q_k(T) = \int_0^\infty y(\tau) g_k(T - \tau) \, d\tau \qquad (2.39)$$

One may identify the following functions corresponding to the
definition of orthonormal functions and their evaluation Eqs. (2.33)-
(2.37):

$$y(t) = f(t)$$
$$g_k(T - t) = \phi_k(t) w(t) \qquad (2.40)$$

Then from (2.37):

$$c_k = q_k(T) \qquad (2.41)$$

From physical realizability conditions y(t) = 0 for t < 0.   This
implies $g_k(t) = 0$ for t < 0, and applied to (2.39) yields:

$$q_k(T) = \int_0^T y(\tau) g_k(T - \tau) \, d\tau \qquad (2.42)$$

This implies that the identification of c_k corresponds to the
functions:

$$y_1(t) = y(t) \quad 0 \leq t \leq T$$
$$y_1(t) = 0 \qquad t > T \qquad (2.43)$$

This means that, in order to recover complete information about
y(t), infinite identification time, T → ∞, is needed.   This, however,
is practically impossible, but T may be chosen large enough to

contain enough information about $y(t)$.

By change of variables in Eq. (2.40),

$$g_k(t) = \phi_k(T - t)w(T - t) \tag{2.44}$$

The spectrum analyzer can be implemented by choosing arbitrarily a set of n functions $\phi_k(t)$, $k = 1$, $\ldots$, n, with poles $s = \alpha_i$, $i = 1$, $\ldots$, n.

$$g_k(t) = r_{k1}e^{-\alpha_1 t} + \cdots + r_{kk}e^{-\alpha_k t} \tag{2.45}$$

where r_{ki} are evaluated from (2.37) to satisfy the orthogonality property with an appropriate weighting function $w(t)$,

$$r_{ki} = \sqrt{2(a + \alpha_i)} \quad i = 1, \ldots, n$$

$$w(t) = e^{-\alpha t} \tag{2.46}$$

Then the impulse responses of the spectrum analyzer may be computed by

$$g_k(t) = r_{k1}e^{-(\alpha_1 + 2a)T}e^{(\alpha_1 + 2a)t} + \cdots + r_{kk}e^{-(\alpha_k + 2a)T}e^{(\alpha_1 + 2a)t} \tag{2.47}$$

or in the frequency domain

$$G_k(s) = \frac{r_{k1}e^{-(\alpha_1 + 2a)T}}{s - \alpha_1 - 2a} + \cdots + \frac{r_{kk}e^{-(\alpha_k + 2a)T}}{s - \alpha_k - 2a} \tag{2.48}$$

It can be observed that, because of the reversal in real time to $T - t$, the filters composing the analyzer are unstable. However, this does not create a problem because the filters operate only in the finite interval $0 \leq t \leq T$. Therefore the filters must be clamped after T sec to avoid instability.

For further simplicity of implementation the filters in the analyzer can be built in a cascaded form by writing

$$G_1(s) = \frac{\sqrt{2}(a + \alpha_1) e^{-(\alpha_1 + 2a)T}}{s - \alpha_1 - 2a} = T_1(s)$$

$$G_2(s) = T_1(s) \cdot T_2(s) \tag{2.49}$$

$$\cdots \cdots \cdots \cdots \cdots \cdots \cdots$$

$$G_n(s) = T_1(s)T_2(s), \ldots, T_n(s)$$

The following example illustrates the method and its hardware implementation given in Fig. 2.8.

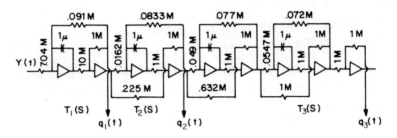

FIG. 2.8.   Hardware Implementation of the Spectrum Analyzer
            for y(t).

Example

An unknown response function $y(t)$ is to be identified.

Select $\alpha_n = n$, $n = 1, 2, 3$, $a = 5$, $T = 0.5$ sec.   Then,

$$\phi_1(s) = \sqrt{12}\ \frac{1}{s + 1}$$

$$\phi_2(s) = \sqrt{14}\ \frac{s - 11}{(s + 1)(s + 2)} = \sqrt{14}\left[\frac{13}{s + 2} - \frac{12}{s + 1}\right]$$

$$\phi_3(s) = \sqrt{16}\ \frac{(s - 11)(s - 12)}{(s + 1)(s + 2)(s + 3)} = \sqrt{16}\left[\frac{78}{s + 1} - \frac{182}{s + 2} + \frac{105}{s + 3}\right]$$

The spectrum analyzer is then composed by the cascaded $T_1(s)$, $T_2(s)$, and $T_3(s)$ filters:

$$G_1(s) = \frac{0.0142}{s - 11} = T_1(s)$$

$$G_2(s) = -0.0632\ \frac{s - 13.9}{(s - 11)(s - 12)} = -4.45\ \frac{s - 13.9}{s - 12}\ T_2(s)$$

$$= T_2(s)T_1(s)$$

$$G_3(s) = 0.1\frac{(s - 12.9)(s - 18.3)}{(s - 11)(s - 12)(s - 13)} = -1.58\ \frac{(s - 12.9)(s - 18.3)}{(s - 13)(s - 13.9)}$$

$$T_2(s)T_1(s) = T_3(s)$$

$T_1(s)$, $T_2(s)$, and $T_3(s)$ may be constructed with active elements, and the function $y(t)$ may be reconstructed by

$$y(t) = c_1\phi_1(t) + c_2\phi_2(t) + c_3\phi_3(T) = r_1e^{-t} + r_2e^{-2t} + r_3e^{-3t}$$

where $c_1 = q_1(T)$, $c_2 = q_2(T)$, $c_3 = q_3(T)$, and r_i are defined accordingly from (2.46). Fig. 2.8 shows the details of the design.

Table 2.2   Comparison of Identified Coefficients for Various
Terminal Times T.

	T = ∞	T = 0.5 sec	T = 0.1 sec
c_1	0.315	0.314	0.210
c_2	-0.0284	-0.0218	0.1095
c_3	0.00466	0.00746	0.036

The effect of the identification time on the accuracy of the
coefficients is demonstrated in Table 2.2. It is obvious that after
T = 0.5 sec the coefficients are almost close to the optimum
estimates (T = ∞). However, this is only an approximation since only
three terms of the series were included. If y(t) were a step, the
optimum $f_3(t)_{T = \infty}$ is within 1% for $0 \le t \le 0.5$, and $f_3(t)_T = 0.5$
is within 3% in the same range, while the $f_3(t)_T = 0.5 = 0.1$ gives a
poor approximation.

The polynomial representation used for this problem is by no
means unique. Instead of selecting $g_k(t)$ by (2.45), one may select
other orthonormal functions from Table 2.1 defined over the same
interval of definition of y(t), i.e., Leguerre polynomials. Such
modifications are discussed in detail in [2.37].

The method presented here deals with linear systems only. A
generalization due to Wiener to characterize nonlinear systems as
well as discussed briefly in the next section.

2.3.2. The Wiener Characterization for Nonlinear Systems

The identification of a process implies finding the input-
output relation of the process. For a linear system this is
characterized by the impulse response. For a nonlinear system,
however, the relation is much more complicated. It is usually an
unknown function of the input and the nonlinearities, and is extremely
difficult to be identified mathematically.

Wiener proposed a model for a nonlinear stable system which was
modified for on-line identification in [2.55]. Because of the

dependence of the model on the input signal and the system structure,
the method suggests separate models for the input signal and the
structure of the system coupled together through the last expansion.
The input signal is approximated by a Laguerre expansion, with
coefficients from Table 2.1, and is depicted in a block diagram in
Fig. 2.9:

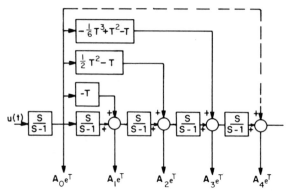

FIG. 2.9.   Laguerre Network for Identification of Coefficients
of a Polynomial Expansion.

$$\hat{u}(t) = A_0 L_0(t) + A_1 L_1(t) + \cdots + A_n L_n(t) \qquad (2.50)$$

Then the output may be approximated, according to Wiener, by a
multivariable finite Hermite series expansion, with coefficients
also given in Table 2.1:

$$\hat{y}(t) = \sum_{a=0}^{Na} \sum_{b=0}^{Nb} \cdots \sum_{k=0}^{Nk} c_{a,b,\cdots,k} \, H_a(A_0) H_b(A_1) \cdots H_k(A_n)$$

$$(2.51)$$

where $c_{a,b,\ldots,k}$ are the identifying coefficients of the expansion
and $H_i(A_j)$, is the ith Hermite polynomial, a function of the Laguerre
coefficients A_j. The unknown factors in this expansion are the
Laguerre coefficients A_i and the identifying coefficients $c_{a,b,\ldots,k}$.
     The identifying coefficients of the Hermite expansion may be
computed by the following expression whenever the system is excited
by Gaussian noise:

$$c_{a,b,\ldots,k} = (2\pi)^{n/2} \lim_{T \to \infty} \frac{1}{2T} \int_{-t}^{T} y(t) H_a(A_0) H_b(A_1)$$

$$\cdots H_k(A_n) e^{-\left[\frac{A_0^2}{2} + \frac{A_n^2}{2}\right]} dt \tag{2.52}$$

or by some similar relation if excited otherwise.

For on-line implementation and in the case of an arbitrary but measureable input a gradient search may be used to determine $c_{a,b,\ldots,k}$ by minimizing the square error criterion

$$I^2 = (y(t) - \hat{y}(t))^2 \tag{2.53}$$

which is unimodal in $c_{a,b,\ldots,k}$.

The proposed scheme would generate the following coefficients of the expansion:

$$\frac{dc_{a,b,\ldots,k}}{dt} = -k \frac{\partial I^2}{\partial c_{a,b,\ldots,k}} = +2kI\ H_a(A_0) H_b(A_1) \cdots H_k(A_n) \tag{2.54}$$

The analog diagram of the identification scheme is illustrated in Fig. 2.10 for two Laguerre polynomials and two Hermite polynomials.

The shortcomings of the method arise from the necessity to use a large number of terms to model the process accurately. The complexity of such a task is obvious from the two-plus-two-term expansion of the example of Fig. 2.10. Therefore, inspite of its elegance, the method has found only limited applications.

2.4 FUNCTIONAL APPROACH FOR SELF-ORGANIZING CONTROL

A general approach to S.O.C. problems can be generated when the system to be controled is represented in a functional form [2.33]. Such an approach was originated by Kulikowski's pioneering work and was extended by Sarachic [2.50], Zaborsky and Humphrey [2.63], Pearson [2.45-2.49], and others.

Since the mathematics involved in the presentation of this method are beyond the level of this book, only a superficial

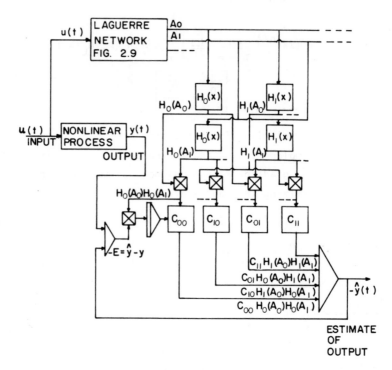

FIG. 2.10.   On-line Nonlinear System Modeling, Using Laguerre
             and Hermite Polynomials.

discussion will be given here. The purpose of this presentation is
to account for a different philosophy for S.O.C. and therefore only
the original work of Sarachic and Pearson is discussed. For more
detailed and up-to-date presentation of this material the reader is
referred to Refs. [2.47] and [2.49].

2.4.1. The Continuous-Time Case
        The problem of optimal design of S.O.C. system may be treated
as the problem of minimization of a cost functional,

$$J(u) = \int_{t_0}^{t_f} L(u(t), y(t), y^d(t)) \quad dt \tag{2.55}$$

subject to the constraints defined by the equation of motion of the
system

$$y(t) = A u(t) \tag{2.56}$$

where $y^d(t)$ is the desired output, $y(t)$ is the actual output of the system, $u(t)$ is the unknown control input to minimize $J(u)$, and $A[\cdot]$ is an unknown nonlinear integrodifferential operator characterizing the unknown dynamics of the plant.

The necessary and sufficient conditions for the minimization of $J(u)$ are

$$dJ(u,h) = \int_{t_o}^{t_f} h(t) \nabla J(u) \, dt = 0 \tag{2.57}$$

$$d^2 J(u,h) > 0$$

where $d^n J(u,h)$ is the nth Frechet differential of $J(u)$ defined by

$$d^n J(u,h) = \left. \frac{\partial^n J(u+ ,h_1+ \cdots + _n h_n)}{\partial \gamma_1 \cdots \partial \gamma_n} \right|_{\gamma_1 = \cdots = \gamma_n = 0} \tag{2.58}$$

Condition (2.57) implies

$$\nabla J(u) = 0$$

which depends implicitly on the unknown operator $dA*(u, \partial G/\partial y)$, adjoint to the differential (Frechet) operator $dA(u,h)$. Pearson [2.45] proposed to identify this operator by input-output measurements and generate sequentially a controller which converges to the optimal by:

$$u_{n+1}(t) = u_n(t) - \gamma \nabla(u_n) \tag{2.59}$$

In order that this controller be implemented on-line, i.e., $dA*(u_n, \partial G/\partial y)$ be identified and used to obtain $\nabla(u_n)$, the following assumptions are necessary:

1.   The process should be *infinite time*, $t_f = \infty$.

2.   The process should reach a *steady state* before the identification starts.

3.   The minimization procedure must be redefined on a *per-interval basis*, i.e.,

$$J_n'(u) = \int_{t_n}^{t_n+T} L(u,y,y^d) \, dt \quad t_n = hT \tag{2.60}$$

where T is a fixed time interval of the subprocess, repeated for the generation of the optimizing sequence (2.59).

The procedure for such a minimization, even under the above strict assumptions is hard to implement.

2.4.2. The Discrete-time Case

The complexity of the problem is considerably reduced if the discrete version of the problem is considered as Sarachic did in Ref. [2.50], where:

$$J(u) = \sum_{k=0}^{N-1} L(y(k), y_d(k), u(k)) \qquad\qquad (2.61)$$

$$Y^{N-1} = B[U^{N-1}] \quad 0 \le k \le N-1 \qquad\qquad (2.62)$$

In this discrete-time equivalent formulation of (2.55) and (2.56), $y(k)$ and $u(k)$ are the scalar output and input to the system, respectively, at the discrete time $k\Delta t$, while $Y^{N-1} = \{y(0),\ldots,y(N-1)\}^T$ and $U^{N-1} = \{u(0),\ldots,u(N-1)\}^T$ represent in a vector form the respective time sequences and $B[\cdot]$ is a discrete-time nonlinear operator. Minimization of (2.61) implies the selection of the optimal sequence U^{*N-1} such that:

$$J(u^*) = 0$$

One way to generate the control sequence sequentially is given by:

$$U_{n-1}^{N-1} = U_n^{N-1} - \gamma \ \nabla J(U_n^{N-1}) \quad n = 1, 2, \ldots \qquad\qquad (2.63)$$

which should converge to the optimal U^{N-1*} as $n \to \infty$. As with the continuous case, ∇J must be identified from input-output records which are provided periodically. In this case the optimization of (2.61) must be considered as an interval optimization over $0 \le K \le N-1$ and repeated as the identification progresses. In order to avoid dependence on initial conditions, the identification must be performed after the operator $B[\cdot]$ has attained its steady-state value.

The gradient $\nabla J(u)$ is computed from

$$\nabla J(u) = \frac{\partial L}{\partial u} + \frac{\partial B}{\partial u}^T \frac{\partial L}{\partial y} \qquad\qquad (2.64)$$

where the matrix $\partial B/\partial u$ may be approximated in terms of its Gateaux derivative defined by

$$\frac{\partial B}{\partial u_i} = \lim_{\gamma \to 0} \left[\frac{B(u+\gamma\varepsilon_i) - B(u)}{\gamma} \right] \cong \frac{Y_i^{N-1} - Y^{N-1}}{\gamma}$$

and Y_i^{N-1} is the response of the system to an input sequence $U^{N-1} + \gamma\varepsilon_i$, and $\varepsilon_i^T = (0,0,\ldots,1 \text{ (ith term)},\ldots,0)$. It takes $(N + 1)$ iterations of the process to evaluate the gradient of B that is used in (2.63) to iterate the control sequence.

The identification described above cannot be used for systems containing an integrator, since an input with nonzero average value causes the output to grow out of bounds, thus steady state can never be established. This can be handled by using inputs with zero average values such as

$$(U^{N-1}{}^T, 0, -U^{N-1}{}^T, 0)$$

This scheme consumes considerable amount of time for the identification since only one out of $(N + 2)$ evaluations of the process is used for the improvement of the controller.

2.4.3. The Fixed-Control Configuration

In many S.O.C. systems the form of the controller is prespecified by the designer with the provision of a parameter vector K left free to be adjusted by the S.O.C. policy. A modified gradient scheme is proposed [2.46], in which the adjustable parameter vector β is held constant over time intervals of nonzero length in order to allow for a realistic assessment of the system's performance with respect to the current value of β before switching to a new value. The discrete-time adjustments of β are performed by the use of an approximation $S(K)$ of the true gradient $\nabla J(K)$ of the *per-interval* performance index

$$J(K) = \int_{t_i}^{t_{i+1}} L(y(t), y^d(t), t) \ dt \tag{2.65}$$

subject to the differential equations of motion of the system:

$$\dot{x} = f(x, r(t), u(x(t), \beta, t), \alpha(t), t) \quad x(0) = x_0 \tag{2.66}$$

$$y = Hx$$

where $x(t)$ is the n-state vector, $y(t)$ is the vector output, $y^d(t)$ is the unknown desired output, $u(x(t),\beta,t)$ is the control input of known structure, possibly depending on the state x, the adjustable parameter β, $r(t)$ is the input to the system, and $\alpha(t)$ is a vector of unknown system parameters. The scheme is given in the block diagram of Fig. 2.11.

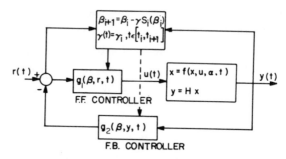

FIG. 2.11.   Block Diagram of the Fixed Control Configuration
             S.0.Control System by Functional Approach Methods.

The algorithm for updating the parameters β at the end of every fixed interval $t_i \leq t \leq t_{i+1}$ is given by

$$\beta_{i+1} = \beta_i - \gamma S_i(\beta_i) \tag{2.67}$$

The vector $S(\beta_i)$ representing an approximation of the gradient vector $\nabla J(\beta)$, which is not directly available, can be computed by the following procedure:

$$S(\beta) = \int_{t_i}^{t_{i+1}} p^T(t) \frac{\partial L}{\partial Y} (y, y^d, t) \, dt \tag{2.68}$$

where the approximate sensitivity matrix $P(t)$ of the output vector $y(t)$ to variation of β is obtained by

$$\dot{v}(t) = \hat{Q}(t) v(t) + \hat{R}(t) \qquad v(t_i) = 0 \tag{2.69}$$

$$P(t) = H v(t)$$

where the matrices $\hat{Q}(t)$ and $\hat{R}(t)$ are defined as the Jacobian and sensitivity matrices of the system, respectively, with the unknown parameter vector α replaced by a given estimate $\hat{\alpha}$:

$$\hat{Q}(t) = \frac{\partial f}{\partial x} (\hat{x}(t), \hat{r}(t), \beta, \hat{\alpha}, t) \qquad (2.70)$$

$$\hat{R}(t) = \frac{\partial f}{\partial \beta} (\hat{x}(t), \hat{r}(t), \beta, \hat{\alpha}, t)$$

If estimates of α are not available, reasonable values of these quantities have been used [2.49] to generate $S(\beta)$, i.e., $\hat{\alpha} = 0$.

Equations (2.69) must be solved on-line at every interval with zero initial conditions to generate $S(k)$. The parameter γ and the intervals (t_i, t_{i+1}) must be specified to satisfy the theoretical condition for *error correctiveness* of the algorithm [2.48].

Such a condition is satisfied if:

$$\text{Min } S^T(\beta) \nabla Q(t) > 0 \qquad (2.71)$$

in which case the algorithm is error corrective. Pearson and his colleagues have further improved this algorithm; their recent research results can be found in Ref. [2.47].

The method presents many difficulties for implementation, especially if it is generalized to include systems operating in a stochastic environment [2.49], [2.50].

2.5   MODEL REFERENCE ADAPTIVE CONTROL SYSTEMS

The next and last school of thought on S.O.C. system to be discussed carries the name of *Model Reference Adaptive* by its creators. Conceptually it represents a Performance Adaptive approach since it updates the control parameters directly to improve the system's performance, and has been attracting considerable attention by researchers since stability has been considered as part of the algorithm and Lyapunov functions have been used to assist the design.

The Model Reference Adaptive Control was first suggested by Whitaker and his colleagues at the MIT Instrumentation Laboratories in Cambridge, Mass., as a solution to the adaptive autopilot problem. The original work [2.57], as well as its modifications [2.58], is more experimental than theoretical. Since then it has been modified on a more theoretical basis by Dressler [2.12] and later by Hsia and Vimolvanich [2.25] who used state variables to represent the plant.

Finally Parks [2.44] redesigned the Model Reference Adaptive algorithm
using Lyapunov functions to account for the stability along with
adaptation of the controller. Since then Monopoli and his colleagues
[2.17], [2.18], [2.31], [2.39], [2.40], Landau [2.29], and Shen and
his colleagues [2.54] have been improving the method with their
numerous contributions. As before, only the algorithms which lead
the major philosophical impact on the method are briefly discussed in
the sequel. For more detailed and up-to-date presentations the reader
is referred to the bibliography at the end of this chapter.

2.5.1. The MIT Model Reference Adaptive Control System

It is appropriate to describe the MIT model reference adaptive
control system as the first Performance Adaptive S.O.C. in the
sequence [2.43], [2.44]. The performance index is defined as the ac-
cummulated error square between the desired output generated by passing
the input of the system through a carefully predesigned model which
satisfies all the required specifications for the plant and the actual
output of the plant:

$$J = \int_0^T ||y^d(t) - y(t)||^2 \ dt \qquad\qquad (2.72)$$

In the case of a linear system written in a transfer-function
form

$$Y_d(s) = M(s)R(s)$$
$$Y(s) = \left[I + G_p(s,\alpha)G_c(s,\beta)G_F(s,\beta)\right]^{-1} G_p(s,\alpha)G_c(s,\beta)R(s) =$$
$$= H(s,\alpha,\beta)R(s) \qquad\qquad (2.73)$$

where

$M(s)$ is the model transfer matrix,

$G_p(s,\alpha)$ is the plant transfer matrix with unknown parameters
vector α, and $G_c(s,\beta)$, $G_F(s,\beta)$ are the transfer matrices of the
fixed controllers with β the vector of adjustable parameters.

The whole system is represented in Fig. 2.12. For an exact
adjustment $M(s)$ should be of the same order as $H(s)$. However, it
has been conjectured that if the order of the model is smaller than

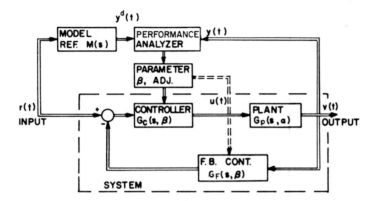

FIG. 2.12. Block Diagram of a Generalized Model Reference
Adaptive Control System.

that of $H(s)$, then the model reference acts in the best way to match
the response of the system. However, this may eventually lead to
pole-zero cancellation between plant and controllers. Therefore,
special consideration is required when the plant is inherently
unstable.

Since the minimization of $J(\beta)$ must be performed on-line, the
performance evaluation over the whole interval is meaningless.
Therefore, a *per-interval* performance criterion is specified by

$$J_p(\beta) = \frac{1}{2} \left|\left| y^d(t) - y(t) \right|\right|^2 \tag{2.74}$$

for a sufficiently small Δt. Such a performance index is dependent
on time and it must be minimized with respect to β by using a search
technique only after it assumed its steady-state value or for
practical purposes when it changes with time much slower than the
rate of the search technique used. In this latter case one obtains
a tracking system. The standard gradient search algorithm, may be
used to generate estimates of the parameters, i.e.,

$$\dot{\beta}(t) = -\gamma \frac{\partial J}{\partial \beta} = +\gamma \frac{\partial y}{\partial \beta}^T c(t) \tag{2.75}$$

where $e(t) \triangleq y^d(t) - y(t)$ $\tag{2.76}$

The sensitivity matrix $\partial y/\partial \beta$ of the output y with respect to a variation of β may be obtained by differentiating the closed-loop transfer matrix $H(s,\alpha,\beta)$ with respect to β. From (2.73), we obtain

$$\frac{\partial y(s)}{\partial \beta} = \frac{\partial H(s,\alpha,\beta)}{\partial \beta} R(s) \tag{2.77}$$

However, since the parameter vector α is not known, the sensitivity matrix cannot be computed exactly. Instead various approximations must be used.

In the original work by Whitaker and his colleagues [2.43], the performance of the system is assumed to be sufficiently close to the performance of the model. It is also assumed that the order of the model is the same as the order of the plant. In this case, the sensitivity matrix may be obtained from the model. In general Osburn, Whitaker, and Kezer [2.43] propose for a single-input/single-output system the following expression for the components of the sensitivity vector:

$$\frac{\partial e}{\partial \beta_i} = M(s)G_{Mi}(s)v_i \tag{2.78}$$

where $M(s)$ is the transfer function of the model, $G_{Mi}(s)$ is the backward transfer function obtained by the partial differentiation of the system and referred to the model, and v_i are the signals at the inputs of the adjustable parameters of the system assumed measurable.

The method has been further improved at MIT [2.58].

Example

The method is illustrated by a model reference adaptive control system for a roll angle control of an aircraft given in Fig. 2.13.

$$H(s) = \frac{Y(s)}{R(s)} = \frac{\beta_1 G_s(s)G_A(s)}{1 + \beta_1 G_s(s)G_A(s)(1 + \beta_2 G_G(s))}$$

$$G_S = \frac{\beta_1}{1 + T_s S} \qquad G_A = \frac{K_A}{S(1 + T_A S)} \qquad G_G = \beta_2 GS$$

$$J = e^2(t) = (y^d - y)^2 \qquad y^d = u_{-1}(t) - u_{-1}(t - \sigma)$$

$$\frac{\partial e}{\partial x_i} = H(s)G_{Mi}(s)v_i \quad i = 1, 2, \qquad G_{M1} = \frac{1}{\beta_1}, \; G_{M2} = 1$$

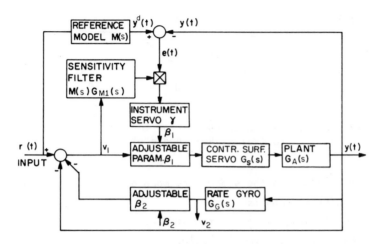

FIG. 2.13.   Block Diagram of Model Reference Adaptive Control
for Roll Angle Control of an Aircraft.

and

$$\frac{\partial e}{\partial \beta_1} = H(s)\frac{1}{\beta_1} v_1 \qquad\qquad \frac{\partial e}{\partial \beta_1} \simeq M(s)\frac{1}{\beta_1} v_1$$

or

$$\frac{\partial e}{\partial \beta_2} = H(s)v_2 \qquad\qquad \frac{\partial e}{\partial \beta_2} \simeq M(s)v_2$$

where the sensitivity functions have been approximated by the
respective functions in the model.

Then the parameter estimation algorithm (2.76) is defined by

$$\frac{d\beta_1}{dt} = -\gamma\frac{\partial e}{\partial \beta_1} e(t) = -\gamma\left[M*\frac{1}{k_1} v_1\right] e(t)$$

$$\frac{d\beta_2}{dt} = -\gamma\frac{\partial e}{\partial \beta_2} e(t) = -\gamma\left[M* v_2\right] e(t)$$

where * stands for convolution of the time functions it associates.
The following assumptions have been made:

1.   The condition that the model is the same order as the system is
       relaxed (second-order model, second-order system).

2.   Initial conditions are zero.

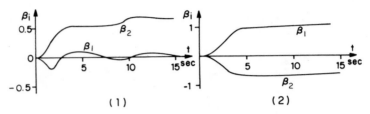

FIG. 2.14.   Parameter Adaptation for Model Reference Adaptive
Control of Roll Angle of an Aircraft.

3.   The changes in the system are much slower than the ones in the
adjusting mechanism.
4.   The signals v_1, v_2 are measurable.
5.   The overall system is stable.
The properties of parameter adaptation of the roll angle control
problem are given in Fig. 2.14 for two different flight conditions.
The results are very satisfactory because of the simplicity of the
model, the justifiable approximations, and the availability of the
measurements v_i.   Any attempt to apply the method to higher-order
systems is plagued with unsurmountable difficulties.

2.5.2.   Dressler's Model Reference Adaptive Control

The assumptions made for the MIT model reference adaptive
control system involved in obtaining the sensitivity functions for
the adjustment of the parameters are too restrictive for a general-
purpose S.O.C. system.   This lead Dressler [2.12] to relax the strict
minimization condition on the index of performance and consider
instead an *error corrective* approach to improve the system's
performance.   This idea is similar to the one explored by Pearson
in [2.48],   and suggests a stability analysis for the convergence
of the algorithm.

The approach described here deals with a linear system with an
unknown parameter vector α, and with a fixed configuration for a
controller with adjustable parameter β that has been incorporated
in the differential equations of the system:

$$\dot{x}(t) = F_s(\alpha, \beta, t)x(t) + B_s(\alpha, \beta, t)r(t)$$

$$y(t) = h^T x(t) \tag{2.79}$$

where F_s and B_s are the system's matrices of dimensions nxn and nxr, respectively, containing the unknown and adjustable parameters, and h is an n-dimensional known vector. Assume that the desired performance is formulated in terms of a set of differential equations *of the same order as the plant* called the "model reference," which is considered as an implicit characterization of a performance criterion:

$$\dot{z}(t) = F_D z(t) + B_D r(t)$$

$$y_D(t) = h^T z(t) \tag{2.80}$$

where F_D and D_D are known nxn and nxr matrices. Define the output error

$$e(t) = y(t) - y^d(t) = h^T(x(t) - z(t)) \tag{2.81}$$

This error can be made equal to zero, $e(t) \equiv 0$ at the steady state, if the plant matrices are made to match the model matrices $F_s(t) = F_D$ and $B_s(t) = D_D$ by properly adjusting the vector β to account for the changes of the unknown parameters $\alpha(t)$.

However, setting the error equal to zero is not always necessary since the error correction requires only an error reduction at every step of the process.

The following decomposition of the system matrices is assumed at this point:

$$F_s(t) = F_D + \varepsilon F_\delta(t, \beta, \alpha)$$

$$B_s(t) = B_D + \varepsilon B_\delta(t, \beta, \alpha) \tag{2.82}$$

where F_δ and B_δ contain the adjustable and unknown parameters and are considered as perturbations of the desired matrices F_D and B_D, and ε is a scalar constant. Using (2.82), the system's response is written by the method of successive approximations:

$$x(t) = \Phi_D(t - t_0)x(t_0) + \int_{t_0}^{t} \Phi_D(t - \tau)B_D r(\tau) \, d\tau$$

$$+ \, \varepsilon \int_{t_0}^{t} \Phi_D(t - \tau) \cdot \left[B_\delta(\tau)r(\tau) + F_D(\tau) \left[\Phi_D(\tau - t_0)x(t_0) \right. \right.$$

$$\left. \left. + \, \varepsilon \int_{t_0}^{\tau} \Phi_D(t - \sigma)B_D r(\sigma) d\sigma \right] \right] \, d\tau + o(\varepsilon) \qquad (2.83)$$

Assuming that the system has reached steady state, in which case the influence of the initial condition on the system's response are negligible, and that F_δ, and B_δ are continuous over the interval $t_0 \le t < \infty$, the output error (2.81) is written as:

$$e(t) = h^T \int_{t_0}^{t} \Phi_D(t - \tau)[\varepsilon B_\delta(\tau)r(\tau) + \varepsilon F_\delta(\tau)y(\tau)] \, d\tau + o(\varepsilon)$$

$$(2.84)$$

Equation (2.84) defined the functional dependence of the parameters β to the error $e(t)$ through F_δ and B_δ.

It is reasonable to assume that ε is small enough so that

$$o(\varepsilon) = \frac{f(\varepsilon)}{\varepsilon} \approx 0$$

in which case the model closely approximates the systems, and that the parameter $\alpha(t)$ changes slowly relative to the adjustable parameter β.

Define

$$\Delta e(t) = e(t + \Delta t) - e(t) \quad \Delta t > 0 \qquad (2.85)$$

Then the *error-corrective* condition reads:

$$e(t) \cdot \Delta e(t) \le 0 \qquad (2.86)$$

implying reduction of the error at t, whenever satisfied, $\Delta e(t)$ may be rewritten as

$$\Delta e(t) = w(t) + \Delta_2 e(t) \qquad (2.87)$$

where $w(t)$ is a function independent of the adjustable parameters Δ and $\Delta_2 e(t)$ represents the second-order terms of the error expansion [2.12] :

$$\Delta_2 e(t) = \frac{\Delta t^2}{2} \left[\sum_{i=1}^{n} \sum_{j=1}^{n} a_i f_{ij}(t)y_j(t) + \sum_{m=1}^{n} \sum_{\ell=1}^{r} a_m b_{m\ell} r_\ell(t) \right] \qquad (2.88)$$

where f_{ij} and $b_{m\ell}$ are elements of F_δ and B_δ, respectively, and a_k is an element of $\phi_D^T(\Delta t)h$; $\Delta_2 e(t)$ contains the terms with the adjustable parameters β.

Instead of (2.86), consider now the following special error-corrective condition:

$$e(t)\, \Delta_2 e(t) \leq 0 \qquad\qquad (2.89)$$

This is satisfied if

$$\dot{f}_{ij}(t) = -\gamma_{ij} a_i y_j(t) e(t)$$
$$\dot{b}_{n\ell}(t) = -\mu_{m\ell} a_m r_\ell(t) e(t) \qquad\qquad (2.90)$$

where γ_{ij} and $\mu_{m\ell}$ are positive constants chosen large enough so that (2.89) will be predominant term over $e(t)w(t)$ in (2.86), so that the error-corrective condition is satisfied.

The convergence of the scheme is guaranteed by the stability of the system of the combined Eqs. (2.80) and (2.90). A stability analysis is therefore necessary to define the regions of stability. Lyapunov's direct method [2.15] is most appropriate for such stability analysis.

Example

The following example is presented to illustrate Dressler's error-corrective method. Let the following differential equation describe the plant and its controller:

$$\dot{x}(t) = [-\beta(t) - f_1(t)]\, x(t) + r(t)$$

where $f_1(t)$ is the unknown plant parameter and $\beta(t)$ the adjustable control parameter. Let the following model describe the desired performance of the system:

$$\dot{z}(t) = -f_2\, z(t) + r(t) \qquad f_2 > 0$$

and $e = x - z$

The system equation may be rewritten as

$$\dot{x} = [-f_2 + f(t)]\, x + r(t)$$

where $f(t) = -\beta(t) - f_1(t) + f_2$

The parameter adjusting algorithm (2.90) yields:

$$\dot{\beta}(t) = \gamma_F' z(t) e(t) \qquad \gamma_F' > 0$$

The system is depicted in Fig. 2.15.

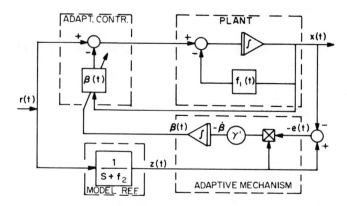

FIG. 2.15.   Block Diagram of a First Order Example of Dressler
            Model Reference Error Corrective Algorithm.

For stability purposes one may consider $e(t)$ and $f(t)$ as the
states of the system.   Let

$$w(t) = \begin{bmatrix} f(t) \\ e(t) \end{bmatrix}$$

Substituting into the previous equations and assuming that
$\dot{f}_1 = 0$, one gets

$$\dot{w}(t) = \begin{bmatrix} 0 & -\gamma_F' y(t) \\ y(t) & -f_2 \end{bmatrix} w(t) + \begin{bmatrix} 0 \\ f(t)e(t) \end{bmatrix}$$

which is a nonlinear differential equation.

Stability investigation of the above equation yields the
number of equilibrium states (one in this case), their stability
situation (asymptotically stable in this case, and the regions around
them for which the system is asymptotically stable, i.e., that the
algorithm converges.

A quadratic Lyapunov function gives sufficient conditions for
the existence of such regions:

$$V(w) = \gamma_F' e^2 + f^2 > 0$$

$$\dot{V}(w) = w^T \dot{w} = -2 \gamma_F' e^2 (f_2 - f(t))$$

It may be shown that $\dot{V} \le 0$ in the region bounded by the ellipse
$V = f_2^2$.

Further discussion on the stability investigation may be found
in the original work of Dressler [2.12]. However, the realization that
convergence of the search algorithm can be incorporated with the
stability investigation of the system suggests a method for designing
model reference-adaptive control systems which will be discussed in
the next section.

2.5.3. The Lyapunov Redesign of Model Reference Adaptive Control

The purpose of the model reference adaptive control algorithm is
to adjust the control parameters dynamically to force the square
error between the model and actual output to converge to zero,
provided that the overall system is stable. Since the reference
model is always assumed to be stable, the above two conditions
governing the dynamics of the error and the parameter adjustment
must converge uniformly to the desired values as $t \to \infty$. This
observation may be used to repose the problem as an asymptotic
stability problem and use Lyapunov's direct method [2.15] for its
investigation. This approach is very important in minimizing the
instantaneous square error criterion between the model and the
actual outputs which forms a nonstationary function in time. The
original work is credited to Parks [2.44], while some generalizations
have been made by Winsor and Roy [2.61], and certain limitations
have been pointed out. Since then Monopoli, Lindorff, Gilbart, and
others [2.17], [2.18], [2.31], [2.39], [2.40] have considerably improved
the method by using more general Lyapunov functions and applying them
to more general systems. Lately Landau [2.29] has used the concept
of hyperstability to implement the method.

A generalized Lyapunov redesign of the Model Reference Adaptive
Control method, was suggested by Parks and was considered as a
performance adaptive S.O.C. technique by Saridis and Kitahara [2.51].
This method is presented here, using Parks' problem is treated as an
illustrative example. A block diagram of the system is given in
Fig. 2.16.

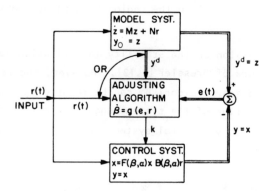

FIG. 2.16.   Lyapunov Design Model Reference Adaptive Control
              System.

Even though the method is applicable to nonlinear systems for
which a Lyapunov function can be generated, it will treat a linear
system with an unknown parameter vector α:

$$\dot{x} = A(\alpha)x + D(\alpha)u \quad x(t_0) = x_0$$

$$y = x \tag{2.91}$$

where $x(t)$ is the state vector available for measurement, $u(t)$ is
the control input, and A and D are matrices of appropriate
dimensions. Defining a controller to be a linear function of the
states and the input $r(t)$ with adjustable parameters β,

$$u = C(\beta)x + E(\beta)r(t) \tag{2.92}$$

the closed-loop system resulting from the combination of (2.91) and
(2.92) is represented by the following system equation:

$$\dot{x} = F(\beta,\alpha)x + B(\beta,\alpha)r \quad x(t_0) = x_0$$

$$y = x \tag{2.93}$$

where $F(\beta,\alpha) \triangleq A(\alpha) + D(\alpha)C(\beta)$ and $B(\beta,\alpha) \triangleq D(\alpha)E(\beta)$.

This system is required to follow as closely as possible the
model system which is linear stable and of the *same order* as the
plant.

$$\dot{z} = Mz + Nr \quad z(t_0) = z_0$$

$$y^d = z \tag{2.94}$$

It must be noticed that all the states of both the model and the systems are available. In case that they are not, a set of filters similar to the ones described in Sec. 2.2.2 that can generate the state vector may be used.

Define the error vector

$$e(t) \underset{=}{\Delta} z - x = y^d - y$$

The differential equation that governs the error becomes

$$\dot{e} = Me + [M - F(\beta,\alpha)] x + [N - B(\beta,\alpha)] r$$

or

$$\dot{e} = Me + \Delta Fx + \Delta Br \tag{2.95}$$

where

$$\Delta F \underset{=}{\Delta} M - F(\beta,\alpha) = \{\delta f_{ij}\} \quad \Delta B = N - B(\beta,\alpha) = \{\delta b_{ij}\} \tag{2.96}$$

It is obvious that the matrices ΔF and ΔB contain the adjustable parameters and therefore they can be varied with time according to the differential equations that will force the system to track the model. Actually, in order that $x(t)$ matches the states of the model $z(t)$, i.e., $y(t) \to y^d(t)$, or equivalently the error $e(t)$ converges to zero, the elements δf_{ij} and δb_{ij} of the matrices ΔF and ΔB, respectively, must converge to zero as $t \to \infty$.

It is realistic to assume that the changes of the tracking parameters δf_{ij} and δb_{ij} are governed by a set of differential equations with state variables δf_{ij} and δb_{ij} in the augmented state space $\Omega_A = \{e, \delta f_{ij}, \delta b_{ij}\}$. Asymptotic stability of the augmented system of differential equations will guarantee perfect tracking of the desired model output, inspite of unequal initial conditions $x_0 \neq z_0$, as well as stable performance of the overall system provided that the model has been designed to be asymptotically stable.

Asymptotic stability in the augmented state space may be investigated by considering a Lyapunov function

$$V(e, \delta f_{ij}, \delta b_{ij}) = e^T Qe + \sum_{i=1}^{m} \sum_{j=1}^{m} \frac{1}{v_{ij}} \delta f_{ij}^2 + \sum_{i=1}^{m} \sum_{j=1}^{m} \frac{1}{w_{ij}} \delta b_{ij}^2 > 0 \tag{2.97}$$

where $Q = Q^T > 0$, $v_{ij} > 0$, $w_{ij} > 0$, $i = 1, \ldots, m$, and $j = 1, \ldots, m$. The total derivative of V along the system trajectories is

$$\dot{V}(e, \delta f_{ij}, \delta b_{ij}) = e^T[M^TQ + QM]e + 2e^TQ\Delta Fx + 2e^TQBr$$

$$+ 2 \sum_{i=1}^{m} \sum_{j=1}^{m} \frac{1}{v_{ij}} \delta f_{ij} \delta \dot{f}_{ij} + 2 \sum_{i=1}^{m} \sum_{j=1}^{r} \frac{1}{w_{ij}} \delta b_{ij} \delta \dot{b}_{ij} \tag{2.98}$$

But

$$e^TQ\Delta Fx = (e_1 q_{11} + \cdots + e_m q_{m1})(\delta f_{11} x_1 + \cdots + \delta f_{1m} x_m) + \cdots$$

$$+ (e_1 q_{1m} + \cdots + e_m q_{mm})(\delta f_{m1} x_1 + \cdots + \delta f_{mm} x_m) \tag{2.99}$$

and

$$e^TQBr = (e_1 q_{11} + \cdots + e_m q_{m1})(\delta b_{11} r_1 + \cdots + \delta b_{1r} r_r) + \cdots$$

$$+ (e_1 q_{1m} + \cdots + e_m q_{mm})(\delta b_{m1} r_1 + \cdots + \delta b_{mr} r_r)$$

By choosing δf_{ij} and δb_{ij} such that,

$$\delta \dot{f}_{ij} = -v_{ij} \left[\sum_{k=1}^{m} e_k q_{ki} x_j + \delta f_{ij} \right] \quad i = 1, \ldots, m \tag{2.100}$$

$$\delta \dot{b}_{ij} = -w_{ij} \left[\sum_{k=1}^{m} e_k q_{ki} r_j + \delta b_{ij} \right] \quad j = 1, \ldots, r$$

the derivative $\dot{V}$ becomes

$$\dot{V} = e^T[M^TQ + QM]e - \sum_{i=1}^{m} \sum_{j=1}^{m} \delta f_{ij}^2 - \sum_{i=1}^{m} \sum_{j=1}^{r} \delta b_{ij}^2 \tag{2.101}$$

Since the model (2.94) is assumed to be designed asymptotically stable, there always exists a positive definite matrix $P = P^T > 0$ such that

$$M^TQ + QM = -P < 0 \tag{2.102}$$

Since the other two terms in (2.101) are also negative, it has been shown that $\dot{V}$ is negative definite and the augmented system is asymptotically stable

$$V > 0 \quad \dot{V} < 0 \tag{2.103}$$

Therefore (2.100) defines a convergent algorithm for the adjustment of the parameters δf_{ij} and δb_{ij}. However, this has not yet given an explicit algorithm for the adjustable parameters β. In order to do that the following equations must be solved for β:

$$\frac{\partial \delta f_{ij}(\beta,\alpha)}{\partial \beta} \dot{\beta} = -v_{ij} \left[\sum_{k=1}^{m} e_k q_{ki} x_j + \delta f_{ij}(\beta,\alpha) \right] \quad i,j = 1, \ldots, m$$

$$\frac{\partial \delta b_{ij}(\beta,\alpha)^T}{\partial \beta} \dot{\beta} = -w_{ij} \left[\sum_{k=1}^{m} e_k q_{ki} r_1 + \delta b_{ij}(\beta,\alpha) \right]$$

$$i,j = 1, \ldots, m, r \qquad\qquad (2.104)$$

Equations (2.104) must be solved simultaneously to yield a set
of differential equations in $\dot{\beta}$:

$$\dot{\beta} = - \left[\frac{\{\delta f_{ij}^T\}}{\{\delta b_{ij}^T\}} \right]^{-1} \left[\frac{\left\{ v_{ij} \left[\sum_{k=1}^{m} e_k q_{ki} x_j + \delta f_{ij}(\beta,\alpha) \right] \right\}}{\left\{ w_{ij} \left[\sum_{k=1}^{m} e_k q_{ki} r_j + \delta b_{ij}(\beta,\alpha) \right] \right\}} \right] \quad \begin{array}{l} \beta(t_0) = \beta \\ \\ (2.105) \end{array}$$

In order for this to be possible the inverse of the matrix in (2.105)
must exist.

Therefore two conditions must be met in order that Eqs. (2.104) have
a solution:

1.   There should be at least $m^2 + m \times r$ available adjustable
      parameters β to solve the simultaneous $m^2 + m \times r$ Eqs. (2.104).

2.   Equations (2.104) must be independent of the unknown parameters
      ω.

The first condition imposes the stricter restrictions since (2.100)
is not the unique set of equations that satisfy Eqs. (2.104).   It
is therefore possible to define only ℓ equations, where ℓ is the
dimension of k to satisfy (2.100).   The second condition, however,
limits considerably the systems for which the method is applicable.
The accessibility of all the states also imposes another restriction
to the method.

Example: Parks' system
      One of the systems that satisfy the above conditions is the
first-order system described by Parks [2.44] given in Fig. 2.17.
The gain K_c is there to compensate the unknown parameter K_v
influenced by the environment.

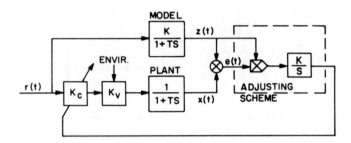

FIG. 2.17.   Park's Example of the Lyapunov Redesign Model
             Reference Adaptive Control Method.

The system equations are

<u>Model</u> $\dot{z} = -\dfrac{z}{T} + \dfrac{K}{T}\, r$ $e \triangleq -x,$ K_v = unknown parameter

<u>Plant</u> $\dot{x} = -\dfrac{x}{T} + \dfrac{K_c K_v}{T}\, r$ K_c = adjustable parameter

and the error equation are given by

$$\dot{e} = -\frac{1}{T}e + \frac{(K - K_c K_v)}{T}\, r = -\frac{1}{T}e + \frac{\delta f r}{T} \quad \delta f \triangleq K - K_c K_v$$

Define the Lyapunov function

$$V = e^2 + \lambda \delta f^2 \quad \lambda > 0$$

Then $\dot{V} = 2e\dot{e} + 2\lambda\delta f\dot{\delta f} = 2e\left(-\dfrac{e}{T} + \dfrac{\delta f}{T}\, r\right) + 2\lambda\delta f\dot{\delta f}$

Letting $\dot{\delta f} = -\dfrac{er}{\lambda T} - \delta f$

$$\dot{V} = -2\,\frac{e^2}{T} - 2\lambda\delta^2 f$$

This guarantees convergence of the algorithm. However, to
implement this algorithm one observes that

$$\delta f \triangleq K - K_c K_v \quad \dot{\delta f} = -K_v \dot{K}_c$$

Substituting back, one has

$$-\dot{K}_c K_v = -\frac{er}{\lambda T} - (K - K_v K_c)$$

or $\dot{K}_c = +\dfrac{er}{\lambda K_v T} - \dfrac{K}{K_v} + K_c$

This cannot be realized for any value of λ since it depends on the unknown parameter K_v. However, Parks, makes the assumption that δf is changing faster than e and r, and goes to zero as $t \to \infty$. Then

$$\dot{\delta f} = - \frac{er}{\lambda T}$$

gives a rate of change for f which makes $\dot{V} < 0$ with respect to e. Then

$$\dot{V} = - \frac{2e^2}{T}$$

In this case the equation for adjustment of K_c is written as

$$\dot{K}_c = \frac{er}{\lambda K_v T} = cer \qquad K_c(0) = 0$$

where for $K_v > 0$, and $\lambda K_v T$ is set equal to the arbitrary number c which is independent of K_v. More discussion on the conditions for recovery of the adjustable parameters are given in [2.62] and elsewhere in the literature. Acceleration of the algorithm may be obtained by considering $\ddot{V}$, [2.36] .

2.6 DISCUSSION

A very concise presentation of certain deterministic Self-Organizing Control algorithms has been given with the main emphasis on general methodologies representing various schools of thought of approaching the problem, rather than on accounting for their numerous applications. As such, the sample of techniques is rather small and their presentation has not been very thorough. However, certain conclusions may be drawn regarding the nature of S.O.C.s that have influenced the development of new algorithms for stochastic systems.

Four classes of S.O.C. have been considered, each one containing an important new idea for treating systems with unknown dynamics. The first one was the use of a search technique like the gradient method, for parameter identification to be used on-line to update a controller which was designed as if the unknown parameters were known.

The second new idea was the use of a polynomial expansion of the measurable state variables with adjustable coefficients to represent an unknown plant and/or even the structure of a controller as an approximation of the effective control.

The third idea is the use of a functional approach to represent a general form of an unknown system, and use functional analysis to obtain the desired optimal control.

Finally, the model tracking idea was introduced to generate a performance adaptive S.O.C. based on stability considerations. This answers the designers question of how to design a performance index to meet desired specifications and provides a method that guarantees stability of the system at the cost of sacrificing transient optimality.

With no intention of reducing the merits of the original methods, these ideas may be found combined in the new class of S.O.C. algorithms for stochastic systems which are described in this book. From another point of view, they serve as the founding ideas which qualify and unite the various methods under the name of "self-organizing control systems."

A detailed presentation of stochastic S.O.C. algorithms is the purpose of this book and will be presented in Part III. Part II is devoted to the development of background material like optimal estimation and search techniques, stochastic and dual optimal control, and parameter identification, necessary for the quantitative study of S.O.C. The reader who is properly versed in this material may skip directly to Part III without any loss in continuity.

2.7 REFERENCES

[2.1]   Andeen, R. E., "A General Adaptive Approach Based Upon Power
        Density Spectrum Moments," 3rd Congress of the International
        Federation of Automatic Control (IFAC) Paper 41B, London,
        (1966).

[2.2]   Anderson, G., Booland, R.., and Cooper, G. R., "Use of
        Crosscorrelation in Adaptive Systems," *Proc. Natl. Electronics
        Conf.*, II, p. 34 (1959).

[2.3]   Åström, K. J., and Wittenmark, B., "On Self-Tuning Regulators,"
        Automatica, 9, 185-199 (1973).

[2.4]   Beck, M. S., and Gough, N. E., 1967, "Some Further Contributions
        to A Model Reference Adaptive Control System using a Cost
        Function Criterion for Driers and other Dead-time Processes -
        The use of a Dynamic Model," *Int. J. Control,* 6, (5), 481-493
        (1967).

[2.5]   Caress, E. J., and Williams, T. J., *Non-linear and Adaptive
        Control Techniques,* Control Engineering Chicago, Ill., April
        1974.

[2.6]   Choe, R. W., and Nikoforuk, P. N., "Model reference approach
        in the control of plants with parameter uncertainties," Paper
        No. 2.3, Preprints of Technical Papers. IFAC Kyoto Symposium
        on Systems Approach to Computer Control, 1970.

[2.7]   Davies, W. D. T., *System Identification for Self-Adaptive
        Control.* Wiley-Interscience, New York, 1970.

[2.8]   Donalson, D. D., and Kishi, P. H., "Review of Adaptive control
        Systems Theories and Techniques", in *Modern Control Systems
        Theory* , (C. T. Leondes, ed.) New York, McGraw-Hill, 1965.

[2.9]   Donalson, D. D., and Leondes, C. T., "A Model Referenced
        Parameter Tracking Technique for Adaptive Control Systems,"
        IEEE Trans. Appl. Ind. , 82, 241-262 (1963).

[2.10] Draper, J., and Li, Y. T., "Principles of Optimalizing
        Control Systems and an Application to the Internal Combustion
        Engine," *ASME,* Publication, 160 (1951).

[2.11] Dreifke, G. E., and Hougen, J. O., "Experimental Determination
        of System Dynamics by Pulse Methods," Preprints, *Joint Automatic
        Control Conference,* 1963.

[2.12] Dressler, R. M., "An Approach to Model-Referenced Adaptive
        Control Systems," *IEEE Trans. Automatic Control,* AC-12, (1),
        75-80 (1967).

[2.13] Dymock, A. J., Meredith, J. F., Hall, A., and White, K. M., "Analysis of a Type of Model-Reference Adaptive Control System," *Proceedings of IEEE*, 112, (4), 743-753 (1965).

[2.14] Eveleigh, V. W., *Adaptive Control and Optimization Techniques*, McGraw-Hill, New York, 1967.

[2.15] Gibson, J. E., *Non-linear Automatic Control*, McGraw-Hill, New York, 1962.

[2.16] Gibson, J. E., "Self-optimizing or Adaptive Control Systems," Proc. IFAC Congress, Moscow, 1960, p. 147.

[2.17] Gilbart, J. W., and Monopoli, R. V., "A Modified Lyapunov Design for Model Reference Adaptive Control Systems," Proc. Seventh Annual Allerton Conference on Circuit and System Theory, Monticello, Ill., Oct. 1969.

[2.18] Gilbart, J. W., Monopoli, R. V., and Price, C. F., "Improved Convergence and Increased Flexibility in the Design of Model Reference Adaptive Control Systems," 9th IEEE Symposium on Adaptive Processes, Austin, Texas, 1970.

[2.19] Graham, D., and Lathrop, R. C., "The Synthesis of Optimum Transient Response, Criteria and Standard Forms," *Trans. AIEE*, 72, 273-288 (1953).

[2.20] Grammaticos, A. J., and Horowitz, B. M., "The Optimal Adaptive Control Law for a Linear Plant with Unknown Input Gains," *Int. J. Cotrol*, 12, 337-346, 1970.

[2.21] Hawkins, J. K., "Self-Organizing Systems - A Review and Commentary," *Proc. Inst. Rubber Ind.*, 49, 31-38 (1961).

[2.22] Hoberock, L., and Kohr, R., "An Experimental Determination of Differential Equations to Describe Simple Nonlinear Systems," Preprints JACC, 1966, pp. 616-623.

[2.23] Horowitz, I. C., "Optimum linear Adaptive Design of Dominant Systems with Large Parameter Variation," *IEEE Trans. Automatic Control*, AC-14, 261-269 (1968).

[2.24] Horrocks, T., "Investigations into Model-Reference Adaptive Control Systems," *Proc. IEEE London 111*, (11), 1894-1906 (1964).

[2.25] Vimolvanich, V., and Hsia, T. C., "System Identification Using a Learning Model," *Proc. Natl. Electron. Conf.*, 24, 92-97 (1968).

[2.26] Jacobs, O. L. R., "Design of a Single Input Sinusoidal Perturbation Extremum Control System," *Proc. IEEE (London)*, 115, (1), 212-217 (1968).

[2.27] Kokotovic, P. V., Medanic, J. V., Vuskovic, M. E., "Sensitivity Method in Experimental Design of Adaptive Control Systems," 3rd IFAC Paper 45B, London, 1966.

[2.28] Kumar K. S. P., and Sridhar, R., "On Identification of Control Systems by Quasilinearization," *IEEE Trans. on Automatic Control,* AC-92, 151-153 (1964).

[2.29] Landau, I. D., "A Hyperstability Criterion for Model Reference Adaptive Control Systems," *IEEE Trans. Automatic Control,* AC-14, 5, 552-554 (1969).

[2.30] Lee, R. C. K., *Optimal Estimation Identification and Control,* M.I.T. Press, Cambridge, Mass., 1964.

[2.31] Lindorff, D. P., and Monopoli, R. V., "Control of Time-Variable Nonlinear Multivariable Systems Using Lyapunov's Direct Method," Proc. 1966 JACC, Seattle, Wash., 475-483.

[2.32] Lion, P. M., "Rapic Identification of Linear and Nonlinear Systems," 1966, Preprints JACC, pp. 605-615.

[2.33] Luenberger, D., *Optimization by Vector Space Methods,* Wiley, New York, 1969.

[2.34] Margolis, M., and Leondes, C. T., "A Parameter Tracking Servo for Adaptive Control Systems," IRE, PGAC, November, 1959.

[2.35] Meditch, J., and Gibson, J. E., "On Real-Time Control of Time-Varying Linear Systems," IRE, GAC, July 1962.

[2.36] Mendel, J. M., *Discrete Techniques of Parametric Estimation,* Marcel-Dekker, New York, 1973.

[2.37] Mishkin, E., and Braun, L., *Adaptive Control Systems,* McGraw-Hill, New York, 1961.

[2.38] Moore, R. L., and Schweppe, F. C., "Model Identification for Adaptive Control of Nuclear Power Plants," *Automatica,* 9, 309-318 (1973).

[2.39] Monopoli, R. V., "Lyapunov's Method for Adaptive Control System Design," *IEEE Trans. Automatic Control,* AC-12, (3) 334-335 (1967).

[2.40] Monopoli, R. V., and Gilbart, J. W., "Model Reference Adaptive Control Systems with Feedback and Prefilter Adjustable Gains," Proc. of the Fourth Annual Princeton Conference on Information Sciences and Systems, March 1970, pp. 452-456.

[2.41] Nightingale, J. M., "Parameter-Perturbation Adaptive Control Systems with Imposed Constraints," IEE Monograph No. 518M, May 1962.

[2.42] Nikiforuk, P. N., and Gupta, M. M., "Control of Unknown Plants in Reduced State Space," *IEEE Trans. Automatic Control*, <u>AC-14</u>, (5), 489-495 (1964).

[2.43] Osburn, P. V., Whitaker, H. P., and Kezer, A., "New Developments in the Design of Model Reference Adaptive Control Systems," Institute of Aerospace Sciences, Paper No. 61-39, Jan. 1961.

[2.44] Parks, P. C., "Lyapunov Redesign of Model Reference Adaptive Control System," *IEEE Trans. Automatic Control*, <u>AC-11</u>, (3), 362-367 (1966).

[2.45] Pearson, A. E., "Adaptive Optimal Control of Nonlinear Processes," Preprints JACC 1963, p. 80. (1963).

[2.46] Pearson, A. E., "Analysis of an Adaptive Control Algorithm for Linear Systems," *Proc. Natl. Electron. Conf.*, <u>24</u>, 88-91 (1968).

[2.47] Pearson, A. E., "An Adaptive Control Algorithm for Linear Systems," *IEEE Trans. Automatic Control*, <u>AC-14</u>, (5), 497-503 (1969).

[2.48] Pearson, A. E., "A Modified Gradient Procedure for Adaptation in Nonlinear Control Systems," Proc. of 1969 Joint Automatic Control Conference, Boulder, Colo., pp. 596-602.

[2.49] Pearson, A. E., and Noonan, F., "On the Model Reference Adaptive Control Problem," Proc. of 1968 Joint Automatic Control Conference, Ann Arbor, Mich., pp. 538-545.

[2.50] Sarachic, P. E., "An Approach to the Design of Nonlinear Discrete Self-Optimizing Systems," 3rd Symposium on Discrete Adaptive Processes Proc. Natl. Electronics Conf., 20, 617-619, (1964), p. 128.

[2.51] Saridis, G. N., and Kitahara, R. T., "Computational Aspects of Performance-Adaptive Self-Organizing Control Algorithms," TR-EE 71-41, School of Electrical Engineering, Purdue University, Lafayette, Ind., 1971.

[2.52] Saridis, G. N., and Ricker, D., "Analog Methods for On-line System Identification Using Noisy Measurements", *Simulation*, 11, (5), 241-248 (1968).

[2.53] Shackcloth, B., and Butchart, R. L., "Synthesis of Model Reference Adaptive Systems by Lyapunov's Second Method," Proc. of 1965 IFAC Symposium on Adaptive Control, Teddington, England, pp. 145-152 (1965).

 2.54 Schalein, H. L. H., Ghonaimy, M. A. R.. and Shen, D. W. C., "Accelerated Model-Reference Adaptation via Lyapunov and Steepest Descent Design Techniques," *IEEE Trans. Automatic Control*, <u>AC-17</u>, (1), 125-128 (1972).

[2.55] Van Trees, H. L., *Synthesis of Optimum Nonlinear Control Systems*, MIT Press, Cambridge, Mass., 1962.

[2.56] Westcott, J. H., *An Exposition of Adaptive Control*, Pergamon Press, New York, 1962.

[2.57] Whitaker, H. P., "Design Capabilities of Model-Reference Adaptive Systems," *Proc. Natl. Electron. Conf.*, 18, 241-249 (1962).

[2.58] Whitaker, H. P., and Rediess, H. A., "A New Model Performance Index for Engineering Design of Flight Control Systems," *AIAA, J. Aircraft, 7*, (1970).

[2.59] Wiener, N., *Time Series*, MIT Press, Cambridge, Mass., 1949.

[2.60] Wilde, D. J., *Optimum Seeking Methods*, Prentice-Hall, New York, 1964.

[2.61] Winsor, C. A., and Roy, R. J., "Design of Model Reference Adaptive Control Systems by Lyapunov's Second Method," *IEEE Trans. Automatic Control*, <u>AC-13</u>, (2), 204 (1968).

[2.62] Zaborsky, J., and Berger, R. L., "An Integral Square Self-Optimizing Adaptive Control," *Trans. AIEE*, 263 256 (1962).

[2.63] Zaborsky, J., and Humphrey, W. L., "Control Without Model or Plant Identification," Preprints JACC 1964, p. 366.

PART II

OPTIMAL STOCHASTIC ESTIMATION, IDENTIFICATION, AND CONTROL

Chapter 3

OPTIMAL ESTIMATION AND RELATED SEARCH ALGORITHMS

3.1   INTRODUCTION

Estimation theory deals with the process of selecting the appropriate value of an uncertain quantity based on the available pertinent information. It seems, therefore, appropriate to discuss briefly certain estimation algorithms and related search techniques, since the purpose of this book is to treat systems with uncertainties.

It is assumed that the measurement error v introduced is a random variable defined over the triple (Ω, F, P), and therefore the estimates produced are also appropriate random variables. Therefore, stochastic methods will be used only to obtain such estimators.

Average values have been long used as *estimates* of a quantity x measured with a measurement error. The average value $\bar{x}$ of x, which may be used as an estimate $\hat{x}$ of x, based on n noisy measurements of x, $\{x_1, \ldots, x_n\}$ is given by

$$\hat{x} = \bar{x} = \frac{1}{n} \sum_{i=1}^{n} x_i \qquad (3.1)$$

while a measure of the accuracy of the estimation is given in the form of *standard deviation* σ_x [3.6]:

$$\sigma_x = \{\frac{1}{n-1} \sum_{i=1}^{n} (x_i - \bar{x})^2\}^{1/2} \qquad (3.2)$$

Irregular quantities, such as the sequence $\{x_1, \ldots, x_n\}$, exhibit the property of statistical regularity for large n (law of large numbers) [3.7]. Therefore, statistical methods based on probabilistic ideas are applicable in the development of estimation techniques. Such techniques will be discussed in the sequel.

Estimation of quantities from a stationary random process is first discussed, and the various optimal estimates and acceptable estimates are established. Optimal estimates are the best estimates which minimize or maximize a given criterion of optimality.

The recursive formulas used to update the estimate of a quantity based on the latest measurement generated, are called "sequential estimation." The estimation of the states of a dynamic system requires, in most cases, a sequential dynamic estimation algorithm to account for the nonstationary node of the states of a dynamic system. All these cases will be discussed in the following sections.

3.2 OPTIMAL POINT ESTIMATION FOR STATIONARY SYSTEMS

Estimation theory deals with the process of making a decision concerning the appropriate value of certain undefined quantities based on the available information. The measurements that provide the available information demonstrate, in general, a statistical regularity. Therefore, it is appropriate to investigate statistical methods in estimation theory. In most of the problems under consideration the estimate of the value of a parameter or a point in the parameter space is sought; therefore, the discussion here will be confined to *point estimation* only.

3.2.1 Optimal and Acceptable Estimators

Three different aspects of estimation theory are presented here in their order of complexity: (1) estimation from a finite set of data, (2) sequential parameter estimation, and (3) sequential dynamic or state estimation. The techniques described as *minimum risk, maximum likelihood, Bayes' rule, stochastic approximation,* or *random optimization* are the ones which are most frequently used in the estimation schemes encountered in the stochastic and S.O.C. algorithms in this book.

Optimal estimation defines the best estimate of a certain quantity, e.g., a parameter, with respect to a criterion of

optimality judiciously chosen and based on all the available information.

Even though optimal estimation is always desirable, such a
result may be either confusing or not feasible. Confusion is
generated when different criteria produce different estimates of the
same quantity as in nonlinear systems with nonsymmetric density
functions, while the feasibility problem arises whenever the solution is difficult to obtain. However, *acceptable estimates* may be
obtained if the following rules are satisfied:

1.   *Unbiased Estimates*.

If the estimate $\hat{x}$ of a parameter vector x, based on subsequent
observations, satisfies

$$E\{\hat{x}\} = E\{x\} = \bar{x} \qquad\qquad (3.3)$$

the estimate $\hat{x}$ is called *unbiased*. Otherwise it may have a positive
or negative bias, $\eta = E\{\hat{x}\} - \bar{x}$.

2.   *Consistent Estimates*.

If the estimate $\hat{x}$ of x converges in probability to x, i.e.,

$$\lim_{N\to\infty} prob[||\hat{x} - x|| \geq \epsilon] = 0 \text{ for arbitrarily small } \epsilon \qquad (3.4)$$

then the estimate $\hat{x}$ is called *consistent*. A consistent estimate is
always unbiased.

3.   *Efficient Estimates*.

An efficient estimate $\hat{x}$ of x is the unbiased estimate of x with
the minimum variance, i.e., $\sigma_{\hat{x}}^2 = E\{||\hat{x} - x||^2\} \leq E\{||y - x||^2\} = \sigma_y^2$ for all other estimates of y of x.

4.   *Sufficient Estimates*.

An estimate $\hat{x}$ is called *sufficient* if it contains all the information in the set of observed values regarding the parameter x to
be estimated. Alternately: if $x_1,\ldots,x_n$ is a finite random sequence
with probability density function $p(x_1,\ldots,x_n,\hat{\theta})$ and $\hat{\theta}$ is an estimate
of θ such that any other estimate y of θ conditioned on $\hat{\theta}$ does not
depend on θ, e.g., $p(y/\hat{\theta})$ is independent of θ, is called *a sufficient
estimate* of θ. Any statistic related to a sufficient estimate is
called *a sufficient statistic*.

Sufficient statistics contain all the information available for the estimation of a parameter, or any other quantity, to be estimated. They are useful in reducing the dimensionality of the estimation problem [3.2].

3.2.2 Minimum Bayes Risk Estimation, Least Squares Fit

Let us construct a simple estimator for a random variable x defined over a space X with an appropriate field F and selected from an ensemble with probability density function $p(x)$ [3.6]. Denote the unknown estimate of x, by $\hat{x}$ to be selected to minimize the following mean-square error criterion:

$$I = E\{(\hat{x} - x)^2\} = E\{x^2\} - 2\hat{x}E\{x\} + \hat{x}^2 \tag{3.5}$$

where E represents the expected value of a random variable, for instance $E\{x\} = \int X x p(x)\, dx$. Such an estimation is equivalent to the least-squares fit algorithm devised by Gauss [3.6], which requires the following necessary condition for minimization:

$$\frac{\partial I}{\partial \hat{x}} = -2E\{x\} + 2\hat{x} = 0$$

which yields

$$x = E\{x\} \triangleq \bar{x} \tag{3.6}$$

Therefore, the best estimate of x in the mean-square sense is the expected value of x.

A direct generalization of the above simple estimation procedure may lead to the *minimum risk* or *Bayesian estimator*.

As generalization of the mean-square estimator one may select a loss function $L(\tilde{x})$ of the extimation error $\tilde{x}$, with the following properties. Let

$$\tilde{x} \triangleq \hat{x} - x \tag{3.7}$$

where $\hat{x}^T$ is the vector optimal estimate of now vector random variable $x^T = [x_1,\ldots,x_m]$ with the probability distribution function

$P(\xi) = \text{Prob}(x < \xi) = \text{Prob}(x_1 < \xi_1, \ldots, x_n < \xi_n).$

Properties 3.1

1.   Assume that $L(\tilde{x})$ satisfies the condition
     $L(\tilde{x}) = 0$, when $\tilde{x} = 0$

$$L(\tilde{x}_2) > L(\tilde{x}_1), \text{ if } \tilde{x}_2 > \tilde{x}_1 \quad 0 \text{ (monotonicity)} \qquad (3.8)$$

$L(\tilde{x}) = L(-\tilde{x})$ (symmetricity)
Note that the above conditions are satisfied under the
stricter condition of convexity of $L(\tilde{x})$ (sufficient condition):

$$L(q_1\tilde{x}_1 + q_2\tilde{x}_2) < q_1 L(\tilde{x}_1) + q_2 L(\tilde{x}_2) \qquad (3.9)$$

for some $q_1, q_2 > 0$; $q_1 + q_2 = 1$
or if $L(\tilde{x})$ is twice differentiable

$$\frac{d^2 L(\tilde{x})}{d \tilde{x}^2} = 0 \qquad (3.9a)$$

2.   Assume that the probability distribution function of the
     random variable $x, P(\xi)$ is symmetric about the point $\bar{x} = E\{x\}$:

$$P(\xi - \bar{x}) = 1 - P(\bar{x} - \xi) \qquad (3.10)$$

Define the risk function $R(\tilde{x})$ as the expected value of the loss
function

$$R(\tilde{x}) \triangleq E\{L(\tilde{x})\} \qquad (3.11)$$

It can be easily shown [3.6] that for a *convex loss function* the
estimate $\hat{x}$ that minimizes $R(x)$ of (3.11) is the expected value of x.

$$\hat{x} = E\{x\} \qquad (3.12)$$

The same result can be obtained if, instead of $L(\tilde{x})$ being convex, it satisfies conditions (3.8) if the distribution function is
symmetric (3.10) and convex for $|x - \bar{x}| > 0$.

Optimal estimates based on past measurements may be obtained
for the random variable x, when a sequence of measurements $Z^m = \{z_1, \ldots, z_m\}$ of a function of the random variable x are available,
defined by the relation

$$z = g(x) \tag{3.13}$$

It is obvious that the best estimate of x should be a function of these measurements

$$\hat{x} = f(Z^m) \tag{3.14}$$

In order to establish the optimal estimate $\hat{x}_z$, conditioned on the available measurement sequence Z^m, the following point distribution function of x and Z^m must be analyzed using the chain rule

$$
\begin{aligned}
P(\xi; \eta_1, \ldots, \eta_m) &\triangleq \operatorname{Prob}(x < \xi; z_1 < \eta_1, \ldots, z_m < \eta_m) \\
&= P(\eta_1, \ldots, \eta_m) P(\xi/\eta_1, \ldots, \eta_m) \\
&\triangleq \operatorname{Prob}(z_1 < \eta_1, \ldots, z_m < \eta_m) \operatorname{Prob}(x < \xi \;/\; z_1 < \eta_1, \\
&\quad \ldots, z_m < \eta_m)
\end{aligned}
\tag{3.15}
$$

where $P(\eta_1, \ldots, \eta_m)$ is the marginal probability distribution function and $P(\xi/\eta_1, \ldots, \eta_m)$ is the conditional distribution function defined by (3.15). By the linearity property of the expectation operator one may write the risk function defined by (3.11)

$$R(\tilde{x}) = E\{L(\hat{x} - x)\} = E_z\{E_{x/z}\{L(\hat{x} - x)/z_1, \ldots, z_m\}\} \tag{3.16}$$

where E_z is the expectation with respect to Z^m, which does not depend on the choice of $\hat{x}$, and $E_{x/z}$ is the conditional expectation of x on the sequence Z^m. Therefore, minimization of $R(\tilde{x})$ with respect to $\hat{x}$ is equivalent to the minimization of the conditional risk $R(\tilde{x}/Z^m)$ for any sequence Z^m, where

$$R(\tilde{x}/Z^m) = E_{x/z}\{L(\hat{x} - x)/z_1, \ldots, z_m\} \tag{3.17}$$

The best estimate of x based on the measurements Z^m can be established through the following theorem, whose proof is given in Ref. [3.6].

Theorem 3.1

Assume that $L(\tilde{x})$ satisfies (3.8) and the conditional distribution function $P(\xi/z_1, \ldots, z_m)$ is

1.   Symmetric about the mean $\bar{x}$

$$P_x[(\xi - \bar{x})/z_1,\ldots,z_m] = 1 - P_x[(\bar{x} - \xi)/z_1,\ldots z_m] \qquad (3.18)$$

2.   Convex for $\xi_1 \leq \xi_2$, i.e.,

$$P_x[\lambda \xi_1 + (1 - \lambda)\xi_2/z_1,\ldots,z_m] \leq \lambda P_x[\xi_1/z_1,\ldots,z_m)$$

$$+(1 - \lambda)P_x(\xi_2/z_1,\ldots,z_m) \quad 0 \leq \lambda \leq 1 \qquad (3.19)$$

Then the random variable $\hat{x}$ which minimizes $R(\tilde{x}/Z^m)$ of Eq. (3.17) is the conditional expectation

$$\hat{x} = E\{x/z_1,\ldots,z_m\} \qquad (3.20)$$

The same results may be obtained by replacing condition 2, Eq. (3.19), by the convexity of $L(\tilde{x})$.

In the case of normal distribution functions

$$P(\xi) = \int_{-\infty}^{\xi} (2\pi\sigma^2)^{-1/2} \exp\left(-\frac{1}{2}\frac{(x-\bar{x})^2}{\sigma^2}\right) dx \qquad (3.21)$$

with $\bar{x} = E\{x\}$, $\sigma^2 = E\{(x-\bar{x})^2\}$, the conditions on $L(\tilde{x})$, Eq. (3.8), may be relaxed and nonsymmetric functions may be considered [3.6].

3.2.3 The Orthogonal Projection Theorem

The optimal estimation problem based on past available measurements can be geometrically reinterpreted through the orthogonal projection Theorem [3.7].

Theorem 3.2

The orthogonal projection $\hat{x}$ of the random vector x belonging to the space X, on the subspace Z spanned by the measurement sequence $Z^m = \{z_1,\ldots,z_m\}$ is the best estimate of x conditioned on the measurements Z^m and its normal component $\tilde{x} = \hat{x} - x$ orthogonal to Z accounts for the estimation error.

The orthogonality to a subspace of measurements Z implies that the following properties of $\hat{x}$ are satisfied:

1. $\hat{x} \in Z$

2. $\tilde{x}$ is orthogonal to all vectors $z \in Z$. This means that
     $E\{\tilde{x} \cdot z\} = 0,\ z \in Z$ \hfill (3.22)

3. $\hat{x}$ is unique in the Hilbert space X. This can be easily
    demonstrated by defining a vector $b = x - a$ where $a \in Z$.
    Let b be orthogonal to any $z \in Z$. Then $a \equiv \hat{x}$; because if
    not, there is at least another $a´ \ne a$, $a´ \in Z$ which $a´ =$
    $\hat{x}$. If so $b´ = x - a´$ and $E\{b´^T z\} = 0$ for any $z \in Z$. Then
    $E\{x^T z\}$ $E\{a´^T z\} = 0$. But by property 2, $E\{x^T a\} = E\{x^T z\} =$
    $E\{a^T z\} = E\{a´^T z\}$ for all $z \in Z$. This implies that $\hat{x} = a =$
    $a´$.

4. $E\{||\tilde{x}||^2\}$ is a minimum where $E\{||\cdot||\}$ is defined as the
    norm of a vector. This property can be established by the
    following argument:
    Let

    $z \in Z$ and $(\hat{x} - z) \in Z$

    then

    $$E\{||x - z||^2 = E\{||x - \hat{x} + \hat{x} - z||^2\} = E\{||x - \hat{x}||^2\}$$
    $$+ E\{||\hat{x} - z||^2\} > E\{||x - \hat{x}||^2\}$$
    $$= E\{||\tilde{x}||^2\} \text{ for all } z \in Z$$

    This establishes $\hat{x} = E\{x/Z^m\}$ as the minimum mean-square
    estimator.

5. Furthermore. it can be established that $\hat{x}$ is also the mini-
    mum variance estimator. This means that the covariance
    matrix $E\{\tilde{x}\tilde{x}^T\}$ of the estimation error $\tilde{x}$ is a measure of
    confidence of the estimation, is a minimum. As in (3.24),
    assume an arbitrary vector $z \in Z$; then

    $$E\{(z - x)(z - x)^T\} = E\{(z - \hat{x} + \hat{x} - x)(z - \hat{x} + \hat{x} - x)^T\}$$
    $$= E\{(y - \hat{x})(y - \hat{x})^T\}$$
    $$+ E\{(z - \hat{x})(\hat{x} - x)^T\} + E\{(\hat{x} - x)(z - \hat{x})^T\}$$
    $$+ E\{(\hat{x} - x)(\hat{x} - x)^T \qquad\qquad (3.23)$$

Invoking the properties of $\tilde{x} = \hat{x} - x$ and $\hat{x} - z$ from 2 and 4,
and that $E\{zz^T\} > 0$, for $z \in Z$ one obtains

$$E\{(z - x)(z - x)^T\} = E\{(z - \hat{x})(z - \hat{x})^T\} + E\{\tilde{x}\tilde{x}^T\}$$

or

$$E\{(z - x)(z - x)^T\} \ge E\{\tilde{x}\tilde{x}^T\} \qquad \forall\ z \in Z \qquad (3.24)$$

which establishes the minimum variance estimator.

The projection theorem point of view for estimation will be further emphasized in the discrete time Kalman filter case.

3.2.4   Linear Mean-Square Estimator

As an example of the mean-square estimation approach, the following problem is considered: Let us try to estimate the parameters x from a set of measurements z linearly related to x by

$$z = Ax + v \tag{3.25}$$

where z is an m-vector of observations, x is an n-vector of the unknown parameters, and v is an m-vector of random error with zero mean $E\{v\} = 0$ and $E\{v^2\} < \infty$;   A is an mxn known nonsingular matrix.

Define the risk function to be minimized

$$R(\tilde{x}) = E\{||z - A\hat{x}||^2\} \tag{3.26}$$

Because of (3.19) it is equivalent to minimize

$$R(\tilde{x}/z) = E_x\{||z - A\hat{x}||^2/z\} = ||z - A\hat{x}||^2 \tag{3.27}$$

since $\hat{x}$ is a function of z only.   This statement establishes the correspondence to the least-squares method.

Consider two cases

1. $m \geq n$.   The minimization of (3.27) yields

$$A^T (z - A\hat{x}) = 0 \tag{3.28}$$

or, because $m \geq n, A^T A$ is an n x n nonsingular matrix, it is positive definite and has an inverse.   Therefore

$$\hat{x} = (A^T A)^{-1} A^T z \tag{3.29}$$

for $m = n$, $\hat{x} = A^{-1} z$

2. $m < n$.   In this case

$$\hat{x} = A^\# z \tag{3.30}$$

where $A^\#$ is the Moore-Penrose generalized inverse of A of rank r.

If

$$A[mxn] = B^{[mxr]} C^{[rxn]}$$

then

$$A^\# = C^T[CC^T]^{-1}[B^TB]^{-1}B^T$$

If instead of the risk (3.26) the weighted risk is used

$$R(x) = E\{(z - Ax)^T Q^{-1}(z - Ax)\} \tag{3.31}$$

where $Q = Q^T > 0$, the best estimate is given for case 1.

$$\hat{x} = (A^TQ^{-1}A)^{-1}A^TQ^{-1}z \tag{3.32}$$

The estimator $\hat{x}$ obtained from (3.32) is unbiased

$$E\{\hat{x} - x\} = (A^TQ^{-1}A)^{-1}A^TQ^{-1}A \ E\{x - x\} + (A^TQ^{-1}A)^{-1}A^TQ \ E\{v\} = 0$$

and the covariance matrix of the mean square estimator is

$$E\{(\hat{x} - x)(\hat{x} - x)^T\} = [(A^TQ^{-1}A)^{-1}A^TQ^{-1}]\Lambda[(A^TQ^{-1}A)^{-1}A^TQ^{-1}]^T \tag{3.33}$$

where

$$\Lambda = E\{vv^T\}$$

In the case that Q is chosen so that

$$Q = \Lambda \tag{3.34}$$

then as the covariance matrix (3.33) corresponds to the minimum variance defined in (3.24) for a linear system (3.25) as in [3.6]

$$E_\Lambda\{(\hat{x} - x)(\hat{x} - x)^T\} = [A^T\Lambda^{-1}A]^{-1} \tag{3.35}$$

Consequently

$$E_\Lambda\{(\hat{x} - x)(\hat{x} - x)^T\} < E_Q\{(\hat{x} - x)(\hat{x} - x)^T\} \quad \forall \ Q \neq \Lambda \tag{3.36}$$

and therefore (3.32) with (3.34) is the *minimum variance estimator*.

In the case that the relation between z and x is not linear, the mean-square estimation method is not directly applicable. A polynomial or Taylor series approximation is proposed in the litera-ture for the solution of this problem [3.6].

3.2.5   Other Bayesian Estimators

In the previous section we have seen the estimators resulting from the minimization of a Bayesian risk function. Two more estimators may be devised by using different performance criteria.

1.   *The most probable estimate.* This corresponds to the mode or the maximum of the *a-posteriori* density function, when it exists, $p(x|z_1,z_2,\ldots,z_m)$

$$\underset{x}{\text{Max}}\ p(x/z_1,\ldots,z_m) \tag{3.37}$$

The *a-posteriori* density function may be calculated by the Bayes rule:

$$p(x/z_1,\ldots,z_m) = \frac{p(x,z_1,\ldots,z_m)}{p(z_1,z_2,\ldots,z_m)} = \frac{p(z_1,\ldots,z_m|x)p(x)}{p(z_1,\ldots,z_m)} \tag{3.38}$$

where $p(z_1,\ldots,z_m|x)$ is the *a-priori* density function.

This approach usually requires *a-priori* information on the density functions and involved computation.

2.   *The minimax estimate of x.* This is the estimate of x that minimizes the maximum possible error $||\hat{x} - x||$. This is the *median* of the density function. In the case of uni-modal symmetric density function, all three estimates coincide.

3.2.6   The Maximum Likelihood Estimator

This method is based on the work of Fisher[3.5]. It relies on the assumption that the joint probabilities of certain measurements $\{z_1,\ldots,z_m\}$ conditioned on the value of the parameter x to be estimated, can be calculated. Then the most likely value of x is obtained by minimizing the "likelihood function," $L(z,x)$

$$L(z,x) = p(z_1,\ldots,z_m|x) \tag{3.39}$$

Consider the following example for demonstration purposes:

$$z = Ax + v$$

where z is the measurement m-vector, x is the unknown parameter n-vector, v is the random noise m-vector with Gaussian probability density $p(v)$ independent of x, and A is an mxn fixed matrix. Find the most likely estimate of x.

Define

$$p(z/x) = \frac{p(z,x)}{p(x)} = \frac{p(x,v)}{p(x)} = \frac{p(z)p(v)}{p(x)} = p(v)$$

$$p(v) = p_v(z - A\hat{x})$$

Now since v is Gaussian with zero mean and covariance matrix, say Λ

$$p(v) = \frac{1}{(2\pi^2|\Lambda|)^{1/2}} \exp\left[-\frac{1}{2} v^T \Lambda^{-1} v\right] = c \exp\left[-\frac{1}{2} ||z - A\hat{x}||^2_{\Lambda^{-1}}\right]$$

Then the likelihood function is

$$L(\tilde{x}) = \ln c - \frac{1}{2} ||z - A\hat{x}||^2_{\Lambda^{-1}}$$

where $\ln(\cdot)$ is the natural logarithm.

The maximum is obtained by

$$\frac{\partial L}{\partial x} = A^T \Lambda^{-1} (z - A\hat{x}) = 0$$

or if m > n

$$\hat{x} = (A^T \Lambda^{-1} A)^{-1} A^T \Lambda^{-1} z \tag{3.40}$$

This is the same result as previously obtained for the minimum variance estimator (3.32). This shows that for Gaussian processes the maximum likelihod method yields the minimum variance mean-square estimator.

It should be noted, however, that the joint density function $p(x,v)$ may not be separable if x is not independent of v.

On the other hand, the maximum likelihood estimator does not always maximize the *a-posteriori* probability (3.37) to yield the most probable estimate. Consider

$$p(x/z) = \frac{p(z/x)p(x)}{p(z)} \tag{3.41}$$

The maximum of $p(z/x)$ with respect to x will correspond to the maximum of $p(x/z)$ only if $p(x)$ does not affect the maximization. This happens if the *a priori* probabilities $p(x)$ are for instance uniform.

3.3 SEQUENTIAL ESTIMATION

The need for sequential estimation arises when the measurement data are obtained <u>on-line</u> as the process evolves and the measurements arrive at the rate of one at a time. Then it is usually desirable to revise the estimate based on n measurements as a new $(n + 1)$ measurement is available without solving the whole problem all over again.

Sequential estimation may be performed with any of the previously discussed methods. Several of them will be demonstrated in the sequel. The least-squares method, which is equivalent to maximum probability estimates for linear systems, will first be demonstrated. Then the stochastic approximation methods will be presented. These are methods that account more for the convergence properties of the algorithm than anything else. They are rather trying to sequentially produce estimates which will be consistent. and will guarantee convergence than minimizing some *a priori* criteria. Finally, random search algorithm will be presented which guarantee a global minimum.

The most probable estimator given by Eq. (3.37) may be first sought. To do this sequentially, the following procedure should be followed [3.27]:

1. Assume that $\hat{x}_n$ is obtained by maximizing $p(x/z_1,\ldots,z_m)$.
2. The measurement z_{m+1} is available.
3. Evaluate either $p(x,z_1,\ldots,z_m,z_{n+1})$ or $p(z_1,\ldots,z_{m+1}/x)$ from the system's equations.
4. Find $\hat{x}_{m+1}$ that maximizes

$$p(x/z_1,\ldots,z_{m+1}) = \frac{p(x,z_1,\ldots,z_{m+1})}{p(z_1,\ldots,z_{m+1})} = \frac{p(z_1,\ldots,z_{m+1}/x)p(x)}{p(z_1,\ldots,z_{m+1})}$$

However, the evaluation of $p(x,z_1,\ldots,z_{m+1})$ or $p(z_1,z_2,\ldots,z_{m+1}/x)$ might be a very difficult task. Therefore, the maximum *a-posteriori* probability estimator is popular only in the cases where the above probabilities can be generated. This is the case with Gaussian random variables which are usually reproducible by a linear operation.

Similar difficulties arise when the maximum likelihood estimator is used for sequential estimation. Therefore, we should look into methods which may yield estimators which do not depend on the probability distribution function of the random variables involved.

3.3.1 Linear Mean-Square Sequential Estimator

The mean-square estimator developed in Sec. 3.2.4 is one that does not depend explicitly on the type of the probability distribution function of the random variables involved. It can be extended to estimate the unknown n-dimensional vector x sequentially. Assume that after m iterations, m measurements $z_1,\ldots,z_m$ are obtained. Define the vector of measurements $Z^{m^T} = [z_1,z_2,\ldots,z_m]^T$. These measurements are related to the vector x by

$$z_1 = a_1^T x + v_1$$
$$\cdots\cdots\cdots\cdots \qquad \text{or} \quad Z^m = A_m x + V^m \qquad\qquad (3.42)$$
$$z_m = a_m^T x + v_m$$

where $A_m = \begin{bmatrix} a_1^T \\ \vdots \\ a_m^T \end{bmatrix}$ is an n x m matrix of known coefficients and $v^{m^T} =$ $[v_1,\ldots,v_m]^T$ is a vector of the past sample of the noise random variable v_i which has zero mean and finite variance, and forms an independent and identically distributed sequence.

Then, after the mth iteration of the algorithm, the best estimate $\hat{x}_m$ of x is given by (3.29) for m > n where n is the dimension of x.

$$\hat{x}_m = (A_m^T A_m)^{-1} A_m^T Z^m \qquad\qquad (3.43)$$

Then a new measurement z_{m+1} is obtained and related to x by

$$z_{m+1} = a_{m+1}^T x + v_{m+1} \tag{3.44}$$

Should the process (3.25)-(3.29) be repeated, one may write in partitioned matrix vector form:

$$Z_{m+1} = \begin{bmatrix} Z^m \\ \hline z_{m+1} \end{bmatrix} = \begin{bmatrix} A_m \\ \hline a_{m+1}^T \end{bmatrix} x + \begin{bmatrix} V^m \\ \hline v_{m+1} \end{bmatrix} = A_{m+1} x + V^{m+1} \tag{3.45}$$

for which the mean-square solution is known to be Eq. (3.29)

$$\hat{x}_{m+1} = (A_{m+1}^T A_{m+1})^{-1} A_{m+1}^T Z^{m+1} \tag{3.46}$$

However, the form of the solution (3.46) is not computationally appealing because of the presence of the continuously growing matrix A_{m+1}, the vector Z^{m+1} and the matrix inversion required to solve the problem. Instead, if the previous estimate $\hat{x}_m$ is used, which contains all the information about x accumulated by the mth iteration, a simpler recursive expression is obtained. For this purpose observe that

$$[A_{m+1}^T A_{m+1}]^T = \{ [A_m^T \vdots a_{m+1}] \begin{bmatrix} A_m \\ \hline a_{m+1}^T \end{bmatrix} \}^{-1} = [A_m^T A_m$$
$$+ a_{m+1} a_{m+1}^T]^{-1} \tag{3.47}$$

Then the matrix $P_m = P_m^T$ is defined in order to obtain a recursive formula to evaluate $(A_{m+1}^T A_{m+1})^{-1}$ given $A_m^T A_m$.

$$P_m^{-1} = A_m^T A_m$$

$$P_{m+1}^{-1} = A_{m+1}^T A_{m+1} = A^T A_m + a_{m+1} a_{m+1}^T = P_m^{-1} + a_{m+1} a_{m+1}^T \tag{3.48}$$

And using the matrix inversion lemma as in Ref. [1.3]

$$P_{m+1} = P_m - P_m a_{m+1} (a_{m+1}^T P_m a_{m+1} + 1)^{-1} a_{m+1}^T P_m \tag{3.49}$$

Substituting this in (3.46) and identifying

$$A_{m+1}^T Z^{m+1} = A_m^T Z^m + a_{m+1} Z_{m+1} \tag{3.50}$$

and with (3.43) one obtains

$$\hat{x}_{m+1} = \hat{x}_m + P_m a_{m+1} (a_{m+1}^T P_m a_{m+1} + 1)^{-1} (z_{m+1} - a_{m+1}^T \hat{x}_m) \tag{3.51}$$

Equations (3.51) with (3.49) give a pair of equations for sequential mean-square estimation of $\hat{x}$. Furthermore, one may observe that in the mean-square estimation, the estimation error $\tilde{x}_m$ is given by

$$\tilde{x}_m \triangleq \hat{x}_m - x = (A_m^T A_m)^{-1} A_m^T z^m - (A_m^T A_m)^{-1} A_m (z^m - v^m) \tag{3.52}$$

or $\tilde{x}_m \triangleq (A_m^T A_m)^{-1} A_m v^m$

Then $\hat{x}_m$ is an unbiased estimator since

$$E\{\tilde{x}_m\} = E\{(A_m^T A_m)^{-1} A_m v^m\} = (A_m^T A_m)^{-1} A_m E\{v^m\} = 0 \tag{3.53}$$

The covariance matrix of $\tilde{x}_m$ is given by

$$\tilde{P}_m \triangleq E\{\tilde{x}_m \tilde{x}_m^T\} = E\{(A_m^T A_m)^{-1} A_m^T v^m v^{m^T} A_m (A_m^T A_m)^{-1}\}$$

$$= (A_m^T A_m)^{-1} A_m^T E\{v^m v^m\}^T A_m (A_m^T A_m)^{-1} = \sigma^2 P_m \tag{3.54}$$

where $E\{v^m v^{m^T}\} = \sigma^2 I$ because v^m has been assumed to be an independent identically distributed sequence of random variables.

The estimation scheme can be written in terms of the covariance matrix

$$\hat{x}_{m+1} = \hat{x}_m + \tilde{P}_m a_{m+1} (a_{m+1}^T \tilde{P}_m a_{m+1} + \sigma^2)^{-1} (z_{m+1} - a_{m+1}^T \hat{x}_m) \tag{3.55}$$

$$\tilde{P}_{m+1} = \tilde{P}_m - \tilde{P}_m a_{m+1} (a_{m+1}^T \tilde{P}_m a_{m+1} + \sigma^2)^{-1} a_{m+1}^T \tilde{P}_m$$

Observe that $\lim\limits_{m \to \infty} \tilde{P}_m = \lim\limits_{m \to \infty} \left[P_0^{-1} + \sum\limits_{i=1}^{m} a_i a_i^T \right]^{-1} = 0.$ This proves

that the estimation is consistent.

3.3.2.   Stochastic Approximation: The Robbins-Monro Algorithm

Stochastic approximation algorithms may be used for parameter estimation.   Actually they were developed to determine critical points of a regression function.

To motivate the stochastic approximation procedure, let us examine a very special problem from statistics.   Suppose that one has to find the true value of y of a quantity, by performing n measurements $\{z_n\}$ corrupted with the measurement error $v = \{v_n\}$ with zero mean $E\{v\} = 0$ and variance $E\{v^2\} = \sigma^2 < \infty$ such that

$$z_i = y + v_i, \quad i = 1,\ldots,n \tag{3.56}$$

The best estimate of y after n measurements, called $\hat{y}_n$, is given as the sample mean of the process

$$\hat{y}_n = \frac{1}{n} \sum_{i=1}^{n} z_i \tag{3.57}$$

For such a process, Kolmogorov's strong law of large numbers, can be used to show that

$$\lim_{n \to \infty} \left\{ \frac{1}{n} \sum_{i=1}^{n} z_i \right\} = y \tag{3.58}$$

Let us now attempt to obtain a recursive formula for the computation of $\hat{y}_{n+1}$ from the sample mean $\hat{y}_n$ after n measurements and a new measurement z_{n+1} performed afterwards.   Using formula (3.57) for n + 1

$$\hat{y}_{n+1} = \frac{1}{n+1} \sum_{i=1}^{n+1} z_i = \frac{1}{n+1} \sum_{i=1}^{n} z_i + \frac{1}{n+1} z_{n+1} \tag{3.59}$$

Substituting now (3.57) in (3.59) we obtain

$$\hat{y}_{n+1} = \frac{n}{n+1} \hat{y}_n + \frac{1}{n+1} z_{n+1} \tag{3.60}$$

or

$$\hat{y}_{n+1} = \hat{y}_n + \frac{1}{n+1} (z_{n+1} - \hat{y}_n) \tag{3.61}$$

Let us analyze now Eq. (3.61). The updated estimate of y, i.e., $\hat{y}_{n+1}$ is equal to the old estimate $\hat{y}_n$ plus a correction term weighted with the quantity $1/_{n+1}$. For n = {1, 2, 3,...} the weighting quanti- ty takes the form of the harmonic sequence 1, 1/2, 1/3,...,1/n,... which decreases the influence of the correction term as n increases in agreement with (3.58). Equation (3.61) represents a special *stochastic approximation* scheme for estimation of y. It does not require special knowledge of the probability distribution or any statistics of the error except for the zero mean for unbiased esti- mate and finite variance. Studying carefully the harmonic sequence, one observes that it converges to zero while its sum diverges, i.e.,

$$\lim_{n \to \infty} \frac{1}{n} = 0 \quad \lim_{n \to \infty} \sum_{k=1}^{n} \frac{1}{k} = \infty \tag{3.62}$$

A heuristic interpretation [3.9], may be given for the above pro- perties of the sequence. The weighting of the correction error should decrease because the strong law of large numbers ensures that the procedure will approach the true value with a large number of iterations. However, correction should be made, no matter how far away one starts the search. This is accomplished because the sum of the sequence is divergent for any starting value $n \leq N < \infty$, and therefore the correction effort is unlimited.

Another important property of the weighting sequence is the fact that

$$\lim_{n \to \infty} \sum_{k=1}^{n} \frac{1}{k^2} < \infty \tag{3.63}$$

This property is important because it implies that individual random errors tend to cancel each other for a large number of iterations.

This is one of the basic requirements for convergence of the general stochastic approximation procedure.

Using the philosophy presented in the above special example, Robbins and Monro [3.38] developed a stochastic approximation algorithm to obtain the root of a regression function from measurements corrupted with noise (see Figs. 3.1 and 3.2). Suppose that the function $y(x)$, where x is a scalar independent variable defined over $a \le x \le b$, is bounded for all $x \in [a,b]$ and has a unique zero for $\theta \in [a,b]$

$$y(\theta) = 0 \tag{3.64}$$

then

$$- \infty < y(x_1)y(x_2) < 0 \text{ for all } : x_1 \in [a,\theta) : x_2 \in (\theta,b] \tag{3.65}$$

If, the measurable output is

$$z(x) = y(x) + v \tag{3.66}$$

where v is a random noise with

$$E\{v\} = 0$$

$$E\{v^2\} < \sigma^2 < \infty \tag{3.67}$$

Then the sequence

$$z_{n+1} - x_n - \gamma_n z(x_n) \tag{3.68}$$

converges to   , with probability one, i.e.,

$$\text{Prob} \{ \lim_{n \to \infty} x_n = \theta\} = 1 \tag{3.69}$$

and in mean square, i.e.,

$$\lim_{n \to \infty} E\{(x_n - \theta)^2\} = 0 \tag{3.70}$$

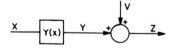

Fig. 3.1.   Block Diagram of a System in the Robbins-Monro and Kiefer-Wolfowitz Procedures.

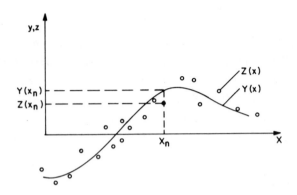

Fig. 3.2. Measurements and Regression Function in the
Robbins-Monro Method.

if the sequence $\{\gamma_n\}$ has the following properties:

1. $\lim\limits_{n \to \infty} \gamma_n = 0$

2. $\lim\limits_{n \to \infty} \sum\limits_{k=1}^{n} \gamma_k = \infty$ (3.71)

3. $\lim\limits_{n \to \infty} \sum\limits_{k=1}^{n} \gamma_k^2 < \infty$

The proof of the Robbins-Monro method will not be presented here.

It is obvious that the harmonic sequence $\{1/n\}$ satisfies all the conditions (3.71) and can be used as a special case of $\{\gamma_n\}$. As a matter of fact, it is the sequence which gives the fastest reduction in the correction term from any sequence with diverging sum of the form $\{1/n^q\}$, $1/2 < q \le 1$. Under the same conditions the method can handle the case of finding the value $x(t) \in [a,b]$ for which $y(x(t)) = A$ by writing (3.68) as

$$x_{n+1} = x_n - \gamma_n (z(x_n) - A)$$ (3.72)

3.3.3   Stochastic Approximation: The Kiefer-Wolfowitz Algorithms

The Robbins-Monro method may be adapted to search for the minimum or the maximum of a unimodal hill defined by $w(x)$ in the presence of noise.   Since $dw/dx = 0$ for such an extremum, one may define

$$y = \frac{dw}{dx} \tag{3.73}$$

and use the previous scheme.   However, the measurement of the derivative of a function in the presence of noise is not always feasible.

Kiefer and Wolfowitz [3.21] proposed a different stochastic approximation scheme to search for the maximum (minimum) of a bounded unimodal, but not necessarily convex, function $w(x)$ defined over $x \, \varepsilon \, [a,b]$ in the presence of random noise v, defined by (3.67). If the measurable output, see Fig. 3.1, is

$$z(x) = w(x) + v \tag{3.74}$$

then from two observations $z(x_n + c_n)$ and $z(x_n - c_n)$ one may calculate the average slope to be used as y, see Fig. 3.3.

$$y = \frac{z(x_n + c_n) - z(x_n - c_n)}{2c_n} \tag{3.75}$$

Then the sequence

$$x_{n+1} = x_n + \gamma_n \frac{[z(x_n + c_n) - z(x_n - c_n)]}{2c_n} \tag{3.76}$$

converges to $\theta = \{\theta / \frac{dw(\theta)}{dx} = 0\}$, with probability one, i.e.,

$$\text{Prob } \{ \lim_{n \to \infty} x_n = \theta\} = 0 \tag{3.77}$$

and in mean square

$$\lim_{n \to \infty} E\{(x_n - \theta)^2\} = 0 \tag{3.78}$$

if

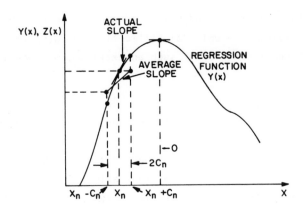

Fig. 3.3.   Average Slope Estimation in the Kiefer-Wolfowitz
           Procedure.

1. $\lim\limits_{n \to \infty} y_n = 0$, $\lim\limits_{n \to \infty} c_n = 0$

2. $\lim\limits_{n \to \infty} \sum\limits_{k=1}^{n} y_k = \infty$ (3.79)

3. $\lim\limits_{n \to \infty} \sum\limits_{k=1}^{n} \left[\dfrac{y_k}{c_k}\right]^2 < \infty$

The original proof of the scheme is not presented here. Various
modifications of the Kiefer-Wolfowitz method which demonstrate
acceleration of convergence have been proposed in the literature.

The normalized version of the method [3.48] uses the following
expression instead of (3.76):

$$x_{n+1} = x_n + y_n \; \text{sgn} \left[\frac{z_n(x_n + c_n) - z_n(x_n - c_n)}{2c_n}\right] \qquad (3.80)$$

with the convention that sgn $[0] = 0$. This form is recommended for
the case where the $w(x)$ is considerably flat in most of the domain
of $x \, \varepsilon \, [a,b]$. Algorithm (3.80) satisfies also the conditions of
Dvoretzky's theorem, which will be presented below, and therefore
converges with probability one and in mean square.

3.3.4 Dvoretzky's Theorems for Stochastic Approximation

Dvoretzky published in 1956 a very general theorem covering most of the known stochastic approximation algorithms and gave the loosest conditions for an estimator based mainly on the convergence properties of the algorithm. Here is the general theorem [3.9].

Theorem 3.3

Let $\alpha_n(r_1,\ldots,r_n)$, $\beta_n(r_1,\ldots,r_n)$, $\zeta_n(r_1,\ldots,r_n)$ be nonnegative measurable functions of real variables $r_1,\ldots,r_n$ satisfying the following conditions:

1. $\alpha_n(r_1,\ldots,r_n)$ are uniformly bounded and

$$\lim_{n \to \infty} \alpha_n(r_1,\ldots,r_n) = 0$$

2.   The $\sum_{n=1}^{n} \beta_n(r_1,\ldots,r_n) < \infty$ uniformly and $\beta_n(r_1,\ldots,r_n)$ converges uniformly for any sequence r_1, $r_2,\ldots$,

3.   The series $\sum_{k=1}^{n} \zeta_k(r_1,\ldots,r_n)$ diverges uniformly for any sequence r_1, $r_2,\ldots$, bounded in absolute value, i.e., for any sequence r_1, $r_2,\ldots$, such that $\sup_{n=1,2,\ldots} |r_n| < c$, c = arbitrary finite number

4.   Let T_1, $T_2,\ldots$, be measurable transformations, satisfying the inequality

$$|T_n(r_1,\ldots,r_n) - \theta| \leq \max\,[\alpha_n, (1+\beta_n)\,|r_n - \theta| - \zeta_n]$$

for any real sequence r_1, $r_2,\ldots$, where θ is a real number. Further, let $x_1,\ldots,x_n$ and y_1, $y_2,\ldots$, be random variables; for $n \geq 1$, let

5. $x_{n+1} = T_n(x_1, x_2,\ldots,x_n) + y_n + g_n(x_1,\ldots,x_n)$ where $g_n(r_1,\ldots,r_n)$ are measurable functions such that:

6.   The series $|g_n(r_1,\ldots,r_n)|$ uniformly converges and its sum if uniformly bounded for any sequence r_1, $r_2,\ldots$;

7. $E\{y_n/x_1,\ldots,x_n\} = 0$ with probability 1;

8. $\sum\limits_{n=1}^{\infty}$ $E\{y_n^2\} < \infty$ and $E\{x_1^2\} < \infty$

Then as $n \to \infty$

1.   Prob $\{ \lim\limits_{n \to \infty} x_n = \theta\} = 1$

2. $\lim\limits_{n \to \infty} E\{x_n - \theta\}^2 = 0$

The proof of this theorem is given in Ref. [3.9]. It represents the
most general case of stochastic approximation algorithms and both the
Robbins-Monro as well as the Kiefer-Wolfowitz algorithm can be considered as a special case of this theorem [3.48]. Dvoretzky presented another theorem which is a special case of the general theorem
3.3 but extends the problem to the multidimensional vector case of
x. This special theorem is also given here without proof. It covers
most of the stochastic approximation algorithms of interest in this
book.

Theorem 3.4

Assume that the vectors $\hat{x}_n$ and y_n are vectors in a normed linear
space Ω and that $T_n(x_1,\ldots,x_n)$ is a measurable transformation
from $\Omega^n = \Omega x,\ldots,x\Omega$ into Ω such that

$$||T_n(x_1,\ldots,x_n) - \theta|| \leq (1 - \gamma_n)||x_n - \theta|| \tag{3.81}$$

with $\theta \in \Omega_\theta$ and instead of 4. (Theorem 3.3).

$$(1 - \gamma_n) > 0 \quad \prod_{n=1}^{\infty} (1 - \gamma_n) = 0, \quad \sum_{n=1}^{\infty} \gamma_n^2 < \infty \tag{3.82}$$

$$x_{n+1} = T_n(x_1,\ldots,x_n) + y_n \tag{3.83}$$

If

$$E\{||\Phi(x_1,\ldots,x_n) + y_n||^2\} < E\{||\Phi(x_1,\ldots,x_n)||^2\} \tag{3.84}$$
$$+ E\{|| y_n ||^2\}$$

for every measurable function $\Phi(x_1,\ldots,x_n)$ over Ω, including T_n
with

$$E\{||x_1{}^2||\} + \sum_{n=1}^{\infty} E\{||y_n||^2\} < \infty \tag{3.85}$$

Then

$$Prob \{ \lim_{n \to \infty} ||x_n - \theta|| = 0\} = 1 \tag{3.86}$$

$$\lim_{n \to \infty} E\{||x_n - \theta||^2\} = 0 \tag{3.87}$$

3.3.5   Acceleration of Stochastic Approximation Algorithms

Stochastic approximation algorithms are usually slow to converge to the true parameter value, since the emphasis of the algorithm is on the convergence and not the speed of convergence. Many
schemes have been proposed in the literature to accelerate stochastic
approximation algorithms.

Kesten in [3.48] proposed an accelerated scheme where y_n is
kept constant until the correction term in (3.76) changes sign and
only then assumes the next value in the $\{y_n\}$ sequence. This scheme
converges with probability one. It has proven to considerably
accelerate the convergence of the Kiefer-Wolfowitz or Robbins-Monro
algorithms and to be one of the most effective schemes.

Another scheme may be generated by minimizing upper bounds on
the variance of the estimation error of the special case of the
Dvoretzky theorem 3.4.

$$V_n{}^2 \geq E\{|| x_n - \theta||^2\} \tag{3.88}$$

An optimal sequence of gains on which satisfies (3.82), can be
generated

$$\gamma_n{}^* = \frac{1}{n + \sigma^2/V_1{}^2} \tag{3.89}$$

which minimizes the upper bound of the error variable [3.43].

$$V_{n+1}{}^2 < \frac{\sigma^2}{n + \sigma^2/V_1{}^2} \tag{3.90}$$

This sequence $\{\gamma_n^*\}$ satisfies the Rao-Cramér inequality which in simple words says that convergence cannot be faster than at $c_1/(n + c_2)$ rate. Since this is a serious limitation on the speed of convergence other methods have been employed to accelerate the algorithm. It has been observed that in the mean-square case a matrix gain has been used instead of the gain γ_n. This idea has been utilized to generate an acceleration scheme with learning capabilities. This algorithm may be implemented as follows [3.42]:
Let

$$x_{n+1} = x_n + \gamma_{n+1} Q(n) f(x_n) \tag{3.91}$$

where γ_n satisfies conditions (3.82) and

$$Q(n) = N \begin{bmatrix} q_1(n) & 0 & \cdots & 0 \\ 0 & q_2(n) & \cdots & 0 \\ 0 & 0 & \cdots & q_N(n) \end{bmatrix}$$

$$0 \leq q_i(n) \leq 1, \quad q_i(0) = \frac{1}{N}, \quad i = 1,\ldots,N$$

$$N \sum_{i=1}^{N} q_i(n) = tr[q(n)] = N, \quad n = 1, 2,\ldots \tag{3.92}$$

A linear reinforcement learning algorithm whose properties are extensively discussed in Ref. [3.42] is used to update at every step the components of matrix $P(n)$ and alter the direction of the search accordingly.

$$q_i(n) = \alpha \, q_i(n - 1) + (1 - \alpha) \, \lambda_i(n) \tag{3.93}$$

where

$$0 < \alpha < 1$$

and

$$\lambda_i(n) = \begin{cases} \epsilon, & f_i(x_n) \cdot f_i(x_{n-1}) \leq 0 \\ \\ \frac{1}{m}, & f_i(x_n) \cdot f_i(x_{n-1}) > 0 \end{cases} \tag{3.94}$$

where m, $0 \leq m \leq N$ is the number of $f(x)$ which did not change sign at the last step.   If all components change sign,

$$\lambda_i(n) = \frac{1}{N} \ , \quad i = 1,2,\ldots,N \ ; \quad m = 0$$

This algorithm was successfully applied to nonlinear system identification.

These are only some of the successful acceleration schemes for stochastic approximation algorithms.   Their main quality is the simplicity of implementation but they are not optimal in any sense and cannot handle multimodal minimization problems.   Such an algorithm will be presented in the next sections.

3.3.6   Random Search Algorithms

Several algorithms have been proposed for the location of the maximum or the minimum of an optimization function in a stochastic environment.   The main difficulty is, however, that most of these methods, like the stochastic approximation, are local techniques and are not applicable to multimodal surfaces.   Several global techniques have been developed in recent years.   Fel'dbaum and Bocharov [3.10] were the first to use a random scheme of selecting the initial point for repeated application of the method of steepest descent to multimodal deterministic hills.   Fu and McMurtry [3.11] used a probabilistic automaton in search of multimodal hills. Kushner [3.24] developed a method for the location of a global maximum or minimum of a function.   He divides his procedure into two parts: (1) the choice of a suitable model for the unknown function, and (2) the choice of a decision policy based upon the conditional mean and variance of the surface conditioned over the past measurements and weighted properly according to a criterion of competitiveness.

However, the most successful and simple procedures for the location of the extremum of a multimodal multidimensional function are the so-called *random search techniques,* due to Matyash and Rastrigin [3.31], [3.37] for the deterministic case, and Gurin and Rastrigin

[3.13], [3.14], for the stochastic case. This method will be presented here, since it has found interesting applications to the S.O.C.
problem.

   Random search may be described as follows:

   Let x be a vector defined over the space Ω_x belonging to the n-
dimensional Euclidean space. The components of x are the parameters
with respect to which the scalar multimodal function Q(x) defined
over Ω_x is to be optimized. The deterministic case considers Q(x)
as a deterministic function of x, while the stochastic case considers
noise-corrupted measurements of Q(x), i.e.,

$$\tilde{Q}(x) = Q(x) + v \tag{3.95}$$

where v is a sample of an independent identically distributed random
process with unknown density function, and $E\{v\} = 0$, $E\{v^2\} < \infty$. In
both cases it is desired to find $x^* \varepsilon \Omega_x$ such that

$$Q(x^*) \leq Q(x) \quad \text{for all } x \varepsilon \Omega_x \tag{3.96}$$

Define ξ an n-dimensional random vector with $E\{\xi\} = 0$ and probability
density function $p(\xi) \neq 0$ for all ξ. An iterative procedure may be
constructed, starting at $x_o \varepsilon \Omega_x$ in such a way that at the ith step,
step one may define a vector $x \varepsilon \Omega_x$ as follows:

$$x_i = x_{i-1} + \xi_i , \quad x \varepsilon \Omega_x \tag{3.97}$$

If the measured values of the function Q and a small $\varepsilon > 0$ satisfy

$$Q(x_{i-1} + \xi_i) \leq Q(x_{i-1}) - \varepsilon \text{ are in the deterministic case} \tag{3.98}$$

or $$\tilde{Q}(x_{i-1} + \xi_i) \leq \tilde{Q}(x_{i-1}) - \varepsilon \text{ in the stochastic case} \tag{3.99}$$

the iteration is considered as a successful step and one may set

$$x_i = x_{i-1} + \xi_i , \quad x_i \varepsilon \Omega_x \tag{3.100}$$

Otherwise, the step is not successful and one may set

$$x_i = x_{i-1} \tag{3.101}$$

and proceed.

Be defining an index function A_i such that

$$A_i = \begin{cases} 1 \text{ if the ith step is successful} \\ \\ 0 \text{ if the ith step is not successful} \end{cases} \qquad (3.102)$$

The algorithm may be written in a compact form

$$A_i = x_{i-1} + A_i \xi_i \qquad (3.103)$$

Matyash showed in Ref. [3.31] that the random sequence $\{x_i\}$ converges in probability to x^*, i.e.,

$$\lim_{i \to \infty} \text{Prob } \{||x_i - x^*|| \le \delta\} = 1 \qquad (3.104)$$

for an arbitrary $\delta > 0$, and x^* corresponding to the global minimum of $Q(x)$. On the other hand Gurin [3.13] proved convergence in probability of (3.104) for the stochastic case. He also used repeated sampling at each point to reduce the variance of the estimates of the function. Gurin and Rastrigin [3.14] compared the convergence of random optimization and the gradient method. They concluded that the gradient method is more efficient for functions of few parameters, and that the random optimization is more efficient for functions of many variables.

An acceleration scheme has been successfully used in parameter identification problems [3.44]. It applies a bias to the search in the direction of successful iterations and also reduces the spread of the search. The algorithm (3.103) is rewritten as

$$x_i = x_{i-1} + A_i \zeta_i \qquad (3.105)$$

where the random vector ζ_i is generated by

$$\zeta_i = c_i + \beta_i + \xi_i \qquad (3.106)$$

The vector β_i is a moving mean vector, providing the appropriate memory to the algorithm for k the member of successful iterations.

$$\beta_k = \sum_{j=1}^{k} (c_j - c_{j-1}) \exp[-(k-j)] \qquad (3.107)$$

The random vector ξ_i is selected from a zero-mean Gaussian distribution overlap with covariance

$$\text{cov}[\xi_i, \xi_j] = \frac{1}{k} P \, \delta_{ij}, \quad P = P^T > 0 \tag{3.108}$$

Such an accelerated random search has given satisfactory results,
e.g., a 10% improvement of convergence on the Rosenbrock curve
described in [3.48].

Gilbert and Saridis used this algorithm for a stochastic optimal control problem with adjustable parameters. In this case one
assumes that the process is of infinite length and that the performance index is split to a sequence of per-interval "sub-goals" which
are asymptotically ergodic in the mean. Then the following theorem
is proved [3.44]:

Theorem 3.5

Let an arbitrary number $\delta > 0$ be given. Then for the

$$x_{k+1} = \begin{cases} \xi_k, & Q(x_k) - Q(\xi_k) > 2\mu \\ \\ x_k, & Q(x_k) - Q(\xi_k) \le 2\mu \end{cases} \tag{3.109}$$

there exists a number $\mu > 0$ such that as $k \to \infty$ the "accrued"
cost

$$\bar{J}(k) = \frac{\sum\limits_{i=1}^{k} [T_i E\{Q(\xi_i)\}]}{\sum\limits_{i=1}^{k} T_i} \quad \text{satisfying the condition, } T_j < T_{j+1},$$

converges in probability

$$\lim_{k \to \infty} \quad \text{Prob } \{\bar{J}(k) - \bar{J}^* < 4\delta\} = 1 \tag{3.110}$$

3.4   STATE ESTIMATION: LINEAR SYSTEMS

Until now, the problem of concern was parameter estimation, that is,
the construction of stationary random processes converging to the true

parameter value. The property of stationary allowed the use of simple
statistical properties to evaluate the estimates. The problem, however,
that arises in control theory and similar disciplines is to estimate
the states of a dynamic system in a stochastic environment. Such
states constitute nonstationary processes that, however, often
possess the Markovian property, i.e., $p(x/z_1, z_2, \ldots, z_n) = p(x/z_n)$,
where z_n is the most recent observation. The previously described
methods must be extensively modified to apply to the problem of
dynamic estimation. The Wiener derivation of the stationary filter
for a dynamic system will first be presented. Then the work of
Kalman and Bucy will provide the dynamic estimator for continuous
and discrete-time systems. The presentation will be from two different points of view to give a better perspective to the problem.
Finally, other state estimators will be briefly discussed to complete the picture of state estimation, which is of major importance
to the stochastic control problem to follow.

3.4.1 The Wiener Filter

    In the United States, the first work on state estimation was
done by Wiener 1942, while similar results were published by
Kolmogorov in U.S.S.R. in 1941. In his pioneering work [3.47],
Wiener used for the first time a variational approach to solve the
state-estimation problem. The problem that Wiener actually solved
is more general, since it gives the best estimates $\hat{x}(t)$ of Markovian
nonstationary signals $x(t)$, corrupted by a stationary noise process
$v(t)$ with known statistical characteristics. Estimates may be obtained for the *prediction problem*, i.e., for values of $\hat{x}(t/t_1)$,
$t > t_1 > 0$, for the future of the process, the *filtering problem*
$\hat{x}(t/t) = \hat{x}(t)$ at the present time, or the *smoothing problem* $\hat{x}(t/t_2)$,
$t_2 > t > 0$, for the past process. Here we are concerned with the
filtering problem only; for the other problems, see Refs. [3.12],
[3.33], [3.37], [3.45]. The best estimates will be obtained with
respect to the mean-square error of the signal

$$R^*(\hat{x}) = \underset{\hat{x}(t)}{\text{Min}}\ E\{||\hat{x}(t) - x(t)||^2\} \tag{3.111}$$

A block diagram for the filtering problem is given in Fig. 3.4.

Assuming that one is looking for the best *linear* estimation, the filter would have an impulse response $h_w(t)$, such that the convolution integral holds

$$\hat{x}(t) = \int_{-\infty}^{\infty} h_w(\tau)x(t-\tau)\ d\tau \tag{3.112}$$

Since $\hat{x}(t)$ depends on $h_w(t)$, the impulse response of the filter may be chosen to minimize the risk function of (3.111)

$$R(\hat{x}) = E\{(\tilde{x})^2\} = E\{[\hat{x}(t) - x(t)]^2\} = E\{\hat{x}^2(t)\}$$
$$- 2\ E\{\hat{x}(t)x(t)\} + E\{x^2(t)\} \tag{3.113}$$

The first two terms in (3.113) are computed to be

$$E\{\hat{x}^2(t)\} = \int_{-\infty}^{\infty} h_2(\tau_1)d\tau_1 \int_{-\infty}^{\infty} h_w(\tau_2)d\tau_2 R_{zz}(t,\tau_1,\tau_2) \tag{3.114}$$

$$E\{\hat{x}(t)x(\tau)\} = \int_{-\infty}^{\infty} h_w(\tau)d\ R_{xz}(t-\tau) \tag{3.115}$$

where $R_{zz}(t,\tau_1,\tau_2)$ is the autocorrelation function of $z(t)$, and $R_{xz}(t,\tau)$ is the crosscorrelation function of $z(t)$ and $x(t)$. A variational approach is used for the functional minimization of (3.113). For this purpose the following variation about the optimal $h_w^*(t)$ impulse response is defined:

$$h_w(t) = h_w^*(t) - \varepsilon\ w(t) \tag{3.116}$$

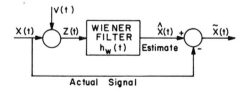

Fig. 3.4.   Wiener Estimator and Error Signal

where $w(t)$ is an "admissible" variation, and ε is an arbitrary parameter.   The necessary condition for optimality of (3.113) is

$$\left.\frac{\partial E\{\tilde{x}^2\}}{\partial \varepsilon}\right|_{\varepsilon \,=\, 0} = 0 \qquad\qquad (3.117)$$

Substituting into the right equation and solving the minimization problem the following expression is obtained:

$$2 \int_{-\infty}^{\infty} d\tau_1 \; w(\tau_1) \left\{\int_{-\infty}^{\infty} d\tau_2 [h^*(\tau_2) R_{zz}(t,\tau_1,\tau_2)]\right\} = R_{zx}(t,\tau_1) \qquad (3.118)$$

Since (3.118) is true for arbitrary $w(\tau)$, the lemma of calculus of variations is applicable and

$$\int_{-\infty}^{\infty} dz \; h_w^*(\tau) R_{zz}(t,\tau) = R_{zx}(t) \qquad\qquad (3.119)$$

This is the celebrated *Wiener-Hopf equation* and its solution yields the best linear estimator for $x(t)$.   Such a solution is not directly obtained.   For $R_{zz}(t,\tau) = R_{zz}(t - \tau)$ and $R_{zx}(t,\tau) = R_{zx}(t - \tau)$ transformations in the frequency domain are applicable, and a method called "*spectral factorization*" may be used to obtain a physically realizable $h_w^*(t)$.   This is not presented here, since the state estimation problem will be treated in more detail in the time domain.   A more general form of (3.119) is given by:

$$\int_{-\infty}^{\infty} dz \; h_w^*(\tau) \; R_{zz}(t - \tau + \alpha) = R_{zx}(t,\alpha) \qquad\qquad (3.120$$

where $\alpha > 0$ for the prediction problem and $\alpha < 0$ for the smoothing problem.

In the case that $z(t)$ is white Gaussian noise

$$R_{zz}(t - \tau) = \delta(t - \tau) \qquad\qquad (3.121)$$

Then (3.120) is given by

$$h_w^*(t + \alpha) = R_{zx}(t,\alpha) \qquad\qquad (3.122)$$

The Wiener estimator and its generalization have created a whole new field in estimation theory.

3.4.2 The Kalman-Bucy Filter: Continuous-Time Case

The work by Wiener in sequential estimation of stochastic processes was extended by Kalman and Bucy for continuous and discrete time processes generated by processing white Gaussian noise through a linear dynamic system [3.18], [3.19]. The original derivation of the continuous-time case which uses Wiener's work will be presented here.

Let a system be described by the vector differential equation

$$\frac{dx}{dt} = F(t)x(t) + G(t)w(t) \qquad x(0) = x_0$$

$$z(t) = y(t) + v(t) = H(t)x(t) + v(t) \tag{3.123}$$

where $x(t)$ is an n-state vector, $F(t)$ an nxn matrix, $G(t)$ an n x m matrix, $H(t)$ an r x n matrix, and $z(t)$ and $y(t)$ are r-vectors where z is the measurable output; $w(t)$ and $v(t)$ are white Gaussian noise processes, and x_0 is also a Gaussian process, with known statistics such that

Means: $E\{w(t)\} = E\{v(t)\} = 0$ $E\{x(0)\} = \bar{x}_0$

Covariances: $E\{w(t)w^T(t - \tau)\} = Q(t)\delta(t - \tau)\}$

$$E\{(x(0) - \bar{x}_0) \cdot (x(0) - \bar{x}_0)^T\} = P_0$$

$$E\{v(t)v(t - \tau)\} = R(t)\delta(t - \tau)\} \quad E\{x(0)v^T(t)\} = 0$$

$$E\{v(t)w^T(t - \tau)\} = 0 \quad E\{x(0)w^T(t)\} = 0 \tag{3.124}$$

Define:

$\hat{x}(t_1/t) \triangleq$ estimate of x at t_1, give all $z(\sigma)$, $0 \leq \sigma \leq t$

$\tilde{x}(t_1/t) \triangleq \hat{x}(t_1/t) - x(t) =$ estimation error

Problem:

Find the estimate-linear filter $A(t,\tau)$, for which

$$\hat{x}(t_1/t) \triangleq \int_0^t A(t_1,\tau)z(\tau) \ d\tau, \quad \hat{x}(0/0) = \bar{x}_o \tag{3.125}$$

such that the risk $R(A)$ is *minimum*:

$$R(A) = E\{||x(t_1) - \hat{x}(t_1/t)||^2\} \tag{3.126}$$

<u>Solution</u>:

The procedure followed by Kalman will be presented here. It is known that the response of (3.127) is given by

$$x(t) = \int_{-\infty}^t \Phi(t,\tau)G(\tau)w(\tau) \ d\tau \tag{3.127}$$

where $\Phi(t,\tau)$ is the state transition matrix and the effect of initial conditions is incorporated in $\int_{-\infty}^o (t,\tau)Gw(\tau) \ d\tau$. It is easily shown that the estimate $\hat{x}(t_1/t)$ is unbiased.

Theorem 3.6

A necessary and sufficient condition that $\hat{x}(t_1/t)$ be a minimum variance estimator for $x(t_1)$ is that the matrix $A(t,\tau)$ satisfies the Wiener-Hopf equation:

$$R_{xz}(t_1 - \sigma) = \int_0^{t_1} A(t_1,\tau)R_{zz}(\tau - \sigma) \ d\tau \tag{3.128}$$

The proof of this theorem for the time invariant case $A(t_1,\tau) = h_w(t)$ was given in the previous section. It has been shown by Kalman in [3.18], [3.19]. The filter equations for the filtering case, are obtained by differentiation of Eq. (3.128) and substitution into (3.125) for $t_1 = t$

$$\frac{d\hat{x}(t/t)}{dt} = \int_0^t \frac{\partial A(t,\tau)}{\partial t} \ z(\tau) \ d\tau + A(t,t)z(t) \tag{3.129}$$

and

$$\frac{d\hat{x}(t/t)}{dt} = F(t)\hat{x}(t/t) + A(t,t)[z(t) - H(t)\hat{x}(t/t)]$$
$$\hat{x}(0/0) = E\{x(0)\} = \bar{x}_o \tag{3.130}$$

The estimation error $\tilde{x}(t/t) = \hat{x}(t/t) - x(t)$ is given by the following differential equation:

$$\frac{d\tilde{x}(t/t)}{dt} = [F(t) - A(t,t)H(t)] \tilde{x}(t/t) - G(t)w(t)$$
$$- A(t,t)v(t) \tag{3.131}$$

One needs to obtain an explicit form of the matrix $A(t,t)$ of the filter. For this purpose an important relation must be established, namely that the estimation error $\tilde{x}(t/t)$ is orthogonal to the estimate $\hat{x}(t/t)$. This very important property may serve as a starting point to develop the state estimation theory from a functional analysis point of view. By using the orthogonal projection theorem 3.2 to show that

$$E\{\tilde{x}(t/t) \, \hat{x}(t_1/t)\} = 0 \quad 0 \leq t_1 \leq t \tag{3.132}$$

one may obtain the same relation defined by (3.128)

$$E\{\tilde{x}(t/t) \, \hat{x}(t_1/t)\} = \int_0^{t_1} \int_0^t d_\sigma A\,(t,\sigma)R_{zz}(\sigma - \tau)A^T(t_1,\tau) \, d\tau$$

$$- \int_0^{t_1} R_{xz}(t - \tau)A^T(t_1,\tau) \, d\tau$$

$$= \int_0^{t_1} R_{xz}(t - \tau)A^T(t_1,\tau)d\tau$$

$$- \int_0^{t_1} R_{xz}(t - \tau)A^T(t_1,\tau) \, d\tau = 0 \tag{3.133}$$

In order to get the relation for $A(t,\tau)$, consider the covariance of the estimation error $E\{\tilde{x}(t/t)\tilde{x}^T(t/t)\} \triangleq P(t)$. The solution of (3.131), assuming that $\psi(t,\tau)$ is its state transition matrix, is:

$$\tilde{x}(t/t) = \psi(t,0)\tilde{x}(0/0) + \int_0^t \psi(t,\tau)[K(\tau)v(\tau) - G(\tau)w(\tau)] \, d\tau$$

$$\tag{3.134}$$

where $A(t/t) \triangleq K(t)$.   Then

$$P(t) = E\{\tilde{x}(t/0)\tilde{x}(t/0)\} = \psi(t,0)E\{\tilde{x}(0/0)\}\psi^T(t,0) + \int_0^t \int_0^t \psi(t,\tau)$$

$$[K(\tau) \ E\{v(\tau)v^T(\sigma)\}]\psi^T(t,\sigma)d\sigma \ d\tau + \int_0^t d\sigma \int_0^t \psi(t,\tau)$$

$$G(\tau)E\{w(\tau)w^T(\sigma)\}G^T(\sigma)\psi^T(t,\tau) \ d\sigma \qquad (3.135)$$

By considering $E\{\tilde{x}(0/0)\tilde{x}^T(0/0)\} = P(0) = P_0$ and (3.124) and differentiating (3.135) with respect to t, one obtains:

$$\frac{dP(t)}{dt} = G(t)Q(t)G^T(t) + F(t)P(t) + P(t)F^T(t)$$
$$- P(t)H^T(t)R^{-1}(t)H(t)P(t) \qquad (3.136)$$

Equation (3.136) is a *matrix Riccati equation* and defines the covariance of the estimation error.

Consider that

$$P(t)H^T(t) = E\{(\hat{x}(t/t) - x(t))(\hat{x}(t) - x(t))^T\}H^T(t)$$
$$= A(t,t)R(t) \qquad (3.137)$$

then

$$K(t) = A(t,t) = P(t)H^T(t)R^{-1}(t) \qquad (3.138)$$

The filter and the system are demonstrated in Fig. 3.5.   The equations for the Kalman-Bucy filter are summarized by

$$\frac{d\hat{x}}{dt} (t/t) = F(t)\hat{x}(t/t) + A(t,t)[z(t) - H(t)\hat{x}(t/t)]; \quad \hat{x}(0/0) = x_0$$

$$A(t,t) = P(t)H^T(t)R^{-1}(t) \qquad (3.139)$$

$$\frac{dP}{dt} = F(t)P(t) + P(t)F^T(t) + G(t)Q(t)G^T(t)$$
$$- P(t)H^T(t)R^{-1}(t)H(t)P(t)$$

If one wishes to obtain estimates for the prediction or the smoothing problem $\hat{x}(t_1/t)$, $t_1 \neq t$, then from the transition properties of
a linear system one may easily obtain

$$\hat{x}(t_1/t) = \Phi(t_1/t) \ \hat{x}(t/t) \qquad (3.140)$$

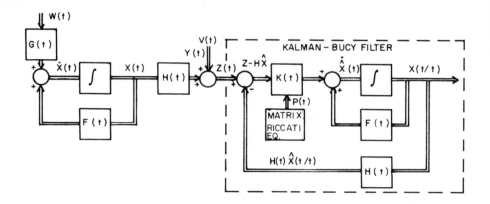

Fig. 3.5.   Continuous-Time System and Kalman-Bucy State
            Estimator

The Kalman-Bucy filter is the best minimum variance estimator for
the linear Gaussian system considered.

3.4.3   The Kalman-Bucy Filter: Discrete-Time Case

The best state estimator for a linear discrete-time system will
be derived next; consider the system

$$x(k + 1) = F(k)\ x(k) + G(k)\ w(k)$$

$$z(k) = H(k)\ x(k) + v(k) \tag{3.141}$$

with random variables $x(0)$, $w(k)$, and $v(k)$ defined by

$$E\{x(0)\} = \bar{x}_0, \quad E\{(x(0) - \bar{x}_0)(x(0) - \bar{x}_0)^T\} = P_0$$

$$E\{w(k)\} = 0 \quad E\{v(k)\} = 0, \ \forall\ k$$

$$E\{w(k)w^T(j)\} = Q(k)\delta_{k_j}, \quad E\{v(k)\ v^T(j)\} = R(k)\delta_{k_j}$$

$$E\{w^T(k)x(0)\} = 0, \quad E\{v(k)x(0)\} = 0, \quad E\{w^T(k)v^T(k)\} = 0$$

and $F(k)$ is an nxn dimensional transition,matrix, $G(k)$ is an nxs gain
matrix, $H(k)$ is an rxn output matrix, $x(k)$ is the n-dimensional state

vector, and $z(k)$ is the r-dimensional output vector; δ_{kj} is the
Kronecker delta, $\delta_{kk} = 1$, $\delta_{kj} = 0$, $k \neq j$.
Define:

> $\hat{x}(k/m)$ = the estimate of $x(k)$ based on the sequence of measure-
> ments
> $Z^m = \{z(1),\ldots,z(m)\}$; $\hat{x}(k/m) = E\{x(k)/Z^m\}$
> $\tilde{x}(k/m) = \hat{x}(k/m) - x(k)$ = the respective estimation error

Problem:

Find the discrete-time filter which yields the best estimate
$\hat{x}(k/n)$ of the states.

Solution:

The solution will be based on the *orthogonal projection theorem*
discussed in Sec. 3.2.3. The theorem will be adapted to fit the
dynamic estimation problem. Consider the space Z_m generated by all
the independent vectors of the sequence $\{z(1),\ldots,z(m)\}$ and the space
X_m generated by the independent vectors of the sequence $\{x(0), x(1),$
$\ldots,x(m)\}$. Since $z(i)$ is a function of $x(i)$ and of dimension $r < n$,
then the space Z_m is a subspace of X_n

$$Z_m \subset X_m$$

Theorem 3.7 - Orthogonal Projection.

Let $\hat{x}(k/m)\epsilon Z_m$ be the orthogonal projection of the state
$x(k)\epsilon X_k$ on the subspace Z_m. Then $\hat{x}(k/m)$ is the best estimate
of $x(k)$ based on Z^m and its normal component $\hat{x}(k/m)$ is orthogonal to Z^m and accounts for the estimation error.

The orthogonality condition implies the following four properties as in 3.2.3:
1. $\hat{x}(k/m)\epsilon Z_m$.
2. $\tilde{x}(k/m)$ is orthogonal to all vectors in Z_m.
3. $\hat{x}(k/m)$ is unique in the Hilbert space X_k.
4.   The estimate $\hat{x}(k/m)$ minimizes the mean-square error cri-
     terion

$$E\{||\tilde{x}(k/m)||^2\} = Min \qquad (3.142)$$

and the error has minimum covariance.

$$E\{\tilde{x}(k/m) \ \tilde{x}^T(k/m)\} = Min \qquad\qquad (3.143)$$

This final property implies that if the state $x(k)$ has a probability density function which satisfies Bayes Theorem 3.1, then the estimate $\hat{x}(k/m)$ is the best possible estimate.

A recursive relation may be obtained for the evaluation of the best estimate $\hat{x}(k + 1/k)$ for the filtering problem under present consideration. This may be generated from the sequential estimator Eq. (3.55), Sec. 3.3.1, based on the estimate $\hat{x}(k/k - 1)$ and the new measurement $z(k)$ from Eq. (3.141). The sequential estimator of Sec. 3.3.1 can be used for this purpose because it is the minimum mean-square error, minimum variance estimator defined by the properties of the orthogonal projection theorem. Equation (3.55) may be modified to read

$$\hat{x}(k/k) = \hat{x}(k/k - 1) + \tilde{P}(k - 1)H^T(k)[H(k)\tilde{P}(k - 1)H^T(k)$$
$$+ R(k)]^{-1}(z(k) - H(k)\hat{x}(k/k - 1))$$

$$\tilde{P}(k/k) = \tilde{P}(k - 1) - \tilde{P}(k - 1)H^T(k)[H(k)\tilde{P}(k - 1)H^T(k)$$
$$+ R(k)]^{-1} H(k)\tilde{P}(k - 1) \qquad\qquad (3.144)$$

where

$$\tilde{P}(k/k) \triangleq E\{\tilde{x}(k/k)\tilde{x}^T(k/k)\} \qquad\qquad (3.145)$$

To complete the recursive relation one should obtain the best estimate of $x(k + 1)$ given Z^k

$$\hat{x}(k + 1/k) = E\{F(k)x(k) + G(k)w(k)/_Z k\} = F(k)\hat{x}(k/k) \qquad (3.146)$$

Then error $\tilde{x}(k + 1/k)$ is also orthogonal on Z_k. Substituting (3.146) into (3.144), the recursive expressions of $\hat{x}(k + 1/k)$ and $\tilde{x}(k + 1/k)$ are obtained.

$$\hat{x}(k + 1/k) = F(k)\hat{x}(k/k - 1) + F(k)\tilde{P}(k - 1)H^T(k)[H(k)\tilde{P}(k - 1)H^T(k)$$
$$+ R(k)]^{-1} (z(k) - H(k)\hat{x}(k/k-1)) \qquad\qquad (3.147)$$

$$\tilde{x}(k + 1/k) = F(k)\tilde{x}(k/k - 1) + F(k)\tilde{P}(k - 1)H^T(k)[H^T(k)\tilde{P}(k - 1)H^T(k)$$
$$+ R(k)]^{-1}(z(k) - H(k)\hat{x}(k/k - 1) - G(k)\ w(k)) \quad (3.148)$$

The covariance of $x(k + 1/k)$, $\tilde{P}(k) \triangleq E\{\tilde{x}(k + 1/k)\tilde{x}^T(k + 1/k)\}$
is derived by

$$\tilde{P}(k) = E\{\tilde{x}(k + 1/k)\tilde{x}^T(k + 1/k)\} = F(k)E\{[\hat{x}(k/k) - x(k)]$$
$$[\hat{x}(k/k) - x(k)]^T\}\ F^T(k) - F(k)E\{[\hat{x}(k/k) - x(k)]w^T(k)\}G^T(k)$$
$$- G(k)E\{w(k)(\hat{x}(k/k) - x(k))^T\}\ F(k) + G(k)E\{w(k)w^T(k)\}\ G^T(k)$$
$$= F(k)\tilde{P}(k/k)F^T(k) + G(k)Q(k)G^T(k) \quad (3.149)$$

Equation (3.149) was obtained by using the covariance of $(\hat{x}(k/k) - x(k))$, namely $\tilde{P}(k/k)$ given in (3.144), and the fact that $w(k)$ is a zero mean process uncorrelated with $(x(k/k) - x(k))$ with covariance $Q(k)$. Therefore, the filter equations become

$$\hat{x}(k + 1/k) = F(k)\hat{x}(k/k - 1) + F(k)\tilde{P}(k - 1)H^T(k)[H(k)\tilde{P}(k-1)H^T(k)$$
$$+ R(k)]^{-1}(z(k) - H(k)\hat{x}(k/k - 1)) \quad (3.150)$$

$$\tilde{P}(k) = F(k)\tilde{P}(k - 1)F^T(k) - F(k)\tilde{P}(k - 1)H^T(k)[H(k)\tilde{P}(k - 1)H^T(k)$$
$$+ R(k)]^{-1}H(k)P(k - 1)F^T(k) + G(k)Q(k)G^T(k) \quad (3.151)$$

Initial conditions may be established if $E\{x_o\}$ and $E\{x_o x_o^T\}$ are known, i.e.,

$$\hat{x}(0/0) = E\{x(0)\} = \bar{x}_o; \quad \tilde{P}(0) = E\{(\hat{x}(0) - \bar{x}_o)(\hat{x}(0) - \bar{x}_o)^T\} =$$
$$P_0 \quad (3.152)$$

A block diagram illustrates in Fig. 3.6 the filter.
As in the continuous case, a predictor can be established based on measurements Z^k by

$$\hat{x}(m/k) = \prod_{i=k}^{m} F(i)\hat{x}(k/k) \quad (3.153)$$

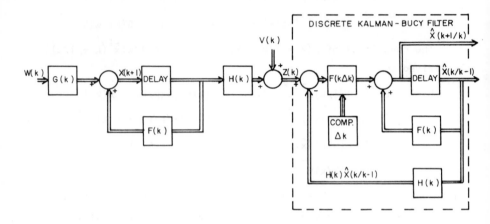

Fig. 3.6.   Discrete-Time Systems and Kalman-Bucy State
            Estimator

3.4.4 Other Optimal State Estimators

The Kalman-Bucy filter is the optimal linear estimator for
linear dynamic systems perturbed by Gaussian independent identically
distributed noise. These properties have been established as an
extension of the Wiener-filter and the minimum mean-square criterion
in the continuous-time case and the orthogonal projection theorem
in the discrete-time case.

However, the other methods described in Sec. 3.2 may be adapted
to produce state estimators optimal in some sense or another. For
linear systems with Gaussian random noises and/or convex criteria
the results are the same as with Kalman-Bucy filter. A good reference for such estimators is the book by Lee [3.27]. A brief discussion will be given in the following section for some of these
estimators.

The recursive *Bayesian or maximum a posteriori estimator* is
given to minimize the *a posteriori* probability density of x(k) given
a sequence of measurement $z^k = \{z(1),\ldots,z(k)\}$

$$\hat{x}(k/k) = \underset{\hat{x}}{\text{Max}}\, p(x(k)/z^k) \tag{3.154}$$

It requires the evaluation of several density functions. However,
if the noises are independent identically distributed Gaussian random variables corresponding to the linear system (3.141), then the
problem is reduced to the minimization of a quadratic form which
yields recursive equations such as (3.150) and (3.151).

The same problem can be formulated as a least-square fit
problem by minimizing the equation errors generated by the approximations of Eq. (3.151)

$$
J = \frac{1}{2} \sum_{i=0}^{n-1} \{||z(i + 1) - H(k)\hat{x}(i + 1/k)||^2_{R(k)^{-1}}
$$

$$
+ ||\hat{w}(i/k)||^2_{Q(k)^{-1}} \} \tag{3.155}
$$

subject to

$$
\hat{x}(i + 1/k) = F(k)\hat{x}(i/k) + G(k)\hat{w}(i/k) \tag{3.156}
$$

given the measurements Z_k. The problem can be treated as a deterministic optimal control problem with control function $\hat{w}(i/k)$, and
variational methods may be applied. The resulting filters are
similar [3.27] to the equations given by (3.150) and (3.151).

Finally, a minimum mean-square procedure may be applied to
generate the filter equations by forming the equation of measurements from (3.151)

$$
z^k = A_k x(k) + v^k \tag{3.157}
$$

where

$$
z^{k^T} = [z_1,\ldots,z_k], \quad A_k^T = [F^{(-k)^T}h,\ldots,h],
$$

$$
v^{k^T} = [v_1,\ldots,v_k] \tag{3.158}
$$

By following the same procedure as in Sec. 3.3.1 the following
a posteriori filter equations are obtained for time-invariant vectors and matrices:

$$
F(k) = F, \quad G(k) = G, \quad H(k) = H
$$

$$\hat{x}(k + 1/k + 1) = F\ \hat{x}(k/k) + \tilde{P}(k + 1/k)H^T[H\tilde{P}(k + 1/k)H^T + R]^{-1}$$

$$[z(k + 1) - HF\ \hat{x}(k/k)]$$

$$\tilde{P}(k + 1/k) = F\tilde{P}(k/k)\ F^T + GQG^T \qquad\qquad (3.159)$$

$$\tilde{P}(k + 1/k + 1) = \tilde{P}(k + 1/k) - \tilde{P}(k + 1/k)$$

$$H^T\ [H\tilde{P}(k + 1/k)H^T + R]^{-1}H\tilde{P}(k + 1/k)$$

These methods are strongly dependent on the linearity of the
system and the Gaussian property of the noises involved for qualifying as the best (most probable) estimate. However, if the noise
is not Gaussian, optimal estimates will still be obtained with
respect ot the pertinent criteria. They are usually functions of
the first and second moments of the random processes involved.

3.4.5   Parameter Adaptive State Estimation

State estimation cannot be directly performed by a Kalman-Bucy
filter if there are uncertainties in the dynamic system to be estimated. Magill [3.30] and Lainiotis [3.25] proposed a solution to
the problem by using a Bayesian approach to obtain the state estimate
of the system, in spite of the uncertainty which is confined in the
parameter vector $\theta \in \Omega_\theta$. The presentation here will follow
Lainiotis's derivation for continuous- as well as discrete-time
systems. These are described in terms of two theorems [3.25].

Theorem 3.8 - Partition theorem

State estimation is sought for the system

$$\frac{dx}{dt} = F(\theta,t)x + G(\theta,t)w(t) \qquad\qquad (3.160)$$

$$z(t) = H(\theta,t)x + v(t)$$

where $\{v(t)\}$ and $\{w(t)\}$ are independent zero mean white Gaussian
random processes with covariance matrices $R(t)$ and I respectively.
The initial state $x(t_0)$ is a random variable independent of
$\{v(t)\}$ or $\{w(t)\}$ for $t > t_0$ and it is Gaussian distributed with
mean and variance $\hat{x}(t_0|\theta,t_0)$ and $P(t_0|\theta,t_0)$. The model is com-
pletely specified by the parameter vector θ, which is unknown
and it is considered a random variable with *a priori* probability
density $p(\theta)$.

Given the observation record $Z^t = \{z(\tau), \tau \epsilon (t_0, t)\}$ the optimal mean-square estimate $\hat{x}(t/t)$ of $x(t)$ is given by:

$$\hat{x}(t/t) = \int_{\Omega_\theta} \hat{x}(t/\theta, t) \ p(\theta/t) \ d\theta \tag{3.161}$$

where $\hat{x}(t/t) \underset{=}{\Delta} E\{x(t)/Z^t)\}$ and $\hat{x}(t/\theta, t) = E\{x(t)/\theta, Z^t\}$ evaluated by Kalman filter Eq. (3.139) for every parameter vector θ.

The *a posteriori* probability of θ, given Z^t, is:

$$p(\theta/t) \underset{=}{\Delta} p(\theta/Z^t) = p(\theta) \ \exp\left\{-\int_{t_0}^t \hat{x}(\tau/\theta, \tau) H^T(\theta, \tau) R^{-1}(\tau) z(\tau) \ d\tau\right.$$

$$\left. -\frac{1}{2} \int_{t_0}^t ||H(\theta, \tau) \ \hat{x}(\tau/\theta, \tau)||^2_{R^{-1}(\tau)} \ d\tau\right\}$$

$$\exp\left\{-\int_{t_0}^t \hat{y}^T(\tau/\tau) R^{-1}(\tau) z(\tau) \ d\tau\right.$$

$$\left. -\frac{1}{2} \int_{t_0}^t ||\hat{y}(\tau/\tau)||^2_{R^{-1}(\tau)} \ d\tau\right\} \tag{3.162}$$

where $\hat{y}(\tau/\tau) = E\{y(\tau)/Z^t\} = \int_{\Omega_\theta} H(\theta, \tau) \hat{x}(\tau/\tau) p(\theta/\tau) \ d\tau$

The conditional state error-covariance matrix $P(t)$, needed for the evaluation of a Kalman-Bucy estimator is given by:

$$P(t) = E\{\tilde{x}(t/t)\tilde{x}^T(t/t)/Z^t\} = \int_{\Omega_\theta} P(t/\theta, t)[\hat{x}(t/\theta, t) - \hat{x}(t/t)]$$

$$[\hat{x}(t/\theta, t) - \hat{x}(t/t)]^T p(\theta/t) \ d\tau \tag{3.163}$$

where $P(t/\theta, t) = E\{[x(t) - \hat{x}(t/\theta, t)][x(t) - \hat{x}(t/\theta, t)]^T/ Z^t, \theta\}$ and it is evaluated from (3.147) for every θ.

It is noted by Lainiotis that this nonlinear estimation scheme can be partitioned into a linear part consisting of an ordinary Kalman-Bucy filter defined as a function of the unknown parameter vector θ, and a nonlinear part consisting of the adaptive procedure which yields the state estimates as average values over the space of the parameter vector θ.

The same procedure may be established for a discrete-time system and the following theorem is given by Lainiotis [3.25].

Theorem 3.9

The equations for a discrete-time dynamic system with an uncertainty are given by

$$x(k + 1) = F(\theta,k)x(k) + w(k)$$

$$z(k) = H(\theta,k)x(k) + v(k) \qquad\qquad (3.164)$$

where $\{v(k)\}$ and $\{w(k)\}$ are independent zero mean white Gaussian sequences with covariance matrices $R(k)$ and $Q(\theta,k)$, respectively. The initial state $x(0)$ is a random variable independent of $\{v(k)\}$ and $\{w(k)\}$ with mean $\hat{x}(0/\theta,0)$ and covariance $P(0/\theta,0)$. The unknown parameter has *a priori* probability density $p(\theta)$. Given $Z^k = \{z(j), j = 1,2,\ldots,k\}$, the optimal mean-square state estimate $\hat{x}(k/k)$ is given by

$$\hat{x}(k/k) = \int_{\Omega_\theta} \hat{x}(k/\theta,k)p(\theta/k)\ d\theta \qquad\qquad (3.165)$$

where $\hat{x}(k/k) = E\{x(k)/Z^k\}$ and $\hat{x}(k/\theta,k) = E\{x(k)/Z^k\theta\}$.

The *a posteriori* probability of θ given $Z^k = \{z(1),\ldots,z(k)\}$ can be computed from:

$$p(\theta/k) = p(\theta/Z^k) = \left|P_z(k/\theta,k - 1)\right|^{-1/2} \exp\left\{-\tfrac{1}{2}||z(k)\right.$$

$$\left. - H(\theta,k)\hat{x}(k/\theta,k)||^2_{P_z^{-1}(k/\theta,k - 1)}\right\}p(\theta/k - 1)$$

$$/ \int_{\Omega_\theta} p(\theta/k - 1)\left|P_z(k/\theta,k - 1)\right|^{-1/2} ||z(k)$$

$$- H(\theta,k)\hat{x}(k/\theta,k)||^2_{P_z^{-1}(k/\theta,k - 1)}]\ d\theta \qquad\qquad (3.166)$$

where the process $(z(k) - H(\theta,k)\hat{x}(k/\theta,k))$ is a conditionally white-noise process with a covariance

$$P_z(k/\theta,k - 1) = H(\theta,k)P(k/\theta,k)H^T(\theta,k) + R(k) \qquad\qquad (3.167)$$

Furthermore, the state-error covariance $P(k/k)$ can be computed from the individual covariances $P(k/\theta,k) = E\{(x(k) - \hat{x}(k/\theta,k))(x(k) - \hat{x}(k/\theta,k))^T/\theta,z^k\}$ computed for each θ:

$$P(k/k) = \int_{\Omega_\theta} \{P(k/\theta,k) + (\hat{x}(k/\theta,k) - \hat{x}(k/k))(\hat{x}(k/\theta,k)$$
$$- \hat{x}(k/k))^T\}p(\theta/k) \; d\theta \qquad\qquad (3.168)$$

The computation of the density function and the statistics
involved is not an easy task, especially when conditioned on θ.
Computations are simplified if θ belongs to a set of finite values.
The proofs of the above theorems are given in [3.25].

Other parameter adaptive procedures may be applicable to the
problem of the state estimation with an unknown parameter vector.

3.5 NONLINEAR ESTIMATION

The preceding section was concerned with the problem of estimation of linear problems with Gaussian random variables. The
problem has acceptable solutions for the stationary, recursive, and
dynamic cases, which may be implemented with only minor difficulties.
However, the more general problem of estimation of nonlinear systems
perturbed by Gaussian or nonGaussian random processes is still
unsolved. In most cases, a linearization by expansion is proposed
as the most realistic approximate solution. A very brief discussion,
without any attempt of derivation of the existing algorithms for
nonlinear estimation is presented here. They are used in certain
S.O.C. algorithms of Chap. 6, but their mathematical derivation is
beyond the scope of this review, [3.6], [3.12], [3.132], [3.39],
[3.49].

3.5.1 Stationary Nonlinear Estimators

Even though in most cases parameters x enter a system linearly,
it is possible that the measurements z may be nonlinear functions of
the parameters as in

$$z = f(x) + v \qquad\qquad (3.169)$$

where z is an m-measurement vector, x is an n-parameter vector, v
is a zero mean finite variance random variable, and $f(\cdot)$ is a

nonlinear function, differentiable with respect to its argument.
According to the popular Gauss-Newton method a risk function may be
defined as

$$R(x) = [z - f(x)]^T Q[z - f(x)], \quad Q = Q^T > 0 \tag{3.170}$$

Assuming that an initial close estimate of $\hat{x}(0)$ is available and
that $f(x)$ is Taylor-series expandable about x_0, the following algorithm may be used for recursive approximation of the minimum mean-
square estimate:

$$\hat{x}(k) = \hat{x}(k - 1) + [\nabla_x^T f(\hat{x}(k - 1))Q\nabla_x f(\hat{x}(k - 1))]^{-1} \nabla_x^T f(\hat{x}(k - 1))$$

$$Q[z - f(\hat{x}(k - 1))] \qquad\qquad \hat{x}(0) = x_0 \tag{3.171}$$

where the function $f(x)$ has been approximated by the linear term of
the Taylor series

$$f(\hat{x}) \approx f(\hat{x}(k)) + \nabla_x^T f(\hat{x}(k)) (x - \hat{x}(k)) \tag{3.172}$$

Higher-order terms may be included in (3.172) to improve the approximation [3.6].

3.5.2   Nonlinear State Estimation

The nonlinear state estimator for a Markov process perturbed by
Wiener random process can be formulated with the aid of the orthogonal projection theorems for both the continuous and discrete-time
cases.   Its solution is not available in either of the two.   Let a
continuous-time system be defined by

$$\frac{dx}{dt} = f(x,t) + w(t)$$

$$\tag{3.173}$$

$$z = h(x,t) + v(t)$$

where x is the n-dimensional state vector, z is an m-dimensional
measurement vector, $f(\cdot,\cdot)$ and $h(\cdot,\cdot)$ are appropriate function
differentiable with respect to their arguments, and w(t) and v(t)
Gaussian random variables with zero-mean and covariance matrices
Q(t) and R(t), respectively, independent and identically distributed
in time of appropriate dimensions.

The so called *extended Kalman filter* [3.12], [3.39] is produced for the estimation of the state x of (3.182) if one obtains the equation for x the conditional mean of x, which represents the best mean-square error estimate, and approximating it by replacing $f(x,t)$ and $h(x,t)$ by the following truncated Taylor series expansions

$$f(x,t) \simeq f(\hat{x},t) + A(\hat{x},t)(x - \hat{x}) \; ;$$

$$h(x,t) \simeq h(\hat{x},t) + C(\hat{x},t)(x - \hat{x}) \qquad (3.174)$$

where the matrices A and C are defined by

$$A(\hat{x},t) = \begin{bmatrix} \dfrac{\partial f_1(\hat{x})}{\partial x_1} & \cdots & \dfrac{\partial f_1(\hat{x})}{\partial x_n} \\ \cdots\cdots\cdots\cdots\cdots \\ \dfrac{\partial f_n(\hat{x})}{\partial x_1} & \cdots & \dfrac{\partial f_n(\hat{x})}{\partial x_n} \end{bmatrix} ; \quad \gamma_1 = \begin{bmatrix} 0 \\ \vdots \\ 1 \\ 0 \\ \vdots \\ 0 \end{bmatrix} \text{ith row} \qquad (3.175)$$

$$C(\hat{x},t) = \begin{bmatrix} \dfrac{\partial h_1(\hat{x})}{\partial x_1} & \cdots & \dfrac{\partial h_1(\hat{x})}{\partial x_n} \\ \cdots\cdots\cdots\cdots\cdots \\ \dfrac{\partial h_m(\hat{x})}{\partial x_1} & \cdots & \dfrac{\partial h_m(\hat{x})}{\partial x_n} \end{bmatrix} ; \quad F_i(x,t) = \begin{bmatrix} \dfrac{\partial^2 h_i(\hat{x})}{\partial x_1^2} & \cdots & \dfrac{\partial^2 h_i(\hat{x})}{\partial x_1 \partial x_n} \\ \cdots\cdots\cdots\cdots\cdots \\ \dfrac{\partial^2 h_i(\hat{x})}{\partial x_n \partial x_1} & \cdots & \dfrac{\partial^2 h_i(\hat{x})}{\partial x_n^2} \end{bmatrix}$$

Then the extended Kalman filter estimator is derived as

$$\frac{d\hat{x}}{dt} = f(\hat{x},t) + \hat{P}(t)C(\hat{x},t)R^{-1}(t)(z(t) - h(\hat{x},t)) \qquad (3.176)$$

$$\frac{d\hat{P}}{dt} = \hat{P}(t)A^T(\hat{x},t) + A(\hat{x},t)\hat{P}(t) + Q(t)$$
$$- \hat{P}(t)C^T(x,t)R^{-1}(t)C(x,t)\hat{P}(t) + \sum_{i=1}^{n} \hat{P}(t)F_i(\hat{x},t)\gamma_i^T R^{-1}(t)$$

$$(z(t) - h(\hat{x},t))$$

The algorithm may be improved by including higher-order terms in the expansions (3.174).

If instead the expansion of $f(\cdot)$ and $h(\cdot)$ is performed about a nominal trajectory, say the previous estimates x'

$$f(x,t) \cong f(x',t) + A(x',t)(x - x')$$
$$h(x,t) \cong h(x',t) + C(x',t)(x - x') \tag{3.177}$$

and then the Kalman filter for linear systems is used, the linearized Kalman estimator will result:

$$\frac{dx}{dt} = A(x',t)\hat{x}(t) + r(t) + \bar{P}(t)C^T(x',t)R^{-1}(t)[z(t) - s(t)$$
$$- C(x',t)\hat{x}(t)]$$

$$\frac{d\bar{P}}{dt} = A(x',t)\bar{P}(t) + \bar{P}(t)A^T(x,t) + Q(t)$$
$$- \bar{P}(t)C^T(x',t)R^{-1}(t)C(x',t)\bar{P}(t) \quad \bar{P}(0) = P_0 \tag{3.178}$$

where

$$r(t) = f(x',t) - A(x',t)x'(t)$$
$$s(t) = h(x',t) - C(x',t)x'(t)$$

It is obvious that for $x' = \hat{x}$ the second algorithm differs from the first by the last term of the matrix Riccati equation in (3.176).

In discrete-time systems the equivalent *extended Kalman filter* is given by

$$\hat{x}(k + 1/k + 1) = f(\hat{x}(k/k) + \hat{P}(k + 1/k + 1)C^T(k + 1)R^{-1}(k + 1)$$
$$(z(k + 1) - C(k + 1)f(\hat{x}(k/k))$$

$$\hat{P}(k + 1/k + 1) = [R(k) + C^T(k + 1)P(k + 1/k)C(k + 1)]^{-1}$$
$$\hat{P}(k + 1/k) \tag{3.179}$$

$$\hat{P}(k + 1/k) = A(k)\hat{P}(k/k)A^T(k) + Q(k) \quad P(0) = P_0$$

where $A(k) = A(\hat{x},t)$, $C(k) = C(\hat{x},t)$ of (3.183).

The *linearized Kalman filter* [3.17], [3.30] for the discrete-time case can be produced by linearizing about a nominal trajectory defined by

$$\hat{x}_0(k + 1) = f(\hat{x}_0(k)) \quad \hat{x}_0(0) = \bar{x}_0$$
$$\hat{x}(k|k) = \hat{x}_0(k) + \delta\hat{x}(k|k) \tag{3.180}$$

Then the estimation equations are given by

$$\hat{x}(k/k) = \hat{x}_0(k) + \delta\hat{x}(k/k)$$

$$\delta\hat{x}(k + 1/k + 1) = A(k)\delta\hat{x}(k/k) + P(k + 1/k + 1)C^T(k)R^{-1}(k)$$

$$(z(k + 1) - C(k)A(k)\delta\hat{x}(k/k)) \qquad (3.181)$$

$$P(k + 1/k + 1) = P(k + 1/k) - P(k + 1/k)C^T(k)$$

$$[C(k)P(k + 1/k)C^T(k) + R(k)]^{-1} \cdot C(k)P(k + 1/k)$$

$$P(k + 1/k) = A(k)P(k/k)A^T(k) + Q(k)$$

and

$$\delta z(k + 1) = z(k + 1) - h(\hat{x}_0(k + 1)) \qquad (3.182)$$

and $A(k)$ and $C(k)$ are defined as before. The application of these two algorithms will be demonstrated in Chap. 6 for joint parameter identification and state estimation of certain parameter adaptive S.O.C. algorithms.

3.6 DISCUSSION

Estimation, the subject of this chapter, is essential for the analysis and control of stochastic systems. The most fundamental techniques for estimation have been presented, but certain of them have been singled out as especially suitable for the S.O.C. problem. The criteria for selection were the simplicity of implementation and in many cases the generality of the method with regard to the class of distribution functions of the stochastic process. It has been explained in Chap. 1 that the uncertainties involved in a S.O.C. system may include ignorance of the distribution of the random processes; therefore estimators which do not depend on them are more desirable for implementation. What is sometimes sacrificed is the optimality of the estimator, which usually imposes very strict mathematical conditions limiting the applicability of these methods. The stochastic approximation method and random search algorithms satisfy these requirements for unimodal and multimodal surface estimation respectively.

State estimation, as discussed from the Kalman-Bucy filter
point of view, presents only a limiting aspect of the problem since
it yields optimal estimates only for linear systems and Gaussian
random processes. If the process is not Gaussian, the estimator is
still minimum in the mean-square sense but not necessarily most
probable. In the case of a nonlinear system linearization techniques
have been recommended in order to use the popular Kalman-Bucy filter.
This approach, however, has limited value for small deviations from
the point of linearization and no guarantee of so-called "approximate
optimality" is made. Further discussion on such linearized Kalman
filters, popular with the investigators of space problems, can be
found in Chap. 6 where the method has been used for parameter identification as well [3.40].

However, an idea, which has been explored only recently, is the
implementation of a parameter-adaptive observer [3.12]. An observer
is an estimator with structure sufficiently defined to guarantee
consistent estimation. Otherwise the parameters involved with the
estimator are adjustable and may be chosen to minimize some criterion.
A parameter-adaptive estimator can be specified to improve its coefficients as the process evolves. Such an idea, although perhaps
heuristic, is appealing because of the degree of freedom available
to the estimator and the simplicity of its implementation.

No examples have been presented in this chapter because only
background material has been covered. All these methods are used
in the rest of this book and will provide the appropriate illustra-
tion of the methods as applied to the S.O.C. problem.

3.7 REFERENCES

[3.1]   Albert, A. E., and Gardner, L. A., *Stochastic Approximation and Nonlinear Regressions*, MIT Press, Cambridge, Mass., 1967.

[3.2]   Aoki, M., *Optimization of Stochastic Systems*, Academic Press, New York, 1967.

[3.3]   Blum, J., "Multidimensional Stochastic Approximation Method," *Ann. Math. Stat.*, 25 400-407 (1954).

[3.4]   Bryson, A. E., and Ho, Y. C., *Applied Optimal Control*, Blaisdell, Waltham, Mass., 1969.

[3.5]   Bucy, R. S., and Joseph, P. D., *Filtering for Stochastic Processes with Applications to Guidance*," Interscience, New York, 1968.

[3.6]   Deutz, R., *Estimation Theory*, Prentice Hall, New York, 1965.

[3.7]   Doob, J. L., *Stochastic Processes*, Wiley, New York, 1953.

[3.8]   Dupac, V., "A Dynamic Stochastic Approximation Method," *Ann. Math. Stat.*, 36 (6) 1695 (1966).

[3.9]   Dvoretzky, A., "On Stochastic Approximation," Proc. 3rd Berkeley Symposium of Math. Stat. and Prob., 1, University of California Press, Los Angeles, 1956.

[3.10] Fel'dbaum, A. A., and Bochorov, V., "An Automatic Optimizer for the Search of Smallest Several Minima," *Automation Remote Control*, 23 (3) (1962).

[3.11] Fu, K. S., and McMurtry, G., "A Variable Structure Automaton used as a Multimodal Searching Technique," *IEEE Trans. Automatic Control*, AC-11 (3), 379-387 (1966).

[3.12] Gelb, A. (ed.), *Applied Optimal Estimation*, MIT Press, Cambridge, Mass., 1974.

[3.13] Gurin, L. S., "Random Search in the Presence of Noise," *Eng. Cybernetics*, 3, 252-260 (1966).

[3.14] Gurin, L. S., and Rastrigin, L. A., "Convergence of the Random Search Method in the Presence of Noise," *Automation Remote Control*, 25 (9), 1505-1511 (1965).

[3.15] Joseph, P. D., and Tou, J. T., "On Linear Control Theory," *AIEE Trans. Appl. Ind.*, 80, 193-196 (1961).

[3.16] Kailath, T., and Frost, P. A., "An Innovation Approach to Least Squares Estimations," Parts I, II, *IEEE Trans. Automatic Control*, AC-13 (6), 655-660 (1968).

[3.17] Kailath, T., Kozin, F., Bucy, R., and Wonham, W. M., *Stochastic Problems in Control*, ASME Publications, New York, 1968.

[3.18] Kalman, R., "A New Approach to Linear Filtering and Predictions Problems," *Trans. ASME J. Basic Eng.*, 82, 34-35 (1960).

[3.19]   Kalman, R., and Bucy, R., "New Results in Linear Filtering and Prediction Theory," *Trans. ASME J. Basic Eng.*, 83-95 (1961).

[3.20]   Kelly, H., "Method of Gradients," in *Optimization Techniques* (G. Leitman, ed.), Academic Press, New York, 1963.

[3.21]   Kiefer, J., and Wolfowitz, J., "Stochastic Estimation of the Maximum of a Regression Function," *Ann. Math. Stat.*, $\underline{23}$, 462 (1952).

[3.22]   Kushner, H. J., *Stochastic Stability and Control*, Academic Press, New York, 1967.

[3.23]   Kushner, H. J., "Hill-climbing Methods for Optimization of Multiparameter Noise Disturbed Systems," Preprints JACC, 1962.

[3.24]   Kushner, H. J., "A New Method of Locating the Max Point of an Arbitrary Multipeak Presence of Noise," Preprints JACC, 1963.

[3.25]   Lainiotis, D. G., "Optimal Adaptive Estimation: Structure and Parameter Adaptations," *IEEE Trans. Automatic Control*, $\underline{AC-16}$, 160-170 (1971).

[3.26]   Lanning, J. H., Jr., and Battin, R. H., *Random Processes in Automatic Control*, McGraw-Hill, New York, 1956.

[3.27]   Lee, R. C. K., *Optimal Estimation Identification and Control*, MIT Press, Cambridge, Mass., 1964.

[3.28]   Loeve, M., *Probability Theory*, Van Nostrand, New York, 1963.

[3.29]   Luenberger, D., *Optimization by Vector Space Methods*, Wiley, New York, 1969.

[3.30]   Magill, D. T., "Optimal Adaptive Estimation of Sampled Stochastic Processes," *IEEE Trans. Automatic Control*, $\underline{AC-10}$ (4), 434-439 (1965).

[3.31]   Matyash, J., "Random Optimization," *Avtomakima i Telemek-hanika,* $\underline{26}$ (2) (1965).

[3.32]   McGarty, T. P., "Stochastic Systems and State Estimations," Wiley-Interscience, New York, 1974.

[3.33]   Meditch, J., *Optimal Estimation and Control*, McGraw-Hill, New York, 1969.

[3.34]   Mehra, R., "On the Identification of Variances and Adaptive Kalman Filtering", *IEEE Trans. Automatic Control*, $\underline{AC-15}$ (2), 175-189 (1971).

[3.35]   Nahi, N. E., *Estimation Theory and Applications*, Wiley, New York, 1968.

[3.36]   Papoulis, A., *Probability Random Variables and Stochastic Processes*, McGraw-Hill, New York, 1965.

[3.37] Rastrigin, L. A., "The Convergence of the Random Search Method in the Extremal Control of a Many-Parameter System," *Automation Remote Control*, 24 (11), 1337-1342 (1963).

[3.38] Robins, H., and Monro, S., "A Stochastic Approximation Method," *Ann. Math. Stat.*, 22, 400 (1951).

[3.39] Sage, A. P., and Melsa, T. L., *Estimation Theory with Applications to Communications and Control*, McGraw-Hill, New York, 1971.

[3.40] Sage, A. P., and Melsa, T. L., *System Identification*, Academic Press, New York, 1971.

[3.41] Sakrison, D., "A Continuous Kiefer-Wolfowitz Procedure for Random Processes," *Ann. Math. Stat.*, 35 (2), 591 (1964).

[3.42] Saridis, G. N., "Learning Applied to Successive Approximation Algorithms," Preprints JACC 1968; *IEEE Trans. on Systems, Man, and Cybernetics*, SMC-6 (2), 97-108 (1970).

[3.43] Saridis, G. N., Nikolic, Z. T., and Fu, K. S., "Stochastic Approximation Algorithm for System Identification Estimation and Decomposition of Mixtures," *IEEE Trans. on Systems, Man, and Cybernetics*, SMC-5 (1), 8-15 (1969).

[3.44] Saridis, G. N., "Expanding Subinterval Random Search for System Identification and Control," *IEEE Trans. Automatic Control*, (to appear).

[3.45] Sorenson, H. W., and Stubberud, A. R., "Nonlinear Filtering by Approximating A posteriori Density," *Int. J. Control*, 18, 33-51 (1968).

[3.46] Sridhar, R., and Detchmendy, D., "Sequential Estimation of States and Parameters in Noisy Nonlinear Dynamic Systems," Preprints JACC, 1965.

[3.47] Wiener, N., "The Extrapolation Interpolation and Smoothing of Stationary Time Series," Wiley, New York, 1949.

[3.48] Wilde, D., *Optimal Seeking Methods*, Prentice-Hall, New York, 1963.

[3.49] Wong, E., *Stochastic Processes in Information and Dynamical Systems*, McGraw-Hill, New York, 1971.

[3.50] Wong, E. and Zakai, M., "On the Convergence of Ordinary Integrals to Stochastic Integrals," *Ann. Math. Stat.*, 36, 1560-1564 (1965).

Chapter 4

THE STOCHASTIC OPTIMAL CONTROL

4.1   INTRODUCTION

Optimality has been the dominating feature of modern control
theory. Therefore considerable effort has been devoted to the
development of optimal policies for systems operating in a stochastic
environment, as described in numerous references. However, the
transition from the straightforward deterministic optimal formulation
to the stochastic one has always been puzzled by conflicting
interpretations of the results obtained. The stochastic optimal-
control problem has the additional feature of state estimation,
necessary for the open-loop or the feedback solution of the problem.
The different use of estimation in the derivation of the stochastic
optimal control produces different solutions to the problem which
were not present in the deterministic case with complete information
about the states available.

It was not until Fel'dbaum first discovered and formulated the
various stochastic optimal control problems [4.11], that Dreyfus
[4.10] and, more recently, Tse and Bar-Shalom [4.5][4.22], have
elaborated on the various stochastic optimal control problems, and
gave a very comprehensive interpretation of their properties.

The stochastic optimal control problem will be presented
briefly in this chapter. Since from our point of view the problem
has been subdivided to the one with irreducible and the one with
reducible uncertainties, only the algorithms which qualify for the
first case will be treated here. This will include the open-loop
and closed-loop control when the separation theorem is applicable.
The stochastic optimal control with reducible uncertainties in its

most general form, called *Dual Control* by Fel'dbaum, will be discussed
in the next chapter. The latter problem will also be treated from the
self-organizing point of view in the last part of this book.
Discussion of some other algorithms will also be given whenever
appropriate.

4.2 STOCHASTIC OPTIMALITY, A REVIEW

The classification of dynamic control systems with uncertainties
operating in a stochastic environment has been based on the
utilization of information regarding the system and the random
processes employed by the controller. The same principle will be
applied to the stochastic optimal control problem. Such a
classification is in agreement with the work of Fel'dbaum [4.11] and
the more recent interpretation by Bar-Shalom and Tse [4.28]. The
result is that different optimal solutions will result for the
different categories of the classification since each one of them
defines a different control problem, a case not yet encountered in
deterministic optimal control.

The stochastic optimal control may be classified in one of the
following three categories, according to the utilization by the
controller of information regarding the process uncertainties:

1.   *Open-loop control* with no use of subsequent measurements of the
      output of the system.

2.   *Passive feedback control* where subsequent output measurement
      are fed back to the controller without further anticipation
      to reduce the uncertainties of the system.

3.   *Active feedback or dual control* where the subsequent output
      measurements are fed back to the controller with anticipation
      of future measurements to reduce the uncertainties of the system.

The first two categories possess irreducible uncertainties and
are going to be treated in this chapter under the title of
stochastic optimal control. They contain the open-loop policies
[4.28], most of the open-loop feedback policies [4.28], and the

closed-loop policies for which the separation theorem is valid and
the information processing is decoupled from the controller.

The last category where active reduction of the uncertainties
takes place will be considered in Chap. 5 under the title of *dual
control*. It may be viewed as corresponding to the optimal solution
of the parameter-adaptive S.O.C. problem.

The stochastic optimal control problem has been formulated for
continuous-time continuous-state, discrete-time continuous-state,
and discrete-time discrete-state problems. In all cases, the
systems treated here are assumed to involve stochastic processes
which satisfy the Markovian property for continuous and discrete
time systems, respectively:

$$p(y(t)|z(\sigma)); \quad t_0 \le \sigma \le t_1 < t) = p(y(t)|z(t_1)) \tag{4.1}$$

or $\quad p(y(k)|z(1),\cdots,z(k-1)) = p(y(k)|z(k-1))$

where $y(\cdot)$ is a vector representing a variable of the system, and
$z(\cdot)$ a measurement vector. The systems treated in this chapter are
described by the following set of differential equations for the
continuous-time process

$$dx(t) = f(x(t),u(t),\theta,t) \, dt + g(\dot{x}(t),\theta,t) \, dW(t); \; x(t_0) = x_0$$

$$z(t) = h(x(t),\theta,t) + v(t) \tag{4.2}$$

where $x(t)$ is an n-dimensional state vector with x_0 initial value,
$E\{x_0\} = \bar{x}_0(\theta), \text{cov}(x_o) = P_0(\theta), z(t)$ is an r-dimensional output
vector, $u(t)$ a m-dimensional control input vector, $w(t) = dW(t)/dt$
and $v(t)$ are independent and identically distributed random variables
of appropriate dimensions with probability density function $p(w(t),\theta)$
and $p(v(t),\theta)$, respectively, and $f(\cdot,\cdot,\cdot)$, $h(\cdot,\cdot)$ are appropriate
functions differentiable and bounded with respect to their arguments;
θ is a reducible parameter of the system and/or the random processes.

For the discrete-time process

$$x(k+1) = f(x(k),u(k),\theta,k) + g(x(k),\theta,k)w(k) \quad x(0) = x_0$$

$$z(k) = h(x(k),\theta,k) + v(k) \tag{4.3}$$

The variables in this case are defined as in the continuous-time case with the continuous-time variable t replaced by the discrete instant k.

4.3 OPEN-LOOP CONTROL POLICY

The first category corresponds to the stochastic subclass of Fel'dbaum's control with complete information [4.11] and Bar-Shalom and Tse's open-loop policy also described in [4.1]. A controller may be designed and constructed for either system (4.2) or (4.3) with θ known as a preprogrammed function of time, depending only on the known statistics of the system, the initial state x_0 and process $w(t)$ and measurement noise $v(t)$ for the duration of the process $t_0 \leq t \leq T$ or $0 \leq k \leq N$, respectively. This controller u^* may be designed to minimize a performance criterion:

$$u^*(p(x_0),p(w),p(v),t): \quad J(u^*) \triangleq \operatorname{Min} E\left\{ \int_{t_0}^{T} L(x,u,t)\ dt \right\} \quad (4.4)$$

or respectively

$$u^*(p(x_0),p(w),p(v),k): \quad J(u^*) \triangleq \operatorname*{Min}_{\{u\}} E\left\{ \sum_{l=1}^{N} L(x(k),u(k-1)) \right\} \quad (4.5)$$

Such a controller can be designed before the process starts because it does not depend on any current measurements of the process but only on the unreducible statistics of the noise processes and the initial state. If the density functions of the random processes are known *a-priori* then the problem of obtaining u^* is the one of obtaining expectations and defining the respective optimal-control function.

4.4 PASSIVE FEEDBACK CONTROL

This is a large category of control systems that lies between open-loop policies and dual control which corresponds to the globally optimal solution [4.10]. The definition of this area has

created many problems because it includes several aspects of
optimality dependent on the definition of the problem, which confused
many researchers.

In view of the reducible uncertainties involved in the
stochastic control problems, the passive feedback control problem can
be best defined by pointing out its difference from the dual ontrol
problem. For a system of the form (4.2) or (4.3) with irreducible
uncertainties, i.e., θ known, the passive optimal-feedback control u^*
can be defined as

$$u^*(t,\theta = \theta, z(\sigma), u(\sigma); \ t_0 \leq \sigma \leq t):$$
$$J(u^*) \triangleq \underset{\{u(t)\}_t^T}{Min} \left\{ \int_t^T L(x,u,\alpha) \ d\alpha / z^t, u^t; \ \theta = \theta \right\} \tag{4.6}$$

or, respectively, for the discrete-time case

$$u(k,\theta = \theta, z^k, u^k); \quad J(u^*) \triangleq \underset{\{u\}_k^{N-1}}{Min} E \left\{ \sum_{i=k}^{N-1} L(x(i+1), u(i))/z^k, u^k; \right.$$
$$\left. \theta = \theta \right\} \tag{4.7}$$

This formulation of the optimal control corresponds to the selection
of the open-loop control at time t or k only from the sequence of
open-loop controls derived for the process starting at time t or k
with initial state x(t) or x(k), respectively, after the Z and u^*
sequences have been known to the controller. This is the reason why
this control has been called by Dreyfus and Tse *open loop feedback
optimal* [4.10], [4.11]. It performs by using the information about
the states passively, and it does not involve any reduction of any
uncertainty or makes the assumption that such information will be
available in the future.

In contrast, the dual control is formulated in such a way
that there is active use of the information about the state to
reduce the ignorance of the system uncertainties θ and estimate the
states of the system in the present and the future. This naturally
results from the use of Bellman's *principle of optimality* [4.6],
and the backward procedure of dynamic programming with a probabilistic

description of the control which is updated as the process evolves, called *pure random strategy*. The transfer to a *mixed strategy*, where the exact value of the control sequence is obtained, yields in most cases a control which is globally optimal.   The details of the formulation of the dual control problem are given in Chap. 5. However, for comparison purposes the optimal control u^* may be thought of as a sequence minimizing in the continuous-time a limiting form $k \to t$ as in the discrete-time case

$$u^*(k,\hat{\theta}_k),z^k,u^k): \quad J(u^*) = \underset{u(0,\hat{\theta}_0)_1}{\text{Min}E} \left\{ L(x(1),u(0) + \cdots \right.$$

$$+ \underset{u(N-1,\hat{\theta}_{N-1})}{\text{Min}} E \left\{ L(x(N),u(N - 1)/_Z N-1,u^{N-1} \right\} /_Z N-2,u^{N-2} \cdots \right\} \qquad (4.8)$$

where the estimates $\hat{\theta}_i$ are obtained at the ith step by an optimal estimation procedure.   The expression (4.8) may be rewritten in the familiar form of dynamic programming for the continuous system

$$u^*(t,\hat{\theta},\hat{x}(t)): \quad \frac{\partial V}{\partial t}(t,\hat{x},\hat{\theta}) + \underset{u(t,\hat{\theta},\hat{x})}{\text{Min}} E\{ L(x,u,t) + L_u V(t,\hat{x},\hat{\theta}) \Big|$$

$$\Big|_{\hat{\theta}=E\{\theta/z(t)\}}, \quad \hat{x}(t) = E\{x(t)/z(t)\}\} \qquad (4.9)$$

or for the discrete-time system:

$$u^*(k): \quad V_{N-k}(k) = \underset{u(k,\hat{\theta})}{\text{Min}E} \{ L(x(k + 1),u(k),k) + V_{N-k-1}(k + 1) \Big|$$

$$\Big| z^k,u^{k-1},\theta=\hat{\theta}_k \} \qquad (4.10)$$

where $V_0(N) = J(u^*)$, $V_N(0) = 0$, and L_u is the differential generator . defined for the process by the appropriate stochastic calculus [4.14], [4.30].

It is obvious that the two formulations (4.6), (4.7) and (4.9), (4.10) yield different controls and different minimum values of performance since they correspond to different problems.   However, for the same performance criterion and θ known, the dual   ntrol solution should yield a global minimum, because it utilizes the information about the system most effectively.   This property has been demonstrated by Dreyfus in a simple example in [4.10].   It suggests that dual control gives a "better minimum" than the passive optimal feedback or the open-loop optimal.   Actually the latter is

the "worst" of all, as will be demonstrated later in this chapter
in a special case. However, there is a class of processes for
which the dual and the passive optimal solution yield the same
results. These systems are called by Fel'dbaum *neutral systems* for
which the property of separation of the optimal control from the
optimal estimation process holds. Certain authors [4.7], [4.27] call
the rules that govern neutral systems the *certainty equivalence*
principle for which the optimal feedback control is defined as
$u^*(\hat{x}(k),k)$, where $\hat{x}(k)$ is the optimal state estimate performed
separately. Under the properties of neutral systems the deterministic
control law $u^*(...)$ sees the system as deterministic with states $\hat{x}(k)$
and does not require the knowledge of subsequent feedback or any
reduction of the system uncertainties in order to be globally optimal.
In contrast to other authors, since the classification of stochastic
optimal problem is based on the reduction of uncertainties, neutral
systems are treated here as passive optimal feedback systems to be
discussed in the sequel.

A typical case of a neutral system is the Linear Quadratic
(performance) Gaussian (LQG) system which is treated in the following
sections.

Both the continuous- and discrete-time versions are popular,
because they possess interesting properties as well as the only
complete solutions to the stochastic optimal control problem. Certain
other open-loop feedback optimal-control solutions that may be
interpreted as parameter adaptive self-organizing will be treated in
Chap. 7 of this book.

4.5 THE LINEAR QUADRATIC GAUSSIAN OPTIMAL CONTROL

The control problem of a known linear dynamic process operating
in a Gaussian stochastic environment, and having to minimize a
quadratic performance criterion (LQG), possesses an optimal solution
which does not require active processing of information about the
system and therefore has a passive feedback form. This important
property, which satisfies the condition of *neutrality* defined by

Fel'dbaum [4.11], was discovered by Tou for discrete time systems
[4.26] and was generalized to continuous-time systems by Wonham
[4.30], among others. It is presented in the form of the following
separation theorem which earned its name from the ability of such
systems to perform the state estimation and optimal control
separately:

Theorem 4.1

> In linear systems with quadratic performance criteria and
> subjected to Gaussian inputs, the stochastic optimal feed-
> back controller is synthesized by cascading an optimal esti-
> mator (Kalman Bucy filter) with the deterministic optimal
> control.

The separation of estimation and control considerably simplifies
the implementation of the stochastic optimal feedback controller.
This appealing idea has led to a wide collection of attempts in the
literature to approximate the solution of the general *nonneutral*
problem by using the separation principle. Such solutions, however,
in spite of their simplicity, do not guarantee a good approximation
to the optimal solution unless otherwise proven.
The optimal solution to the LQG problem was first established
for the discrete-time case [4.26], and pioneered investigation in the
area. The continuous-time case, as presented by Wonham [4.30], will
be first outlined in the sequel, and then the discrete-time case will
be briefly discussed from Meditch's excellent presentation in Ref.
[4.16]. Since the purpose of this part of the book is to summarize
the existing results on the stochastic optimal-control problem, the
presentation will be brief and no example will be presented. Detailed
discussions and extensions to more general neutral systems are
provided in the appropriate references [4.1], [4.18], [4.20], [4.24]
etc.

4.5.1.   The Continuous-Time Optimal LQG

The derivation of the optimal control for the continuous-time
LQG problem can be obtained as a special case of the stochastic
Hamilton-Jacobi-Bellman equation (4.9) where the differential generator

L_u, required by the stochastic process involved, can be explicitly defined [4.8, 4.9, 4.30]. The system under consideration is given by

$$\frac{dx}{dt} = F(t)x(t) + B(t)u(t) + G(t)w(t), x(t_0) = x_0 \qquad (4.11)$$

$$z(t) = H(t)x(t) + v(t)$$

where x is the n-dimensional state vector, u is the m-dimensional control vector, z is the r-dimensional output vector, w(t) and v(t) are independent and identically distributed Gaussian random vectors of appropriate dimensions, and F(t), B(t), G(t), and H(t) are matrices of appropriate dimensions bounded for all t $\epsilon[t_0,T]$. The Gaussian random processes have the following properties:

$$E\{x_0\} = \bar{x}_0 \quad E\{(x_0 - \bar{x}_0)(x_0 - \bar{x}_0)^T\} = P_0$$

$$E\{w(t)\} = E\{v(t)\} = 0$$

$$E\{w(t)w^T(\tau)\} = Q(t)F(\tau-t) \quad E\{v(t)v^T(\tau)\} = R(t)\delta(\tau-t)$$

$$E\{w(t)v^T(\tau)\} = E\{w(t)x_0^T\} = E\{v(t)x_0^T\} = 0 \quad \forall t,\tau \qquad (4.12)$$

The system (4.11) is assumed to be completely controllable and completely observable. Then a feedback control is sought to minimize the performance criterion

$$J(t_0,x_0;u(t)) \triangleq 1/2 \ E \int_{t_0}^{T} [x^T(s)M(s)x(s) + u^T(s)N(s)u(s)] \ ds \qquad (4.13)$$

where the weighting matrices $M(s) = M^T(s) \geq 0$ and $N(s) = N^T(s) > 0$ are of appropriate dimensions, over the interval $t_0 \leq s \leq T$.

Since the *separation theorem* 4.1 will be invoked for optimality, the best estimates of x(t), given the measurements z(t), must be obtained for the system (4.11). This estimate was found to be $\hat{x}(t/t)$ such that

$$\hat{x}(t/t) = E\{x(t)/z(t)\} \qquad (4.14)$$

and satisfies the appropriate equations of the Kalman-Bucy filter (3.140) and (3.147) in Sec. 3.4.2

$$d\hat{x}(t/t)/dt = F(t)\hat{x}(t/t) + P(t)H^T(t)R^{-1}(t)[z(t) - H(t)\hat{x}(t/t)]$$

$$+ B(t)u^*(t), x(t/t_0) = \bar{x}_0 \qquad (4.15)$$

$$\frac{dP}{dt} = F(t)P(t) + P(t)F^T(t) + G(t)Q(t)G^T(t)$$

$$- P(t)H^T(t)R^{-1}(t)H(t)P(t), \quad P(t_0) = P_0$$

The optimal feedback control $u^*(t)$ must be a function of $\hat{x}(t/t)$, according to the separation theorem

$$u^*(t) = u^*(t,\hat{x}(t/t)) \tag{4.16}$$

This control is obtained using Bellman's dynamic programming formulation [4.6], [4.30]. The optimal return function $V(t,\hat{x})$, is defined via the usual embedding procedure

$$V(t,\beta) = E\left\{1/2 \int [x^T(s)M(s)x(s) + u^T(s)N(s)u(s)] \, ds/_{\hat{x}(t/t) = \beta}\right\} \tag{4.17}$$

$$= \underset{u(t)}{\text{Min}} \, E\{1/2 \int [x^T(s)M(s)x(s) + u^T(s)N(s)u(s)] \, ds/_{\hat{x}(t/t)=\beta}\}$$

$$+ 1/2 \, q(t)$$

Then, the return function $V(t,\beta)$, must satisfy the stochastic Hamilton-Jacobi-Bellman equation (4.9) or the equation of dynamic programming

$$\frac{\partial V}{\partial t}(t,\beta) + \underset{u}{\text{Min}} \, E\{1/2[\hat{x}^T M\hat{x} + u^T Nu + L_u V(t,\hat{x})]_{\hat{x}(t/t) = \beta}\} = 0$$

where the differential generation L_u of this process is defined [4.30] as

$$L_u V(t,\beta) = 1/2 \, \text{tr} \, [H^T(t) \, P(t)\frac{\partial^2 V}{\partial \beta^2} P(t)H^T(t)] + (F\beta + Bu(t))^T \frac{\partial V}{\partial \beta} \tag{4.18}$$

where $\text{tr}A = \sum_{i=1}^{n} a_{ii}$ = the trace of matrix A.

The complete derivation of the equation of dynamic programming and the definition of the differential operator, being beyond the scope of this book, have been omitted. The interested reader is referred to Wonham [4.30].

In view of (4.18) the equation of dynamic programming is written as

$$\frac{\partial V}{\partial t} (t,\beta) + 1/2 \ \underset{u}{\text{Min}}\{\beta^T M\beta + u^T Nu + H(t)P \frac{\partial^2 V}{\partial \beta^2} P(t)H^T(t)$$

$$+ 2(F\beta+Bu)^T \frac{\partial V}{\partial \beta}\} = 0 \qquad\qquad (4.19)$$

Assuming that $V(t,\beta)$ is a quadratic in β,

$$V(t,\beta) = 1/2\beta^T S(t)\beta + 1/2 \ q(t) \qquad\qquad (4.20)$$

the minimization in (4.19) yields

$$u^*(t,\beta) = -N^{-1}B^T \frac{\partial V}{\partial \beta} = -N^{-1}(t)B^T(t)S(t)\beta \qquad\qquad (4.21)$$

Substituting back to (4.19) and equating coefficients in the resulting quadratic form, one obtains

$$\frac{dS(t)}{dt} = S(t)B(t)N^{-1}(t)B^T(t)S(t) - M(t) - S(t)F(t)$$
$$\qquad - F^T(t)S(t) \qquad\qquad\qquad S(T) = 0 \qquad (4.22)$$
$$\frac{dq(t)}{dt} = -tr[H(t)P(t)S(t)P(t)H^T(t)] - tr[M(t)P(t)] \quad q(T) = 0$$

Equations (4.21) and the first of (4.22) constitute the deterministic optimal control with $\hat{x}(t) = \beta$. The difference is in the term $1/2q(t)$, which is not necessary for the evaluation of u^*. The justification of the separation theorem is obvious. In terms of the Kalman-Bucy estimates given by (4.15), the optimal control is given by

$$u^*(t,\hat{x}(t/t)) = -N^{-1}(t)B^T(t)S(t)\hat{x}(t/t) \qquad\qquad (4.23)$$

$$\frac{dS(t)}{dt} = S(t)B(t)N^{-1}(t)B^T(t)S(t) - M(t) - S(t)F(t) - F^T(t)S(t)$$

$$S(T) = 0$$

This control does not require the active use of information about the process. This problem has been active by Wonham for a slightly more general class of systems than LQG, [4.14], [4.31].

4.5.2.   The Infinite Continuous-time Problem

The natural extension of the stochastic feedback optimal control problem solved in the preceding section, is the case where $(T - t_0) \to \infty$, or the duration of the process becomes infinite. This is equivalent for searching for the steady-state feedback optimal control. In order for the problem to have a meaning, all the

time-varying matrices in (4.11), (4.12), and (4.13), namely F(t),
B(t), G(t), H(t), Q(t), R(t), and M(t), N(t), respectively, are
replaced by their time-invariant equivalents F, B, G, H, Q, R, and
M, N. However, this restriction is not sufficient for the solution
of the infinite-time problem. The reason is that, due to the
randomness of the variables in the performance criterion (4.12),
the expected value of its integrand is always finite, thus leading
the integral to attend an infinite value. Since multiplication of
the performance criterion by any quantity not containing u(t) does
not change the optimization results, the following performance
criterion, which attains a finite value in the limit, may be used
to solve the infinite-time problem:

$$J(t_0, x_0; u(t)) = \lim_{(T-t_0) \to \infty} E \left\{ \frac{1}{2(T-t_0)} \int_{t_0}^{T} [x^T(s)Mx(s) + u^T(s)Nu(s)] \, ds \right\}$$

(4.24)

The solution of the infinite-time problem depends on three
conditions:

1.   Stability.
2.   Existence of an admissible control.
3.   Optimality.

Wonham gives a thorough discussion in Ref. [4.30]. In brief,
stability of the optimal solution is possible, in other words the
system (4.11) under the minimization of (4.24) is bounded for all
time, if a certain function $V(\hat{x}(t))$ exists and has a bounded
stochastic derivative. The existance of an admissible optimal
control is possible if a certain inequality is satisfied. Finally,
optimality of the solution is guaranteed if there exists a
steady-state version of the Hamilton-Jacobi-Bellman equation (4.19)
for a return function $V(\hat{x})$ defined by (4.20) for constant S and q.

The optimal feedback control $u^*(\hat{x}(t/t))$ is given, as a natural
extension of (4.22)

$$u^*(\hat{x}(t/t)) = N^{-1}B^T S \hat{x}(t/t)$$

(4.25)

$$SF + F^T S + M - SBN^{-1}B^T S = 0$$

The state estimates $\hat{x}(t/t)$ are obtained from the steady-state
Kalman-Bucy filter.

4.5.3. Optimal Feedback vs. Open-loop Control

In Sec. 4.4 it was claimed that the open-loop solution of the stochastic optimal control problem would yield a higher value of the performance criterion than the one resulting from a passive feedback optimal control. A demonstration of this argument is given by Wonham in [4.30] and Dreyfus in [4.10]. Wonham's argument will be recapitulated here in order to substantiate the difference in the solution of the open-loop vs. the feedback optimal control.

For the simplified case, where the initial state x_0 is fixed and the states are available for direct measurement, e.g., $H(t) \equiv [$ and $v(t) \equiv 0$ for the systems (4.11), Wonham [4.30] has shown that

$$J(u^*_{OL}) - J(u^*_{FB}) = \int_{t_0}^{T} tr[G(s)A(s)G(s)Q(s)] \, ds \qquad (4.26)$$

where $J(u^*_{OL})$ is the optimal open-loop value of the performance criterion (4.13), corresponding to the optimal open-loop control $u^*_{OL}(t)$ defined by (4.4) and realized by

$$u^*_{OL}(t) = -N^{-1}(t)B^T(t)S(t) \Phi(t,t_0)x_0 \qquad (4.27)$$

where $\Phi(t,t_0)$ is the state transition matrix associated with the system:

$$\frac{dx}{dt} = [F(t) - B(t)N^{-1}(t)B^T(t)S(t)] x(t) \quad x(t_0) = x_0 \qquad (4.28)$$

On the other hand, $J(u^*_{FB})$ corresponds to the optimal value of the performance criterion obtained by the feedback optimal control $u^*_{FB}(t)$, described by Eq. (4.23), where the states are available for direct measurement

$$u^*_{FB}(t) = -N^{-1}(t)B^t(t)S(t)x(t) \qquad (4.29)$$

The quantity $A(t)$ is defined as

$$A(t) \triangleq \int_{t}^{T} \Psi^T(s,t)S(s)B(s)N^{-1}(s)B^T(s)S(s)\Psi(s,t) \, ds \qquad (4.30)$$

where $\Psi(s,t)$ is the state transition matrix associated with $dx/dt = F(t) \, x(t)$. It is obvious that $J(u^*_{OL}) > J(u^*_{FB})$ and that the difference increases as $||G(t)||^2$ or $Q(t)$ increases, as well as the duration of the process $(T - t_0)$. When the states are not directly available for measurement, a similar expression exists, which establishes the superiority of the feedback solution over the open-loop one.

4.5.4. The Discrete-Time Optimal LQG

The solution of this problem originates with the celebrated work of Joseph and Tou [4.27] and has been elaborated by many authors [4.7], [4.15]. The formulation follows Bellman's dynamic programming approach which guarantees a feedback optimal solution. However, because of the separation Theorem 4.1 the utilization of information is not active, and the optimal control has the deterministic value with the best estimates in the place of the states. The linear dynamic system in discrete-time form is given by

$$x(k + 1) = F(k)x(k) + B(k)u(k) + G(k)w(k)   x(0) = x_0 \qquad (4.31)$$
$$z(k) = H(k)x(k) + v(k)$$

where $x(k)$ is the n-dimensional state bector; $u(t)$ is the m-dimensional control vector; $z(k)$ is the r-dimensional output vector; $w(k)$ and $v(k)$ are independent and identically distributed Gaussian random vectors; and $F(k)$, $B(k)$, $G(k)$, and $H(k)$ are matrices of appropriate dimensions bounded for all $k\epsilon[0,N]$, the duration of the process. The Gaussian random processes involved have the following properties:

$$E\{x_0\} = \bar{x}_0 \quad E\{(x_0 - \bar{x}_0)(x_0 - \bar{x}_0)^T\} = P_0 \qquad (4.32)$$
$$E\{w(k)\} = E\{v(k)\} = 0$$
$$E\{w(k)w^T(k)\} = Q(k)\delta_{kj} \quad E\{v(k)v^T(j)\} = R(k)\delta_{kj}$$
$$E\{w(k)v^T(j)\} = E\{w(k)x_0^T\} = E\{v(k)x_0^T\} = 0 \quad \forall k,j$$

The system (4.31) is assumed to be completely controllable and completely observable. Then a feedback control sequence $U_0^{N-1} = \{u(k)\}$ is sought to minimize the performance criterion

$$J(u) = E\left\{\sum_{i=0}^{N-1}[x^T(i + 1)M(i)x(i + 1) + u^T(i)N(i)u(i)]\right\} \qquad (4.33)$$

The solution of this problem will be obtained in two steps: first,
the optimal solution will be obtained as if the state were available
for measurement; second, the best estimates of the states will be
introduced and will prove to yield the minimum value of the
performance index when the actual states are not available. The last
step is a verification of the separation theorem.

<u>Step 1</u>

The optimal control is obtained by following Bellman's dynamic
programming approach [4.6], [4.15]. The optimal return function
$V_{N-k}(x(k))$ is defined for an N-k stage process starting at x(k), via
the usual embedding procedure

$$V_{N-k}(x(k)) = \min_{u(k),\ldots,u(N-1)} E\left\{\sum_{i=k}^{N-1}[x^T(i + 1)M(i)x(i + 1)\right.$$

$$+ u^T(i)N(i)u(i)]\right\} = E\{\min_{u(k)} E\left\{[x^T(k + 1)M(k)x(k + 1)\right.$$

$$+ u^T(k)N(k)u(k)]/_z k\}$$

$$+ \min_{u(k+1),\ldots,u(N-1)} E\left\{\sum_{i=k+1}^{N-1}[x^T(i + 1)M(i)x(i + 1)\right.$$

$$+ u(i)N(i)u(i)] /_z k\right\} ; \quad V_0(x(N)) = 0 \qquad (4.34)$$

In (4.34) it is assumed that the proper conditions are satisfied in
order to exchange the minimization and expectation operators and
that $z^{k^T} = \{z(1)\cdots z(k)\}$. Define the conditional return function:

$$V_{N-k}(x(k)/z^k) \triangleq \min_{u(k)} E\{[x^T(k + 1)M(k)x(k + 1) + u^T(k)u(k)]$$

$$+ \min_{u(k + 1),\ldots,u(N - 1)} \sum_{i=k+1}^{N-1}[x^T(i + 1)M(i)x(i + 1)$$

$$+ u^T(i)N(i)u(1)] /_z k\} \qquad (4.35)$$

However, because of the *neutrality* of the system $E\{\cdot /_z k\} = E\{E\{\cdot /_z k+1\}/_z k\}$. Therefore, observing the definitions (4.34) for the
return function for the (N-k-1)-stage process starting at x(k + 1),

Eq. (4.35) takes the familiar recursive form;

$$V_{N-k}(x(k)/z^k) = \underset{u(k)}{\text{Min}} E\{x^T(k+1)M(k)x(k+1) + u^T(k)N(k)u(k)$$

$$+ V_{N-k-1}(x(k+1)/_z k+1) \ /_z k\} \tag{4.36}$$

It is obvious that, since

$$V_{N-k}(x(k)) = E_{z^k}\{V_{N-k}(x(k)/_z k)\} \quad \forall K$$

the minimization of (4.36) implies the minimization of (4.34), which is the requirement of the original problem, e.g., minimization of (4.33) when $k = 0$.

The solution of Eq. (4.36) can be implemented to yield the optimal sequence $\{u^*\}_0^{N-1}$ by observing that $V_{N-k}(x(k))$ assumes the following reproducible form:

$$V_{N-k}(x(k)) = E\{x^T(k)S_{N-k}x(k)/_z k\} + q'_{N-k} \tag{4.37}$$

where $P_{N-k} = P_{N-k}^T > 0$ and q'_{N-k-1} is independent of the previous control sequence. Substitution of (4.37) into (4.38) and carrying through the minimization one obtains

$$u^*(k) = - [B^T(k)(M(k) + S_{N-k-1})B(k)+N(k)]^{-1}B^T(k)(M(k)+S_{N-k-1})$$
$$F(k)x(k) \triangleq -A^T_{N-k}x(k)$$

$$S_{N-k} = [F(k)-B(k)A^T_{N-k}]^T(M(k)+S_{N-k-1}) [F(k)-B(k)A^T_{N-k}]$$

$$+A^T_{N-k}N(k)A_{N-k}$$

$$= F^T(k)(M(k) + S_{N-k-1}[I-B(k)[B^T(k)(M(k) + S_{N-k-1})B(k)$$

$$+N(k)]^{-1}B^T(k)(M(k) + S_{N-k-1})]F(k) \quad S_0 = 0 \tag{4.38}$$

$$q'_{N-k} = q'_{N-k-1} + c_{N-k-1}$$

$$c_{N-k-1} = E\{w^T(k)G^T(k)(M(k)+S_{N-k-1})G(k)w(k)/_z k\} = \text{tr} [G^T(k) (N(k)$$

$$+ S_{N-k-1}) G(k)Q(k)]$$

It is obvious that A_{N-k} are the gains of the deterministic optimal feedback control [4.4], e.g., $w(k) \equiv v(k) \equiv 0$, $x_0 = \bar{x}_0$; $H(k) \equiv I$.

Such a control cannot be realized under the assumptions of the present problem. However, by using Eqs. (4.37) and (4.38) and rearranging (4.36) appropriately, one obtains

$$V_{N-k}(x(k)/_Zk) = \underset{u(k)}{\text{Min}}E\{ (u(k) - u^*(k))^T D_{N-k-1}(u(k) - u^*(k))$$

$$+ x^T(k)S_{N-k}x(k) + q'_{N-k-1} + c_{N-k-1}/_Zk\} \qquad (4.39)$$

$$D_{N-k-1} \triangleq B^T(k)(M(k) + S_{N-k-1})B(k) + N(k)$$

From (4.38) it is obvious that the only term affected by the minimization is the first one in (4.39)

$$I(u(k)) = E\{ (u(k)-u^*(k))^T D_{N-k-1}(u(k)-u^*(k))/_Zk\} \qquad (4.40)$$

Step 2

The stochastic optimal feedback solution will be based on the minimization of $I(u(k))$ for all k.

Since the separation theorem 4.1 will be tested and verified, it is natural to define the best estimate of the states $\hat{x}(k/k) \triangleq E\{x(k)/_Zk\}$ and obtain it recursively, as was done in Chap. 3 via the Kalman-Bucy filter equations.

$$\hat{x}(k/k) = F(k - 1)\hat{x}(k - 1/k - 1) + P(K/k)H^T(k)R^{-1}(k)[z(k)$$
$$- H(k)F(k - 1)\hat{x}(k - 1/k - 1)] + P(k/k)P^{-1}(k/k - 1)$$
$$B(k - 1)u(k - 1)$$

$$P(k/k) = P(k/k - 1) - P(k/k - 1)H^T(k)[H(k)P(k/k - 1)H^T(k) +$$
$$+ R(k)]^{-1}H^T(k)P(k/k - 1) \qquad (4.41)$$

$$P(k/k-1) = F(k - 1)P(k - 1/k - 1)F^T(k - 1) + G(k - 1)Q(k - 1)$$
$$G^T(k - 1)$$

Also define the error

$$\tilde{x}(k/k) = x(k) - \hat{x}(k/k) \qquad (4.42)$$

$$E\{\tilde{x}(k/k)\} = 0 \quad E\{\tilde{x}(k/k)x^T(k/k)\} = P(k/k)$$

It was shown in Chap. 3 that $x(k/k)$ is the orthogonal projection of $x(k)$ to the measurement subspace Z_k; e.g., $E\{\hat{x}^T(k/k)\tilde{x}(k/k)/_Zk\} = 0$. Now partition the optimal control $u^*(k)$ of (4.38) to two orthogonal components $\hat{u}^*(k)$ and $\tilde{u}^*(k)$ such that

$$u^*(k) = \hat{u}^*(k) + \tilde{u}^*(k) \tag{4.43}$$

where

$$\hat{u}^*(k) = E\{u^*(k)|_Z k\} = -A^T_{N-k}\hat{x}(k/k) \tag{4.44}$$

$$\tilde{u}(k) = -A^T_{N-k}\tilde{x}(k/k)$$

In view of (4.44), the function to be minimized $I_{N-k}(u(k))$ can be rewritten as

$$I_{N-k}(u(k)) = (u(k) - \hat{u}(k))^T D_{N-k-1}(u(k) - \hat{u}(k)) +$$

$$+ E\{\tilde{u}(k)^T D_{N-k-1} \tilde{u}(k)/_Z k\} = (u(k) - \hat{u}(k))^T D_{N-k-1}$$

$$(u(k) - \hat{u}(k)) + tr[A^T_{N-k}D_{N-k-1}A_{N-k}P(k/k)] \tag{4.45}$$

$$\tag{4.45}$$

In (4.45) it was assumed that u(k) should be a quantity depending only on z^k, that $\hat{u}(k)$ and $\tilde{u}(k)$ are orthogonal, and $E\{\tilde{u}(k)|_Z k\} = 0$ from (4.42). It is also obvious that the second term of (4.45) is independent of u(k) and therefore the optimal control occurs when the first term is zero; for example,

$$u(k) = \hat{u}^*(k) \tag{4.46}$$

establishes the separation theorem for the discrete-time problem, which, in summary, yields the optimal feedback control for the LQG problem depicted in Fig. 4.1

$$u^*(k) = -A^T_{N-k}x(k/k)$$

$$A^T_{N-k} = [B^T(k)(M(k) + S_{N-k-1})B(k) + N(k)]^{-1}B^T(k)(M(k)$$

$$+ S_{N-k-1})F(k)S_{N-k-1}$$

$$= [F(k) - B(k)A^T_{N-k}]^T(M(k) + S_{N-k-1})[F(k) - B(k)A^T_{N-k}]$$

$$+ A^T_{N-k}N(k)A_{N-k} = F^T(k)(M(k) + S_{N-k-1})[1 - B(k)[B^T(k)(M(k)$$

$$+ S_{N-k-1})B(k) + N(k)]^{-1}B^T(k)(M(k) + S_{N-k-1})   S_0 = 0$$

$$\tag{4.47}$$

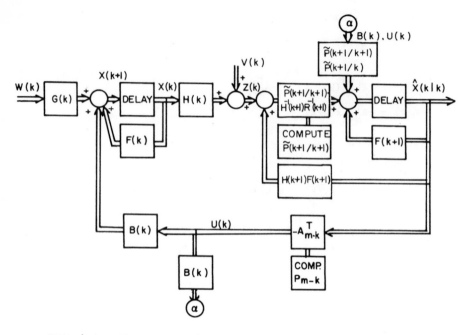

FIG. 4.1.   The Stochastic Optimal Feedback Control for Discrete-Time Systems.

$$q_{N-k} = q_{N-k-1} + tr[A_{N-k}^T[B^T(k)(M(k) + S_{N-k-1})B(k) + N(k)]A_{N-k}P(k/k)]$$
$$+ tr[G^T(k)(M(k) + S_{N-k-1})G(k)Q(k)] = q_{N-k-1}$$
$$+ tr[F^T(k)(M(k) + S_{N-k-1}) B(k)[B^T(k)(M(k) + S_{N-k-1})B(k)$$
$$+ N(k)]^{-1}B^T(k)(M(k) + S_{N-k-1})F(k)P(k/k)] + tr[G^T(k)(M(k)$$
$$+ S_{N-k-1})G(k)Q(k)].$$

$$\hat{x}(k/k) = F(k - 1)\hat{x}(k - 1/k - 1) + P(k/k)H^T(k)R^{-1}(k)[z(k)$$
$$- H(k)F(k - 1)\hat{x}(k - 1/k - 1)]$$
$$- P(k/k)P^{-1}(k/k - 1)B(k - 1)A_{N-k}\hat{x}(k - 1/k - 1)$$
$$\hat{x}(0/0) = x_0 \tag{4.48}$$

$$P(k/k) = P(k/k - 1) - P(k/k - 1)H^T(k) [H(k)P(k/k - 1)H^T(k)$$
$$+ R(k)]^{-1}H^T(k)P(k/k-1) \quad P(0/0) = P_0$$

$$P(k/k - 1) = F(k - 1)P(k - 1/k - 1)F^T(k - 1) + G(k - 1)Q(k - 1)$$
$$G^T(k - 1)$$

4.5.5. The Infinite Discrete-Time Problem

The extension of the problem discussed in the preceding section is, as with the continuous-time systems, the case where the process has infinite duration, e.g., $N \to \infty$. Making similar assumption as in Sec. 4.5.2 all the system matrices in (4.31), (4.32) and (4.33) are replaced with their time-invariant equivalents F, B, G, H, Q, R, M, and N. As in Sec. 4.5.2 the performance criterion (4.33) must also be modified to yield a finite value;

$$J(u) = \lim_{N \to \infty} E\left\{ \frac{1}{N} \sum_{i=0}^{N-1} [x^T(i + 1)M(i)x(i + 1) + u^T(i)N(i)u(i)] \right\} \tag{4.49}$$

As in the continuous-time case, there are three conditions important to the solution of the problem:

1. Stability.
2. Existence of an admissible control.
3. Optimality.

The first is handled by an appropriate Lyapunov function, the second implies the controllability and observability of the system (4.31) along with some other factors, while the third, the optimality is treated by the steady-state form of the Bellman equation of dynamic programming (4.36) and Kalman-Bucy filter (4.41). The result is

$$u^*(k) = -A^T\hat{x}(k/k)$$
$$A^T = [B^T(M + S)B + N]^{-1}B^T(M + S)F \tag{4.50}$$
$$S = (F - B A^T)^T(M + S)(F - B A^T) + A^TNA$$

where $\hat{x}(k/k)$ is the output of a steady-state Kalman-Bucy filter.

4.5.6.   Optimal Feedback vs. Open-loop Controls

A comparison of the optimal feedback to the open-loop controls, similar to the one for the continuous-time case given in Sec. 4.5.3, can be readily presented. In order to avoid complex expressions, w which only complicate the derivative, the comparison is given here for the case that there is no measurement noise, all the states are available for measurement, e.g., $H(k) = I$, $v(k) \equiv 0$ $\forall k$ as in Step 1 of Sec. 4.5.4 and the initial state x_0 is not random.

As in the continuous-time deterministic case, the open-loop control should be designed as if the process noise $w(k)$ was not present at all, in which case the open-loop and the feedback optimal solutions are identical. The open-loop control is thus described by

$$u^*_{OL}(k,x_0) = -A^T_{N-k}\Phi(k)x_0 \tag{4.51}$$

where A^T_{N-k} is the deterministic optimal control gain given in (4.38), and $\Phi(k)$ is the state transition matrix associated with the deterministic closed-loop system:

$$\Phi(k) \triangleq \prod_{i=0}^{k-1} [F(i) - B(i)A^T_{N-i}] \tag{4.52}$$

The optimal feedback control is given by (4.38) also, and is repeated here

$$u^*_{FB}(k,x(k)) = -A^T_{N-k}x(k) \tag{4.53}$$

Comparing u^*_{OL} and u^*_{FB}, one may easily verify that

$$u^*_{FB}(k,x(k)) = u^*_{OL}(k,x_0) - A^T_{N-k}\sum_{i=0}^{k-1} \Psi(k - i - 1)G(i)w(i) \tag{4.54}$$

where $\Psi(j)$ is the state transition matrix associated with the open-loop system

$$\Psi(k - i - 1) = \prod_{j-0}^{k-i-1} F(j) \tag{4.55}$$

Substituting the open-loop u^*_{OL} and feedback u^*_{FB} control in the performance criterion (4.33) to be minimized, and comparing, one gets after considerable algebra

$$J(u_{OL}^*) - J(u_{FB}^*) = \sum_{i=0}^{N-1} tr[G^T(i) \sum_{k=i+1}^{N-1} \Psi(k - i - 1)A_{N-k}N(k)A_{N-k}^T$$

$$\Psi(k - i - 1) \quad G(i)Q(i)] > 0 \tag{4.56}$$

The relation (4.55) is the discrete-time equivalent of (4.26) and Therefore the conclusions are similar, namely, the difference between $J(u_{OL}^*)$ and $J(u_{FB}^*)$ increases with $||G(k)||^2$ or Q(k) or the duration of the process.

In case the measurement of the states is performed indirectly and is corrupted with noise, a similar expression can be obtained which verifies the superiority of the feedback optimal control.

4.6 DISCUSSION

A survey of the stochastic optimal feedback control was presented in this chapter, and the various concepts of stochastic optimality have been discussed. The differences between open-loop and feedback control have been established from the point of view of utilization of the available information about the states and uncertainties of the systems for the general case and the specific LQG problems.

The conclusions are very enlightening, in view of the confusion that still exists among many researchers in the area. The open-loop optimal control is inferior to any feedback optimal control in the value of the performance index as long as process and measurement noise are present and, to quote Wonham [4.30], "No open-loop control can stabilize an inherently unstable system, and feedback is crucial for acceptable long-term operation."

The feedback optimal-control problem was further investigated and it was pointed out that the way that information about the state and the system uncertainties is used changes the optimality of the solution. If the information is used passively, in other words, in the conservative sense of using only the accrued information without anticipation of obtaining more in the future, a control is generated which is inferior to the dual control, resulting from active use of the measurable information about the system.

Dual control is covered in the next chapter. However, the case *neutral systems,* which are indifferent to the active utilization of information, has been treated here. The optimal solution of the linear quadratic Gaussian problem, a special care of this category, has been treated in the continuous-time and discrete-time cases. The neutrality of such systems has been established through the celebrated *separation theorem* which assigns separately the functions of optimal state estimation and optimal control in optimal control systems. This, however, is a special case of stochastic optimal systems, very rarely encountered in reality. The dual optimal solution, as will be seen in the next chapter, is very hard to obtain. Therefore, the various approximations proposed in the modern literature are based on approximation of the passive feedback optimal control, sometimes appearing under the name of *"open-loop feedback optimal"* (OLFO) [4.10]. Some OLFO algorithms which possess a parameter-adaptive property will be discussed in Chap. 7, along with other parameter-adaptive S.O.C. algorithms.

4.7 REFERENCES

[4.1] Aoki, M., *Optimization of Stochastic Control Processes*, Academic Press, New York, 1967.

[4.2] Åstrom, K. J., *Introduction to Stochastic Control Theory*, Academic Press, New York, 1970.

[4.3] Athans, M. (ed.), *IEEE Trans. Automatic Control*, AC-166 (1971) (Special issue on Linear-Quadratic-Gaussian Problem.)

[4.4] Athans, M., and Falb, P., *Optimal Control*, McGraw-Hill, New York, 1966.

[4.5] Bar-Shalom, Y., and Tse, E., "Dual Effect, Certainty Equivalence and Separation in Stochastic Control," *IEEE Trans. Automatic Control*, AC-19, 5 (1974).

[4.6] Bellman, R., *Adaptive Processes - A Guided Tour*, Princeton University Press, Princeton, N.J., 1961.

[4.7] Bryson, A. E., and Ho, Y. C., *Applied Optimal Control*, Blaisdell, Waltham, Mass., 1969.

[4.8] Doob, J. L., *Stochastic Processes*, Wiley, New York, 1953.

[4.9] Dynkin, E. B., *Markov Processes*, New York, Springer-Verlag, 1965.

[4.10] Dreyfus, S. E., *Dynamic Programming and the Calculus of Variations*, Academic Press, New York, 1965.

[4.11] Fel'dbaum, A. A., *Optimal Control Systems*, Academic Press, New York, 1965.

[4.12] Kailath, T., Kozin, F., Bucy, R., and Wonham, W., *Stochastic Problems in Control*, ASME Publication, New York, 1968.

[4.13] Kulikowski, R., "Optimization of Nonlinear Random Control Processes," Preprints, 2nd IFAC, Basel, 1962, p. 69.

[4.14] Kushner, H., *Introduction to Stochastic Control*, Holt, Rinehart, and Winston, New York, 1971.

[4.15] Meier, L., Larson, R. E., and Tether, A. J., "Dynamic Programming for Stochastic Control of Discrete Systems," *IEEE Trans. Automatic Control*, AC-16, 767-775, (1971), (Special issue on Linear Quadratic Gaussian Problem.)

[4.16] Meditch, J., *Stochastic Optimal Linear Estimation and Control*, McGraw-Hill, New York, 1969.

[4.17] Patchell, J. W., and Jacobs, O. L. R., "Separability, Neutrality, and Certainty Equivalence," *Int. J. Control,* 13, 337-342 (1971).

[4.18] Peschon, J., *Disciplines and Techniques of Systems' Control,* Blaisdell, Waltham, Mass., 1965.

[4.19] Raiffa, H., and Schlaifer, R., *Applied Statistical Decision* Theory, M.I.T. Press, Cambridge, Mass., 1972.

[4.20] Root, J. G., "Optimum Control of Non-Gaussian Linear Stochastic Systems with Inaccessible State Variables," *SIAM J. Control,* 7, 317-323 (1969).

[4.21] Sage, A., *Optimum Systems Control,* Prentice-Hall, New York, 1968.

[4.22] Speyer, J., Deyst, J., and Jacobson, D., "Optimization of Stochastic Linear Systems with Additive Measurement and Process Noise Using Exponential Performance Criteria," *IEEE Trans. Automatic Control,* AC-19, 358-366 (1974).

[4.23] Sworder, D. D., *Optimal Adaptive Control Systems,* Academic Press, New York, 1967.

[4.24] Sworder, S. C., and Sworder, D. D., "Feedback Estimation Systems and the Separation Principle of Stochastic Control," *IEEE Trans. Automatic Control,* AC-16, 350-354 (1971).

[4.25] Theil, H., "A Note on Certainty Equivalence in Dynamic Planning," *Econometrica,* 25, 346-349 (1957).

[4.26] Tou, J., *Modern Control Theory,* McGraw-Hill, New York, 1963.

[4.27] Tou, J., and Joseph, P., "On Linear Control Theory," *Pt. II, AIEE Trans.* 80, 18, (1961).

[4.28] Tse, E., and Bar-Shalom, Y., "Concepts and Methods in Stochastic Control," in *Control and Dynamic Systems,* (C. T. Leondes, ed.) Academic Press, New York, 1975.

[4.29] Witsenhausen, H. S., "Separation of Estimation and Control for Discrete-Time Systems," *Proc. IEE,* 59, 1557-1566, (1971).

[4.30] Wonham, W. M., "On the Separation Theorem of Stochastic Control," *SIAM J. Control,* 6, 312-326 (1968).

[4.31] Wonham, W. M., "Random Differential Equations in Control Theory," in *Probabilistic Methods in Applied Mathematics,* (A. T. Barucha-Reid, ed.), Academic Press, New York, 1970.

Chapter 5

THE DUAL CONTROL PROBLEM

5.1 INTRODUCTION

In 1961, A. A. Fel'dbaum wrote four papers in which he formulated his celebrated *Dual Control Problem* treating the general case of systems with incomplete information about its own parameters and operating in a stochastic environment [5.3]. This problem is far more general than the stochastic control problem, since it contains uncertainties reducible by some estimation procedure as the process evolves and new input-output data about the system are available. In this case an optimal control performs the *dual function* of optimizing a given performance criterion and, at the same time, accumulating information regarding the unknown parameters. In the case that all the system and environment parameters are known statistically or deterministically, the dual control problem is reduced to the stochastic control problem discussed in the preceding chapter. It is obvious that the present problem is far more complex than the stochastic control problem of which a complete solution was obtained only for the cases where the separation theorem was applicable. It will be shown that this is the special case of dual control called *neutral systems* that possess the separability property and represent only a very small subclass of the stochastic control problem [5.1], [5.5].

The formulation of the dual control problem closely resembles the formulation of the S.O.C. problem given in Chap. 1, since the accumulation of information is an essential attribute of this particular problem. As a matter of fact, the dual control solution

would represent the optimal solution of the parameter-adaptive
S.O.C. problem. However, it would not generate a solution to the
performance-adaptive S.O.C. problem, since it does not provide any
way to account for structural uncertainties of the plant or
accumulation of information regarding the improvement of the
performance of the system. But, it will be demonstrated that a
solution to the dual control problem is far from being an easy task.
As a matter of fact, due to analytic and computational difficulties,
it is hardly possible to obtain a solution to any but trivial cases.
This is one of the major reasons to resort to S.O.C. as the best
thing next to the optimal solution of the problem.

5.2 THE GENERAL FORMULATION OF THE DUAL CONTROL PROBLEM

The problem that Fel'dbaum formulated originally dealt with a
system without any dynamics surrounded by noisy channels and driven
by a dynamic dual controller, where the parameters with reducible
uncertainties where parameters of the noise vectors.

A different formulation of the dual control problem with
emphasis on the control problem discussed throughout this book is
given here. A discrete-time, nonlinear system depicted in Fig. 5.1

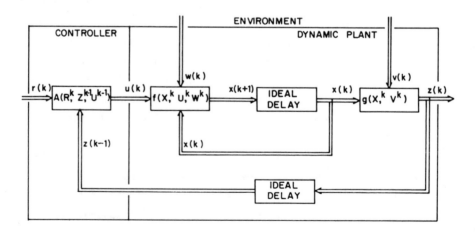

FIG. 5.1. Block Diagram of the Dual Control Problem.

is considered where, for simplicity of presentation, two vectors may
represent either reducible or irreducible uncertainties, or both, of
the system. If one defines the sequence of certain measurements $y(\cdot)$
by $Y^k = \{y(0),...,y(k)\}$, then the general stochastic system can be
formally described by the following difference equations with memory:

$$x(k + 1) = f(x^k, v^k, w^k, k) \quad x(0) = x_0$$

$$z(k) = g(z^k, v^k, k) \tag{5.1}$$

where $x(k)$ is the n-dimensional state vector at time k; $\{u(i)\}$
$i = 0,...,k$ is a sequence of control inputs; and $\{w(i)\}$ and $\{v(i)\}$,
$i = 0,...,k$ are sequences of random variables representing uncertainties of the system. If they represent stochastically known system-
environment interactions, their respective probability density functions
$p(w(i))$ and $p(v(i))$, $i = 0, 2,...,N$ are known and the uncertainties
are not further reducible. This reduces the problem to a nonlinear
stochastic control problem discussed in Chap. 4. However, any of
the vectors $w(i)$ and $v(i)$ may be assigned to represent the unknown
system parameters, or an unknown parameter θ may be implicitly
imbedded in one of the input vectors of the system. In this case,
the respective probability density functions $p(w(i))$, $p(v(i))$, or
$p(\theta)$ may be replaced by the *a-posteriori* density functions $p(w(i)/Z^k)$,
$p(v(i)/Z^k)$, or $p(\theta/Z^k)$. These density functions, which may not
appear explicitly, may be introduced into the problem by considering
the marginal densities of an appropriate dependent variable, say
$y(i)$, i.e., for θ_j:

$$p(y(i)/Z^k) = \int_{\Omega(\theta)} p(y(i)/\theta, Z^k) p(\theta/Z^k) \, d\Omega(\theta)$$

The *a-posteriori* probabilities $p(\theta/Z^k)$ can be learned by the
controller <u>on-line</u>, by using Bayes' rule successively:

$$p(\theta/Z^k) = \frac{p(z(k)/Z^{k-1}, \theta) p(\theta/Z^{k-1})}{\sum_{j=1}^{k} p(z(j)/Z^{j-1}, \theta) p(\theta/Z^{j-1})} \tag{5.2}$$

This additional property of the controller accounts for the name
dual control. The functions $f(\cdot)$ and $g(\cdot)$ are known functions of

their arguments and the initial state x_0 is a random variable with known density function with possibly unknown parameters.

A control law of unknown structure

$$u(k) = A(R^k, z^{k-1}, u^{k-1}) \tag{5.3}$$

is sought to minimize a performance criterion of the form:

$$I = E\{J\} \triangleq E\left\{\sum_{k=0}^{N} J_k(r(k), z(k), u(k), k)\right\} = \sum_{k=0}^{N} I_k \tag{5.4}$$

where the instantaneous performance criterion I_k is defined by

$$I_k = E\{J_k\} = \int_{\Omega(R^k, z^k, u^k)} J_k(r(k), z(k), u(k), k) p(R^k, z^k, u^k)$$
$$d\Omega(R^k, z^k, u^k)$$

$$= \int_{\Omega(R^k, z^{k-1}, u^{k-1})} \rho_k p(R^k, z^{k-1}, u^{k-1}, k) \, d\Omega(R^k, z^{k-1}, u^{k-1}) \tag{5.5}$$

and

$$\rho_k \triangleq E\{J_k(r(k), z(k), u(k), k) / R^k, z^{k-1}, u^{k-1}\}$$

$$= \int_{\Omega(z(k), u(k))} J_k(r(k), z(k), u(k), k) p(z(k), u(k) / R^k, z^{k-1}, u^{k-1})$$
$$d\Omega(z(k), u(k)) \tag{5.6}$$

In Eqs. (5.4) to (5.6) the summability properties of the functions involved have been assumed. The chain rule of probabilities similar to Eq. (5.2) has been used to expand the expressions. R^k is a sequence of known random inputs with given probability density function $p(r(i))$, $i = 0, \ldots, k$, and the appropriate spaces of definition of the various random variable, say $u(i)$, are denoted by $\Omega(u(i))$. $N < \infty$ is the number of stages of the process.

In the above expressions the probability densities $p(w(i))$, $p(v(i))$, or $p(\theta)$ have not appeared explicitly. They occur in the expressions that define the transition probabilities of the plant $T(k)$ probabilistically describing the plant dynamics.

$$T(k) \triangleq p(z(k) / R^k, z^{k-1}, u^k) = p(z(k) / u^k) \tag{5.7}$$

The last equation can be written because u^k represents sufficient statistics for R^k and z^{k-1} as indicated by Eq. (5.3).

The transition probabilities of the plant, called by Bellman
information patterns [5.2], serve to describe the system probabilistically anc can be manipulated by deterministic methods, thus
bypassing the difficulties involving the random nature of the plant
and the required operation in Hilbert spaces. Since the plant is
described by the transition probabilities, it is natural to assume
that the control policy is described by the so-called *decision*
probabilities defined by

$$\Gamma(k) \triangleq p(u(k)|R^k, z^{k-1}, u^{k-1}), \int_{\Omega(u(.))} \Gamma(k) \, d\Omega(u(.)) = 1 \qquad (5.8)$$

Therefore, the controller (5.3) is purely stochastic and it is
completely described by the sequence $\{\Gamma(k)\}$ of the decision probability densities. Such a sequence $\{\Gamma^*(k)\}$, which minimizes the
performance criterion I in (5.4), is called an optimal *pure random*
strategy. The actual implementation of such a controller is rather
difficult and therefore their application is limited. However, in
many cases such as the dual control problem, a deterministic
controller $u^*(k) = A(R^k, z^{k-1}, u^{k-1})$ may be recovered from the decision
probabilities when available structural information of the controller
may provide information pertaining to the minimization of the
performance criterion I.   In such a case the optimal decision
probabilities are reduced to delta functions,

$$\Gamma^*(k) = \delta(u(k) - u^*(k)) \qquad (5.9)$$

and the control strategy becomes a *mixed strategy*. This idea was
proposed by Fel'dbaum [5.4] and will be further explored in the
sequel.

The dependence of the transition probabilities on the probability densities $p(w^k)$ and $p(v^k)$, describing the uncertainties of the
system, can be demonstrated by considering $T(k)$ as marginal
probabilities.

$$T(k) = p(z(k)|u^k) = \int_{\Omega(x^k, w^k, v^k)} p(z(k), x^k, w^k, v^k \, u^k)$$
$$d\Omega(x^k, w^k, v^k)$$

$$= \int_{\Omega(x^k,w^k,v^k)} p(z(k)|x^k,w^k,v^k,u^k)$$

$$p(x^k|w^k,v^k,u^k)$$

$$p(w^k) \ p(v^k) \ d\Omega(x^k,w^k,v^k)$$

$$= \int_{\Omega(x^k,w^k,v^k)} p(z(k)|x^k,v^k) \cdot p(x^k|w^{k-1},u^{k-1})$$

$$p(w^k) \ p(v^k) \ d\Omega(x^k,w^k,v^k) \tag{5.10}$$

The last Eq. in (5.10) was written after considering the
sufficient statistics of each random variable involved. Equation
(5.10) provides a vehicle for the evaluation of the transition
probabilities $T(k)$. The densities $p(z(k+1)|x^k,v^k)$ and $p(x^k|w^{k-1},u^{k-1})$ describing the transitions of the measurement equations and
the dynamic equations of (5.1), respectively. They may be computed
from Eq. (5.1) if the structural form of the system $f(\cdot)$ and $g(\cdot)$ is
known and the form of the density functions of the basic random
variables are known. However, this is a difficult task when the
system is nonlinear and the random processes are of nonreproducible
form. The probability densities $p(z(k)|u^k)$, $p(z(k)|x^k,v^k)$, etc.,
may be simplified considerably if the plant possesses a first-order
Markov property, i.e., $p(z(k)|u(k))$ and $p(z(k)|x(k), v(k))$ respectively.
The first step for the formulation of the optimal solution is
to evaluate the density functions involved in the evaluation of the
performance criteria (4.55) and (4.54), $p(z(k), u(k)|R^k,z^{k-1},u^{k-1})$
and $p(R^k,z^{k-1},u^{k-1})$, respectively.

$$p(z(k),u(k)|R^k,z^{k-1},u^{k-1}) = p(z(k)|R^k,z^{k-1},u^k)$$

$$p(u(k)|R^k,z^{k-1},u^{k-1}) = T(k) \ \Gamma(k)$$

$$\tag{5.11}$$

$$p(R^k,z^{k-1},u^{k-1}) = p(z^{k-1},u^{k-1}|R^k) \ p(R^k) = p(R^k) \ p(z^{k-1},u^{k-1}|R^{k-1})$$

$$= p(R^k) \ p(z(k-1)|R^{k-1},z^{k-2},u^{k-1}) \ p(z^{k-2},u^{k-1}|R^{k-1})$$

$$= p(R^k) \ T(k-1) \ p(u(k-1)|R^{k-1},z^{k-2},u^{k-2})$$

$$p(z^{k-2},u^{k-1}|R^{k-1})$$

$$= p(R^k) \ T(k-1) \ \Gamma(k-1) \ p(z^{k-2},u^{k-2}|R^{k-1})$$

$$= p(R^k) \prod_{i=0}^{k-1} T(i) \prod_{j=0}^{k-1} \Gamma(j) \tag{5.12}$$

where the chain rule has been used and the definitions of $T(i)$ and $\Gamma(i)$ in (5-7) and (5-8) are used. The *a priori* probabilities have been defined as

$$T(0) \triangleq p(z(0)) \tag{5.13}$$

Substituting back to (4.54) and (4.55) one obtains the instantaneous values of the performance criterion I_k

$$I_k = \int_{\Omega(R^k,z^k,u^k)} J_k(r(k),z(k),u(k),k) p(R_k^k) \prod_{i=0}^{k} T(i) \prod_{j=0}^{k} \Gamma(j)$$

$$d\Omega(R^k,z^k,u^k) \tag{5.14}$$

The uncertainties inherent in this system may be introduced through Eqs. (5.2) and (5.10).

5.3 FORMULATION OF THE OPTIMAL SOLUTION

Fel'dbaum applies procedure similar to Bellman's dynamic programming [5.2], [5.4] to formulate the optimal solution to the problem and generate the pure random strategy desired. A mixed strategy is then derived to yield the actual optimal controller. It can be seen from Eq. (5.14) that the instantaneous performance index depends on all the past decision probabilities $\{\Gamma(j)\}$. Therefore, in contrast to what happened in the pure stochastic case, it is expected that every decision depends on all past decisions and this would necessitate the duality of the controller. In order to set up the dynamic programming formulation, define the following functions:

$$\alpha_k(R^k,z^{k-1},u^k) = \int_{\Omega(z(k))} J_k(r(k),z(k),u(k)) p(R^k) \prod_{i=0}^{k} T(i) d\Omega z(k))$$

$$\beta_{k-1}(u^{k-1}) = \prod_{j=0}^{k-1} \Gamma(j) \tag{5.15}$$

Then the instantaneous performance criterion I_k is written as

$$I_k = \int_{\Omega(R^k, z^{k-1}, u^k)} \alpha_k(R^k, z^{k-1}, u^k) \beta_{k-1}(u^{k-1}) \Gamma(k) \; d\Omega(R^k, z^{k-1}, u^k)$$

$$(5.16)$$

As in the case of dynamic programming, the minimization of (5.4) can be imbedded in the more general problem of minimizing the *return function* $V_{N-k}(k)$, starting at time k and having (N - k) states.

$$V_{N-k}(k) \stackrel{\Delta}{=} \min_{\Gamma(k),\dots,\Gamma(N)} \sum_{i=k}^{N} I_i = \min_{\Gamma(k)} [I_k + \min_{\Gamma(k+1),\dots,\Gamma(N)}$$

$$\sum_{i=k+1}^{N} I_i] = \min_{\Gamma(k)} [I_k + V_{N-k-1}(k + 1)] \qquad (5.17)$$

The solution of the problem under consideration will be given for k=0. Moving backwards in time, let k = N

$$V_0(N) = \min_{\Gamma(N)} I_N = \min_{\Gamma(N)} \int_{\Omega(R^N, z^{N-1}, u^N)} \alpha_N(R^N, z^{N-1}, u^{N-1}, u(N))$$

$$\beta_{N-1}(u^{N-1}) \cdot \Gamma(N) \; d\Omega(R^N, z^{N-1}, u^N) \qquad (5.18)$$

Finally, define the Hamiltonian type of function:

$$H_N(R^N, z^{N-1}, u^{N-1}) \stackrel{\Delta}{=} \int_{\Omega(u(N))} \alpha_N(R^N, z^{N-1}, u^{N-1}, u(N)) \Gamma(N) \; d\Omega(u(N))$$

$$(5.19)$$

Under the continuity assumption in Ω, one may write:

$$V_0(N) = \min_{\Gamma(N)} \int_{\Omega(R^N, z^{N-1}, u^{N-1})} \beta_{N-1}(u^{N-1}) H_N(R^N, z^{N-1}, u^{N-1})$$

$$d\Omega(R^N, z^{N-1}, u^{N-1}) = \int_{\Omega R^N, z^{N-1}, u^{N-1}} \beta_{N-1}(u^{N-1})$$

$$\min_{\Gamma(N)} H_N(R, z^{N-1}, u^{N-1}) \; d\Omega(R^N, z^{N-1}, u^{N-1}) \qquad (5.20)$$

Since (5.20) is true for all (R^N, z^{N-1}, u^{N-1}), $\min_{\Gamma(N)} H_N$ implies $\min_{\Gamma(N)} I_N$

and therefore it may replace I_N in the minimization procedure. Hence, the pure random optimal strategy may be defined to minimize H_N.

Assuming that the appropriate conditions are satisfied, one may use
the mean-value theorem [3.28] on (5.19) to write;

$$H_N(R^N, Z^{N-1}, U^{N-1} | u(N) = \bar{u}(N)) = \bar{\alpha}_N(R^N, Z^{N-1}, U^{N-1}, \bar{u}(N))$$

$$\int_{\Omega(u(N))} \Gamma(N) \, d\Omega(u(N))$$

$$= \bar{\alpha}_N(R^N, Z^{N-1}, U^{N-1}, u(N)) \geq \alpha_N^* \qquad (5.21)$$

where $\bar{\alpha}_N(R^N, Z^{N-1}, U^{N-1}, \bar{u}(N))$ is the value of $\alpha_N(R^N, Z^{N-1}, U^N)$ for an
appropriate $\bar{u}(N) \, \epsilon \, \Omega(u(N))$ for which the mean-value theorem is
satisfied, and α_N^* is the minimum value of $\alpha_N(R^N, Z^{N-1}, U^N)$ for an
appropriate optimal controller $u^*(N) \, \epsilon \, \Omega(u(N))$, so that

$$\alpha_N(R^N, Z^{N-1}, U^{N-1}, u(N)) \geq \alpha^*(R^N, Z^{N-1}, U^{N-1}, u^*(N)) = \alpha_N^* \qquad (5.22)$$

for all $u(N) \neq u^*(N)$ and any R^N, Z^{N-1}, U^{N-1}.

One may proceed now to obtain the optimal mixed strategy, which
is available for the dual control problem. This is, of course,
contingent upon the availability of a function $u^*(N)$ that minimizes
α_N from Eq. (5.22). It is claimed that $u^*(N)$ is the optimal controller and the strategy is now mixed. Let

$$u^*(N) = A^*(R^N, Z^{N-1}, U^{N-1}): \quad \underset{u(N) \epsilon \Omega(u(N))}{\text{Min}} \alpha_N(R^N, Z^{N-1}, U^{N-1}, u(N))$$

$$\underline{\underline{\Delta}} \, S_0 \qquad (5.23)$$

Assume that $u^*(N)$ is the optimal controller, then

$$\Gamma^*(N) = \delta(u(N) - u^*(N)) \qquad (5.24)$$

where $\delta(\cdot)$ is a Dirac delta function. It is easy to show that $T^*(N)$
minimizes H_N of (5.19).

$$H_N(R^N, Z^{N-1}, U^{N-1}, T^*(N)) = \int_{\Omega(u(N))} \alpha_N(R^N, Z^{N-1}, U^{N-1}, u(N)) \, \delta \, (u(N)$$

$$- u^*(N)) \, d\Omega(u(N)) = \alpha(R^N, Z^{N-1}, U^{N-1}, u^*(N))$$

$$= \alpha_N^* = \underset{\Gamma(N)}{\text{Min}} \, H_N(R^N, Z^{N-1}, U^{N-1}) \qquad (5.25)$$

The last statement is justified by (5.21) for which α_N^* is a lower bound on H_N and, if attainable, the minimum of H_N. Since $u^*(N)$ renders H_N minimum, it renders I_N minimum as well according to (5.20). This establishes the optimal mixed strategy for a one-stage process.

$$V_0 = \underset{\Gamma(N)}{\text{Min}}\ I_N = \int_{\Omega(R^N,Z^{N-1},U^{N-1})} \beta_{N-1}(U^{N-1})\ S_0^*\ d\ (R^N,Z^{N-1},U^{N-1})$$

$$(5.26)$$

To obtain the general recursive formula for N-k stages, one may proceed *inductively* backward from the one-stage process.

In order to obtain such an expression, one should start from Eq. (5.20) which defines the recursive relation for the return function. The return function $V_{N-k}(k)$ may be inductively obtained from Eq. (5.25) by

$$V_{N-k}(k) = \int_{\Omega(R^k,Z^{k-1},U^{k-1})} \beta_{k-1}(U^{k-1}) \cdot S_{N-k}^*(R^k,Z^{k-1},U^{k-1},u^*(k))$$

$$d\Omega(R^k,Z^k,U^{k-1})\qquad\qquad(5.27)$$

where S_{N-k}^* will be recursively obtained in the sequel. Then Eq. (5.20) may be written as

$$V_{N-k}(k) = \underset{\Gamma(k)}{\text{Min}}\ [I_k + V_{N-k-1}(k + 1)] = \underset{\Gamma(k)}{\text{Min}}\int_{\Omega(R^k,Z^{k-1},U^k)}$$

$$\beta_{k-1}(U^{k-1})\Gamma(k)\ [\alpha_k(R^k,Z^{k-1},U^k) + \int_{\Omega(r(k+1),z(k))}$$

$$S_{N-k-1}^*(R^{k+1},Z^k,U^k,u^*(k + 1))\ d\Omega(r(k + 1),z(k))]$$

$$d\Omega(R^k,Z^{k-1},U^k)\qquad\qquad(5.28)$$

and

$$S_{N-k}^* = \alpha_k(R^k,Z^{k-1},U^k) + \int_{\Omega(r(k+1),z(k))} S_{N-k-1}^*(R^{k+1},Z^k,U^k,u^*(k + 1)$$

$$d\Omega(r(k + 1),z(k))\qquad S_0^* = \alpha_N^*\qquad\qquad(5.29)$$

As it was done for the one-stage process, one may define the Hamiltonian function

$$H_k(R^k, z^{k-1}, U^{k-1}) \triangleq \int_{\Omega(u(k))} S_{N-k}(R^k, z^{k-1}, U^{k-1}, u(K)) T(k) \; d\Omega(u(k))$$

$$(5.30)$$

Then for an appropriate $\bar{u}(k)$, satisfying the conditions of the mean-value theorem

$$H_k(R^k, z^{k-1}, U^{k-1} | u(k) = \bar{u}(k)) = \bar{S}_{N-k}(R^k, z^{k-1}, U^{k-1}, \bar{u}(k))$$

$$\int_{\Omega(u(k))} \Gamma(k) \; d\Omega(u(k))$$

$$= \bar{S}_{N-k}(R^k, z^{k-1}, U^{k-1}, \bar{u}(k))$$

$$\geq S_{N-k}^*(R^k, z^{k-1}, U^{k-1}, u^*(k)) \qquad (5.31)$$

Then the mixed optimal strategy can be defined as follows:

$$u^*(k) = A^*(R^k, z^{k-1}, U^{k-1}): \quad \underset{u(k) \in \Omega(u(k))}{\text{Min}} S_{N-k}(R^k, z^{k-1}, U^k) = V_{N-k}$$

$$\Gamma^*(k) = \delta(u(k) - u^*(k)) \qquad (5.32)$$

It is very easy to show that $u^*(k)$ minimizes the $H_k(R^k, z^{k-1}, U^{k-1})$ and therefore the $(N-k)$-stage performance criterion (5.27).

$$H_k(R^k, z^{k-1}, U^{k-1} | u(k) = u^*(k)) = \int_{\Omega(u(k))} S_{N-k}^*(R^k, z^{k-1}, U^{k-1}, u^*(k))$$

$$\delta(u(k) - u^*(k)) \; d\Omega(u(k))$$

$$= S_{N-k}^*(R^k, z^{k-1}, U^{k-1}, u^*(k))$$

$$= \underset{\Gamma(k)}{\text{Min}} H_k(R^k, z^{k-1}, U^{k-1})$$

$$(5.33)$$

It is obvious from (5.32) that $u^*(k)$ depends on all the past and future optimal decisions $u(i)$, $i = 0,\ldots,N$; furthermore, the learning property of the optimal control is implicitly assumed from the evaluation of α_k in Eq. (5.15). The dependence of the optimal controller on the integral term of the right-hand side of Eq. (5.28) will be explored in the next section. Meanwhile, the optimal mixed policy is completely described by (5.32).

5.4   THE OPTIMAL DUAL CONTROL AND THE NEUTRAL SYSTEM

It has been demonstrated that the general stochastic control problem, where the plant has complete memory of the past, the optimal control depends on all the past decisions, and, in addition it depends to the predicted future, optimal decisions enter the computation of S_{N-k}^* recursively in (5.29). Of course, this property is independent of the learning property of the dual controller which is assumed implicitly in this derivation of the optimal controller and is performed by using (5.2) and (5.10) as the transition probabilities of the plant are evaluated.

The dependence of the optimal controller at time k on the other optimal decisions complicates greatly the process of obtaining the optimal control policy for the system.

However, there is a class of systems for which the right-hand side of the expression (5.28) used to evaluate S_{N-k}^* does not depend on $\{u(j); j = k + 1, \ldots, N\}$ explicitly. In cases like this, the optimal controller can be evaluated as in the case of the linear, quadratic, Gaussian problem discussed in Chap. 4. In this case, the system is called *neutral* [8.4], and covers a much larger class than the linear quadratic Gaussian problem, like all open-loop controls plus several closed-loop cases. It is obvious that neutral problems may find a solution much easier than the general dual control problem, for which, as seen from the formulation in the previous section, solutions are prohibitively difficult and implementation is almost impossible, due to the dependence to future decisions.

A more special case of neutral systems arises when the right-hand-side term of (5.28) is completely independent of U^N. In such a case, the control u(k) may be evaluated directly from α_k and the return function is not needed. Then the present decision is completely independent of the future states or decisions of the system

$$u^*(k) = A^*(R^k, z^{k-1}, u^{k-1}): \quad \underset{u(k) \in \Omega(u(k))}{\text{Min}} \quad \alpha_k(R^k, z^{k-1}, u^{k-1})$$

$$= \underset{u(k) \in \Omega(u(k))}{\text{Min}} \int_{\Omega(z(k))} J_k P(R^k) \prod_{l=0}^{k} T(i) \, dz(k) \qquad (5.34)$$

Such a case will be demonstrated by an example in the next section.

5.5   TWO CASE STUDIES OF DUAL CONTROL

The purpose of this section is to demonstrate the problems involved with the solution of the equations formulated in the previous
sections to determine the optimal dual control, the computational
limitations, and the difficulty of implementation of such a controller.
Two cases are discussed: one of a neutral system, and another with
a nonneutral one. Both cases treat trivial nondynamic first-order
systems, because of the minimum amount of computations required.

5.5.1   Dual Control for a Neutral Plant

The first system to be considered is depicted in Fig. 5.2, and
is described as follows:

<u>Plant</u> $x(k) = u(k) + w(k)$ (5.35)

   $z(k) = x(k) + v(k)$ $v(k) = \mu$ = the unknown parameter

with

$$p(w(k)) = \frac{1}{\sigma_w \sqrt{2\pi}} \exp\left\{-\frac{w(k)^2}{2\sigma_w^2}\right\} \quad p_0(\mu) = \frac{1}{\sigma_\mu \sqrt{2\pi}} \exp\left\{-\frac{\mu^2}{2\sigma_\mu^2}\right\}$$

where $p_0(\mu)$ is the *a priori* density.

<u>Performance Criterion</u>: $I = E\{\sum_{k=0}^{N} (r(k) - z(k))^2\}$, where $r(k)$,

   $k = 0,\ldots,N$, is a deterministic desired
   response.

Then

   $J_k = (r(k) - z(k))^2$ (5.36)

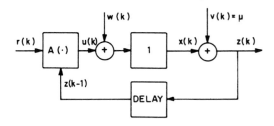

Fig. 5.2.   Dual Control of a Neutral System

The plant equations may be combined to read

$$z(k) = u(k) + \mu + w(k) \tag{5.37}$$

Problem

A controller $u(k) = A(R^k, z^{k-1}, u^{k-1})$, $k = 0, \ldots, N$, is sought to minimize I.

The optimal control will be evaluated backward, as in Sec. 5.3, from the Hamiltonian expression and therefore, S_{N-k}^*. To do this, α_k should be computed as in (5.15). However, since the system is first-order Markov, i.e., $p(z(k)|u^k) = p(z(k)|u(k))$, the part needed for the optimization is

$$\alpha_k = \int_{\Omega(z(k))} (z(k) - r(k))^2 p(z(k)|u(k)) \, d\Omega(z(k)) \tag{5.38}$$

Using (5.2) to introduce μ, consider the marginal probability expressing

$$\alpha_k = \int_{\Omega(z(k),\mu)} (z(k) - r(k))^2 p(z(k)|\mu,u(k)) p(\mu|u(k)) \, d\Omega(z(k),\mu) \tag{5.39}$$

But by Bayes' rule

$$p(\mu|u(k)) = \prod_{i=0}^{k-1} p(z(i)|\mu,u(i)) p_0(\mu) \tag{5.40}$$

and

$$\alpha_k = \int_{\Omega(z(k),\mu)} (z(k) - r(k))^2 \prod_{i=0}^{k} p(z(i)|\mu,u(i)) p_0(\mu) \, d\Omega(z(k),\mu) \tag{5.41}$$

However, from

$$p(z(i)|\mu,u(i)) = p(w(i)) = p_w(z(i) - u(i) - \mu) \tag{5.42}$$

therefore,

$$\alpha_k = \int_{-\infty}^{\infty} \int_{-\infty}^{\infty} (z(k) - r(k))^2 \, P_0(\mu) \prod_{i=0}^{k} P_w(z(i) - u(i) - \mu) \, dz(k) \, d\mu$$

$$= \frac{1}{\sigma_\mu \sigma_w^{k+1} \, 2\pi^{(k/2)} + 1} \int_{\mu=-\infty}^{\infty} \exp\left\{ -\frac{\mu^2}{2\sigma_\mu^2} - \right.$$

$$\left. \sum_{i=0}^{k-1} \frac{(z(i) - u(i) - \mu)^2}{2\sigma_w^2} \right\} \int_{-\infty}^{\infty} \int_{z(k)=-\infty}^{\infty} (z(k) - r(k))^2$$

$$\exp\left\{ \frac{(-z(k) + u(k) + \mu)^2}{2\sigma_w^2} \right\} dz(k) \, d\mu \qquad (5.43)$$

Then

$$\alpha_k = \left[A_k + B_k + \frac{2B_k}{C_k} \left[(u(k) - r(k)C_k + \frac{\sum\limits_{i=0}^{k-1}(z(i) - u(i))}{2\sigma_w^2} \right]^2 \right]$$

$$\exp\left\{ -\frac{\sum\limits_{i=0}^{k-1}(z(i) - u(i))^2}{2\sigma_w^2} + \frac{\left[\sum\limits_{i=0}^{k-1}(z(i) - u(i))\right]^2}{4\sigma_w^4 \, C_k} \right\} \qquad (5.44)$$

where

$$A_k = \frac{\sigma_w^2}{\sqrt{2}\sigma_w^k \sigma_\mu (2\pi)^k \sqrt{C_k}}$$

$$B_k = \frac{1}{2\sigma_w^k (2\pi)^{k/2} (2C_k)^{3/2}}$$

$$C_k = \frac{k}{2\sigma_w^2} + \frac{1}{2\sigma_\mu^2}$$

The optimal control is computed as follows:

Let k = N

$$u^*(N): \quad \underset{u(N)}{\text{Min}} \, \alpha_N$$

which yields

$$u^*(N) = r(N) - \frac{\sum\limits_{i=0}^{N-1} (z(i) - u(i))}{2\sigma_w^2 \, C_N} \tag{5.45}$$

and

$$S_0^* = \alpha_N^* = (A_N + B_N) \, \exp \left\{ - \frac{\sum\limits_{i=0}^{N-1} (z(i) - u(i))^2}{2\sigma_w^2} - \frac{\left[\sum\limits_{i=0}^{N-1} (z(i) - u(i)) \right]^2}{4\sigma_w^4 \, \sigma_\mu^2} \right\} \tag{5.46}$$

Proceeding backward, one should minimize S_1

$$S_1 = \alpha_{N-1} + \int_{-\infty}^{\infty} S_0^* \, dz(N-1) \tag{5.47}$$

and

$$
\begin{aligned}
S_1 = &\left[A_{N-1} + B_{N-1} + \frac{2B_{N-1}}{C_{N-1}} \left[(u(N-1) - r(N-1)) \, C_{N-1} \right. \right. \\
&\left. \left. + \frac{\sum\limits_{i=0}^{N-2} (z(i) - u(i))}{2\sigma_w^2} \right] \right] \exp \left\{ - \frac{\sum\limits_{i=0}^{N-2} (z(i) - u(i))^2}{2\sigma_w^2} \right. \\
&\left. + \frac{\left[\sum\limits_{i=0}^{N-2} (z(i) - u(i)) \right]^2}{4\sigma_w^2 \, \sigma_\mu^2} \right\} + \int_{-\infty}^{\infty} (A_N + B_N) \\
&\exp \left\{ - \frac{(z(N-1) - u(N-1))^2 + \sum\limits_{i=0}^{N-2} (z(i) - u(i))^2}{2\sigma_w^2} \right. \\
&\left. + \frac{\left[\sum\limits_{i=0}^{N-2} (z(i) - u(i)) + (z(N-1) - u(N-1)) \right]^2}{4\sigma_w^4 \, \sigma_\mu^2} \right\} dz(N-1)
\end{aligned}
$$

Since $z(N-1) - u(N-1) = y(N-1)$ and $dz(N-1) = dy(N-1)$, the above quantities may be replaced in the integrand of Eq. (5.46) and make the integral independent of $u(N - 1)$. Therefore, the minimization of S_1 depends only on the minimization of α_{N-1}, e.g.,

$$\underset{u(N-1)}{\text{Min}} \ S_1 \Rightarrow \underset{u(N-1)}{\text{Min}} \ \alpha_{N-1} \Rightarrow u^*(N - 1) = r(N - 1) - \frac{\displaystyle\sum_{i=0}^{N-2}(z(i) - u(i))}{2\sigma_N^2 \, c_{N-1}} \quad (5.48)$$

and inductively

$$\underset{u(k)}{\text{Min}} \ S_{N-k} \Rightarrow \underset{u(k)}{\text{Min}} \ \alpha_k \Rightarrow u^*(k) = r(k) - \frac{\displaystyle\sum_{i=0}^{k-1}(z(i) - u(i))}{2\sigma_w^2 \, c_{k-1}} \quad (5.49)$$

Obviously this is the case of a *neutral system* since the minimization depends only on α_k. It is easily seen that the controller depends only on the past history of the system, i.e., past controls and states and not the future ones, and therefore, it is possible to implement. Finally, the learning part of the controller is performed implicitly by this controller since the estimation of the unknown parameter is included in the expression (5.41).

5.5.2   Dual Control for a Nonneutral Plant

The second system to be considered is depicted in Fig. 5.3, and is described by:

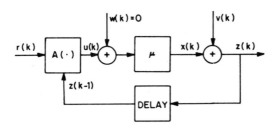

Fig. 5.3.   Dual Control of a Nonneutral System

Plant $x(k) = \mu\, u(k)$ (5.50)

$\qquad\quad z(k) = x(k) + v(k)$ $\mu =$ the unknown parameter

with $p(v(k)) = \dfrac{1}{\sigma\sqrt{2\pi}} \exp\left\{ -\dfrac{v(k)^2}{2\sigma_v^2} \right\}$

$\qquad p_0(\mu) = \dfrac{1}{\sigma_\mu\sqrt{2\pi}} \exp\left[-\dfrac{(\mu - \mu_0)}{2\sigma_\mu^2} \right] =$ the a-priori probability

$\qquad\qquad$ density

Performance Criterion:

$$I = E\left\{ \sum_{k=0}^{N} (r(k) - x(k))^2 \right\} = \sum_{k=0}^{N} I_k \qquad\qquad (5.51)$$

where $r(k)$, $k = 0,\ldots,N$ is a deterministic desired response. Then
$J_k = (r(k) - x(k))^2$.

Problem

$\qquad$ A controller is sought to minimize I:

$$u^*(k) = A(R^k, z^{k-1}, u^{k-1})$$

Solution

$\qquad$ Due to the Markovian property of the system $p(z(k)|U^k) = p(z(k)|u(k))$, the optimal control can be evaluated backward from the Hamiltonian function, as suggested in Sec. 5.3.   The quantity α_k should be evaluated for this purpose

$$\alpha_k = \int_{\Omega(z(k))} J_k\, p(z(k)|u(k))\, dz(k) = \int_{\Omega(x(k),v(k),\mu)} J_k$$

$\qquad\qquad p(v(k))\, p(x(k)|v(k),u(k),\mu)p(\mu|u(k))\, dx(k)\, dv(k)\, d\mu$

$$= \int_{\Omega(x(k),\mu)} J_k\, p(x(k)|u(k),\mu)p(\mu|u(k))\, dx(k)\, d\mu$$

$$\int_{\Omega(v(k))} p(v(k))\, dv(k) = \int_{\Omega(x(k),\mu)} (r(k) - x(k))^2$$

$\qquad\qquad p(x(k)|u(k),\mu)p(\mu|u(k))\, dx(k)\, d\mu \qquad\qquad (5.52)$

However, from the plant Eq. (5.50)

$$p(x(k)|u(k),\mu) = \delta(x(k) - \mu\, u(k)) \tag{5.53}$$

and from the chain rule and the combined plant equation

$$p(\mu|u(k)) = \prod_{i=0}^{k-1} p(z(i))p_0(\mu) = \prod_{i=0}^{k-1} p_v(z(i) - x(i))p_0(\mu) \tag{5.54}$$

Substituting (5.53) and (5.54) into (5.52) one gets

$$\alpha_k = \int_{\mu=-\infty}^{\infty} (r(k) - \mu u(k))p_0(\mu) \prod_{i=0}^{k-1} \frac{1}{\sigma_v \sqrt{2\pi}}$$

$$\exp\left\{ -\frac{(z(i) - \mu u(i))}{2\sigma_v^2} \right\} d\mu \tag{5.55}$$

Letting

$$A_{k-1} = \frac{1}{2\sigma_\mu^2} + \frac{\sum\limits_{i=0}^{k-1} u^2(i)}{2\sigma_v^2} > 0$$

$$B_{k-1} = \frac{\mu_0}{\sigma_\mu^2} + \frac{1}{\sigma_v^2} \sum_{i=0}^{k-1} u(i)\, z(i)$$

$$C_{k-1} = \frac{\mu_0^2}{2\sigma_\mu^2} + \frac{1}{2\sigma_v^2} \sum_{i=0}^{k-1} z^2(i) > 0$$

$$D_{k-1} = \left[\sum_{i=0}^{k-1} z(i)u(i)\right]^2 - \left[\sum_{i=0}^{k-1} u^2(i)\right]\left[\sum_{i=0}^{k-1} z^2(i)\right]$$

$$E_{k-1} = \mu_0^2 \sum_{i=0}^{k-1} u^2(i) - 2\mu_0 \sum_{i=0}^{k-1} z(i)u(i) + \sum_{i=0}^{k-1} z^2(i)$$

The expression for α_k is modified

$$\alpha_k = (\text{const.}) \exp\{-C_{k-1}\} \int_{-\infty}^{\infty} (r(k) - \mu u(k))^2 \exp\{-A_{k-1}\mu^2$$

$$+ B_{k-1}\mu\} \, d\mu = (\text{const.}) \exp\left\{\frac{\sigma_\mu^2 D_{k-1} - \sigma_v^2 E_{k-1}}{4\sigma_v^4 \sigma_\mu^2 A_{k-1}}\right\}$$

$$\frac{1}{2A_{k-1}} \sqrt{\frac{\pi}{A_{k-1}}} \, [u^2(k) + (2A_{k-1}r(k) - 2B_{k-1}u(k))^2] \qquad (5.56)$$

The optimal control may be computed backward as in the preceding case.

Let $k = N$

$$u_N^*: \quad \operatorname*{Min}_{u(N)} \alpha_N \Rightarrow \operatorname*{Min}_{u(N)} \left[u^2(N) + \frac{(2A_{N-1}r(N) - 2B_{N-1}u(N))^2}{2A_{N-1}} \right]$$

or

$$u^*(N) = \frac{2A_{N-1}B_{N-1}r(N)}{A_{N-1} + 2B_{N-1}^2} \qquad (5.57)$$

and $$S_0^* = \alpha_N^* = \frac{r^2(N)\sqrt{\pi A_{N-1}}}{A_{N-1} + 2B_{N-1}^2} \exp\left\{\frac{\sigma_\mu^2 D_{N-1} - \sigma_v^2 E_{N-1}}{4\sigma_v^4 \sigma_\mu^2 A_{N-1}}\right\} \qquad (5.58)$$

The next step is to evaluate S_1 in order to obtain $u^*(N - 1)$. In order to simplify the notation, the sufficient statistics for $u(k)$ may be introduced by the following quantities:

$$L_k = \sum_{i=0}^{k} u^2(i) = L_{k-1} + u^2(k) \qquad k = 1, 2, \ldots, N-1$$

$$M_k = \sum_{i=0}^{k} u(i)z(i) = M_{k-1} + u(k)z(k)$$

In this case, one may rewrite A_{k-1} and B_{k-1} of (5.54)

$$A_{k-1} = \frac{1}{\sigma_\mu^2} + \frac{L_{k-1}}{2\sigma_v^2} \qquad B_{k-1} = \frac{\mu_0}{\sigma_v^2} + \frac{M_{k-1}}{\sigma_v^2}$$

and (5.55)

$$\alpha_k = (\text{const.})\ \exp\left\{\frac{1}{\sigma_v^2}\sum_{i=0}^{k-1} z^2(i)\right\}\left[u^2(k)\phi_{2k}(L_{k-1},M_{k-1})\right.$$

$$\left. - u(k)r(k)\phi_{1k}(L_{k-1},M_{k-1}) + r^2(k)\ \phi_{0k}(L_{k-1},M_{k-1})\right] \qquad (5.59)$$

where

$$\phi_{2k}(L_{k-1},M_{k-1}) = \int_{-\infty}^{\infty} \mu^2 \exp\{-A_{k-1}\mu^2 + B_{k-1}\mu\}\,d\mu$$

$$= \frac{1}{2A_{k-1}}\sqrt{\frac{\pi}{A_{k-1}}}\left[1 + \frac{2B_{k-1}^2}{A_{k-1}}\right]\exp\left\{\frac{B_{k-1}^2}{A_{k-1}}\right\} > 0$$

$$\phi_{1k}(L_{k-1},M_{k-1}) = \int_{-\infty}^{\infty} \mu\ \exp\{-A_{k-1}\mu^2 + B_{k-1}\mu\}\,d\mu$$

$$= 2\sqrt{\frac{\pi}{A_{k-1}}}\ \frac{B_{k-1}}{A_{k-1}}\ \exp\left\{\frac{B_{k-1}^2}{A_{k-1}}\right\}$$

$$\phi_{0k}(L_{k-1},M_{k-1}) = \int_{-\infty}^{\infty} \exp\left\{-A_{k-1}\mu^2 + B_{k-1}\mu\right\}d\mu$$

$$= \sqrt{\frac{\pi}{A_{k-1}}}\ \exp\left\{\frac{B_{k-1}^2}{A_{k-1}}\right\}$$

Then the expressions for α_N and $u^*(N)$ are given from (5.57) in terms of the sufficient statistics

$$\alpha_N = (\text{const.})\ \exp\left\{\frac{1}{2\sigma_v^2}\sum_{i=0}^{N-1} z^2(i)\right\}\left[u^2(N)\phi_{2N}(L_{N-1},M_{N-1})\right.$$

$$\left. - u(N)\ r(N)\ \phi_{1N}(L_{N-1},M_{N-1}) + r^2(N)\ \phi_{0N}(L_{N-1},M_{N-1})\right] \qquad (5.60)$$

Then

$$u^*(N):\quad \alpha_N^* = \underset{U(N)}{\text{Min}}\ \alpha_N \Rightarrow \underset{u(N)}{\text{Min}}\ u^2(N)\phi_{2N}(L_{N-1},\ _{N-1})$$

$$- u(N)r(N)\phi_{1N}(L_{N-1},M_{N-1}) + r^2(N)\phi_{0N}(L_{N-1},M_{N-1})$$

$$u^*(N) = u^*(N,L_{N-1},M_{N-1}) = u^*(N,U^{N-1},Z^{N-1},R^N)$$

$$= \frac{r(N)}{2}\ \frac{\phi_{1N}(L_{N-1},M_{N-1})}{\phi_{2N}(L_{N-1},M_{N-1})} = \frac{2r(N)B_{N-1}A_{N-1}}{A_{N-1} + 2B_{N-1}^2} \qquad (5.61)$$

$$S_0^* = \alpha_N^* = (\text{const.}) \exp\left\{\frac{1}{\sigma_v^2} \sum_{i=0}^{N-1} z^2(i)\right\} r^2(N)\left[\phi_{0N}(L_{N-1},M_{N-1})\right.$$

$$- \frac{1}{4} \frac{\phi_{1N}^2(L_{N-1},M_{N-1})}{\phi_{2N}(L_{N-1},M_{N-1})} = (\text{const.}) \exp\left\{\frac{1}{\sigma_v^2} \sum_{i=0}^{N-1} z^2(i)\right\} r^2(N)$$

$$\sqrt{\frac{\pi}{A_{N-1}}} \exp\left\{\frac{B_{N-1}^2}{A_{N-1}}\right\}\left[\frac{A_{N-1}}{A_{N-1} + 2B_{N-1}^2}\right]$$

Proceeding to evaluate S_1 and $u^*(N-1)$

$$S_1(r(N-1),r(N),U^{N-1},Z^{N-2}) = \alpha_{N-1}(r(N-1),U^{N-1},Z^{N-2})$$

$$+ \int_{-\infty}^{\infty} S_0^*(r(N),U^{N-1},Z^{N-1})\, dz(N-1)$$

or by substituting (5.60) and (5.61)

$$S_1(r(N-1),r(N),U^{N-1},Z^{N-2}) = (\text{const.}) \exp\left\{\frac{1}{2\sigma_v^2} \sum_{i=0}^{N-2} z^2(i)\right\}$$

$$[u^2(N-1)\ \phi_{2N-1}(L_{N-2},M_{N-2})$$

$$- u(N-1)r(N-1)\ \phi_{1N-1}(L_{N-2},M_{N-2})$$

$$+ r^2(N-1)\ \cdot\phi_{0N-1}(L_{N-2},M_{N-2})$$

$$+ r(N) \int_{-\infty}^{\infty} \exp\left\{\frac{1}{\sigma_v^2} \sum_{i=0}^{N-1} z^2(i)\right\}$$

$$\sqrt{\frac{\pi}{A_{N-1}}} (\frac{A_{N-1}}{A_{N-1} + 2B_{N-1}^2})$$

$$\exp\left\{\frac{B_{N-1}^2}{A_{N-1}}\right\}\ dz(N-1)] \qquad (5.62)$$

Since $u(N-1)$ does not depend on $z(N-1)$, the integral in the right-hand side of (5.61) may be integrated with respect to $z(N-1)$ and result in a function $\psi(Z^{N-2},U^{N-2},u(N-1))$.

$$S_1(r(N-1), r(N), U^{N-1}, Z^{N-2}) = (const). \ exp \left\{ \frac{1}{2\sigma_v^2} \sum_{i=0}^{N-2} z^2(i) \right\}$$

$$[u^2(N-1) \ \phi_{2N-1}(L_{N-2}, M_{N-2})$$

$$- u(N-1)r(N-1)\phi_{1N}(L_{N-2}, M_{N-2})$$

$$+ r^2(N-1) \ \phi_{0N-1}(L_{N-2}, M_{N-2})$$

$$+ r(N)\psi(Z^{N-2}, U^{N-2}, u(N-1)] \quad (5.63)$$

and

$$u^*(N-1): \quad \underset{u(N-1)}{Min} \quad S_1(r(N-1), r(N), U^{N-1}, Z^{N-2}).$$

It is obvious that the integral of the right-hand side of (5.62) must be integrated analytically, since it contains unspecified variables such as $u(N-1)$. Such an integration is almost impossible due to the complexity of the integrand. This difficulty is enhanced as one moves inductively backward to obtain

$$u^*(k): \quad \underset{u(k)}{Min} \ S_{N-k}(R^k, U^k, Z^{k-1}) \quad\quad\quad (5.64)$$

This is an obvious case of nonneutral dual control with all the associated difficulties.

5.6 DISCUSSION

The dual control problem was presented here as a generalization of the stochastic control problem, which could be imbedded in the dual control as a special case. It covers the very general case of the control of a dynamic system with reducible uncertainties operating in a stochastic environment. As a matter of fact, the dual control solution could serve as the optimal solution to the parameter-adaptive S.O.C. problem defined in Chap. 1. However, the impossibility of obtaining a solution to this problem for all but trivial cases has been demonstrated by both theory as well as examples.

Thus, the dual control problem as well as the stochastic control problem will be insolvable with the current mathematical methods. However, since the purpose of engineering research is the solution of the problem and not mathematical elegance, the next step is to define an approximation which may yield a solution to the problem, by relaxing one of the most restrictive but less important conditions, and retaining the other properties of the original requirements. Such a condition was found to be the *transient optimality*. The difficulties regarding the computational aspects of the problem may be avoided if the optimality condition is replaced by some condition which will utilize a learning procedure to account for the uncertainty and the dual control during the transient period. In the mean time, an asymptotical optimal solution may still be sought after the learning procedure yields the expected results. Such a procedure is under the jurisdiction of S.O.C. systems to be discussed in the next part of the book.

5.7 REFERENCES

[5.1] Bar-Shalom, Y. and Tse, E., "Concepts and Methods, in Stochastic Control in *Control and Dynamic Systems: Advances in Theory and Application* (C. T. Leondes, ed.), Academic Press, New York, 1975.

[5.2] Bellman, R., *Adaptive Control Processes: A Guided Tour*, Princeton University Press, Princeton, N.J., 1961.

[5.3] Fel'dbaum, A. A., "Dual Control Theory," Parts I and II, *Automation Remote Control*, 21 (9,10) 1240-1249 and 1453-1464 (1960); Parts III and IV, 22 (1,2) 3-16 and 129-143 (1961).

[5.4] Feld'baum, A. A., *Optimal Control Systems*, Academic Press, New York, 1965.

[5.5] Meier, L., 3rd, "Combined Optimal Control and Estimations," 3d Annual Allerton Conference Proceedings, Monticello, Ill., 1961.

[5.6] Tse, E., Bar-Shalom, Y., and Meier, L., 3rd, "Wide Sense Dual Control for Nonlinear Stochastic Systems," *IEEE Trans. Automatic Control*, AC-18 (2), 98-108 (1973).

Chapter 6

A SUMMARY OF ON-LINE PARAMETER IDENTIFICATION

6.1 INTRODUCTION

The attempt to write a summary of existing identification
methods is not a gratifying task; it is impossible to accomplish it
without neglecting some schemes and offending someone. In other
words, so many identification algorithms have been developed for so
many different applications that they would not fit under one title.
Therefore, the task of summarizing them is a demanding one, and the
selection of parameter identification algorithms presented in this
book has been limited mainly to the ones which can be and have been
used in S.O.C.

The selection of the appropriate algorithms, in view of the
above qualifications, has to be limited to on-line schemes since
identification must be performed on-line in order to qualify for
S.O.C. The second factor considered in limiting the spectrum of
identification algorithms in this book is the simplicity of
implementation, which is another of the qualifications of S.O.C.
With these two limitations the list of identification algorithms
narrows down to relatively manageable length.

At the risk of incurring more criticism, this list is further
limited to the following eleven algorithms which are felt to be the
most suitable for S.O.C.

1. Modified on-line crosscorrelation algorithms [6.40].

2. First-order stochastic approximation algorithm [6.47].

3. Second-order stochastic approximation algorithm (extended least
    squares) [6.47].

4. Perturbation input stochastic approximation [6.48].

5.   On-line maximum likelihood algorithm [6.21].

6.   Maximum *a-posteriori* probability algorithm, linearized
     Kalman-filter [6.36].

7.   Extended Kalman filter algorithm [6.36].

8.   Random search algorithm [6.42].

9.   Generalized least squares algorithm [6.7].

10. Instrumental variable algorithm [6.53], [6.54].

11. Fourier analysis algorithm [6.16], [6.17].

These eleven algorithms have been used for various forms of
S.O.C., some by the author and his colleagues, and some by other
researchers with different degrees of success. Due to the limitation
in space, only the first eight will be briefly presented in the
basic form in which they have been developed, so that some comparisons
can be made among them. Numerous variations of these algorithms have
been developed in the literature to yield considerable improvements,
but it was felt that they belong to a detailed discussion of
identification and not to a summary intended to suppliment the theory
of S.O.C.

The above algorithms are designed to identify unknown system
parameters sequentially and they accomplish that asymptotically.
However, the speed of convergence to an acceptable value, the
sensitivity to the initial guess, the simplicity of implementation,
and the size of the computer required are criteria for their evaluation. Specific acceleration and other schemes, developed for each
algorithm separately, may improve their performance for a particular
job. A great amount of research is still in progress in this area.

6.2   PARAMETER VS. STRUCTURAL IDENTIFICATION

The problem of system identification is much more general than
discussed in the previous section; it is the problem of modeling
a process from measurable input-output data when it operates in a
deterministic or stochastic environment. Such a process involves
two separate procedures, namely, the structural identification and
the parameter identification.

The structural identification pertains to the reconstruction of
the mathematical dynamic equations which describe the system. If the
laws governing the system are known, as in most electromechanical
plants, the structure of the system is assumed *a priori*. However,
in cases where these laws are not known, or a smaller-scale model
is desired, an identification procedure is needed to recover the
structure from input-output data.

Wiener [6.51], [6.52] was the first to propose a method for
structural identification of nonlinear systems. It comproses the
approximation of the signal through the system by a finite set of
Laguerre polynomials and the approximation of the system by a set
of Hermite polynomials, the coefficients of which could be identified
through regular parameter identification algorithms. One version of
the method is given in Fig. 6.1. It has the disadvantage that, in
order to obtain a close approximation of the system, a prohibitively

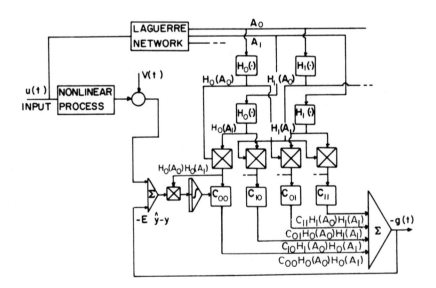

FIG. 6.1.   An On-line Structural Identification of Nonlinear
Systems After Wiener.

large number of polynomials is required with almost unmanageable
computation to be performed. Variations of the idea of series
expansion can be found in the literature [6.55].

   An alternate approach for structural identification has been
proposed by Saridis and Hofstadter [6.44]. According to this method,
a nonlinear system is classified within one of a score of classes
of nonlinear systems with similar properties, via pattern recognition
methods. The input-output crosscorrelation function has been
successfully used as a feature vector. Fig. 6.2 shows a block diagram
of the method which has produced very satisfactory results.

   Parameter identification is performed to obtain the unknown
coefficients of the assumed structure of the system. It may be
performed off-line or on-line, the latter being the only appropriate
one for S.O.C.s. Parameter identification may be further simplified
by the observation that in 90% of the cases parameters enter the
system model linearly. Therefore, they require mostly linear
estimation techniques for their identification. In the few cases
where parameters enter the model nonlinearly, such as chemical or
biological processes where the parameters appear in exponentials,
speical techniques insensitive to nonlinearities must be used.

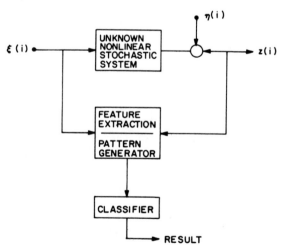

FIG. 6.2. Nonlinear System Classification Method of
           Saridis and Hofstadter.

The parameter identification algorithms can be further classified as applicable to linear systems with Gaussian environments (algorithms 5, 6, 7, and 11), linear systems only (algorithm 1 and 10), and linear or nonlinear systems with any kind of environment (algorithms 2, 3, 4, 8, and 9). Unfortunately their convergence properties are inversely proportional to the requirements of the system.

The first eight algorithms listed in Sec. 6.1 will be discussed in the sequel. They are presented mostly for discrete-time systems since a digital computer must be used for their implementation.

The parameter identification problem is best illustrated by Fig. 6.3, where the unknown system parameters must be identified by input-output data, which means that the records of all the variables are available for measurement.

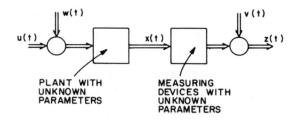

FIG. 6.3.   Systems with Unknown Parameters.

6.3   THE CROSSCORRELATION METHODS

Crosscorrelation methods for linear system identification are among the oldest, and still popular, since they do not require any information about the statistics of the noises involved.

Crosscorrelation was originally developed for continuous-time systems and then used for discrete-time systems [6.33], [6.11]. Their successful implementation is due to the use of the *discrete interval binary noise* (DIBN) composed of a periodic sequence of fixed amplitude pulses with randomly assigned signs. The DIBN is depicted in Fig. 6.4, and represents a most desirable test signal for identification. This test-sequel is injected in the plant causing it to

generate more energy in the plant and undesired perturbations of the
output.   However, this disadvantage is balanced by the speed of
convergence and simplicity of the algorithms.

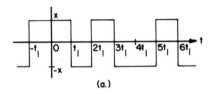

(a.)

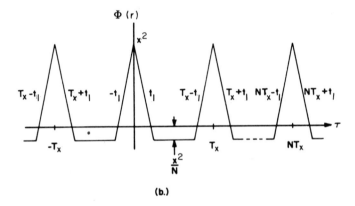

(b.)

FIG. 6.4.   a.   The DIBN Wave Form.
            b.   Auto-correlation Function of Synthesized
                 Periodic DIBN.

6.3.1 Continuous Crosscorrelation

In a linear single-input/single-output dynamic system with
time-varying coefficients, the relation between input and output is
defined through the impulse response $c(t,\sigma)$ which satisfies

$$z(t) = \int_{t_0}^{t} c(t,\sigma)[\, u(\sigma) + w(\sigma)] \ d\sigma + v(t) \qquad (6.1)$$

In Eq. (6.1), $z(t)$ is the scalar output, $u(t)$, the scalar input,
and $w(t)$ and $v(t)$ are scalar zero-mean random variables uncorrelated
between themselves and $u(t)$.   Assigning

$$u(t) = \xi(t) \qquad (6.2)$$

where $\xi(t)$ is an injected random variable, the crosscorrelations of the output and the injected input are given by

$$R_{z\xi}(t,\tau) \triangleq E\{z(t)\xi(\tau)\} = \int_{t_0}^{t} c(t,\sigma)E\{[\xi(\sigma) + w(\sigma)]\xi(\tau)\}\,d\sigma$$

$$+ E\{\xi(\tau)v(t)\} \qquad\qquad (6.3)$$

Using the properties of the random variable involved

$$R_{z\xi}(t,\tau) = \int_{t_0}^{t} c(t,\sigma)R_{\xi\xi}(\tau - \sigma)\,d\sigma \qquad\qquad (6.4)$$

If $\delta(t)$ was a white noise with autocorrelation

$$R_{\xi\xi}(\tau - \sigma) = \delta(\tau - \sigma)$$

where $\delta(\tau - \sigma)$ is a Dirac delta function. Then Eqs. (6.4) can be reduced to

$$R_{z\xi}(t,\tau) \equiv c(t,\tau) \qquad\qquad (6.5)$$

which indicates that the value of the impulse response at time t for an impulse applied at σ can be readily evaluated through its crosscorrelation function. Since white noise cannot be physically realized, the DIBN maybe used instead [6.1]. For $\xi(t)$, wide band relative to $c(t,\tau)$, the autocorrelation function of the DIBN can be approximated by a Dirac delta function

$$R_{\xi\xi}(t - k\tau) \simeq A\delta(t - k\tau)$$

and the identification algorithms for a part of the impulse response can be given by

$$c(t,k\tau) \simeq \frac{1}{A} R_{z\xi}(t,k\tau) \qquad\qquad (6.6)$$

The complete identification can be achieved by a bank of functions generating the crosscorrelation $R_{z\xi}(t,k\tau)$ for k = 1, 2,.... Since only discrete points of the impulse response are available interpolation is necessary.

The block diagram of Fig. 6.5 illustrates impulse-response identifications. It is so designed that the random processes involved are ergodic and ensemble averages are replaced by time averages. Since DIBN is periodic with period T, the averaging filters must integrate only over T.

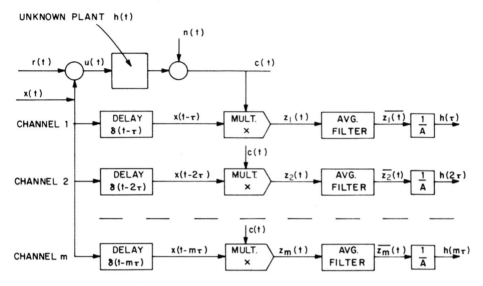

FIG. 6.5. Crosscorrelation Identification.

There are two disadvantages of the crosscorrelation technique:
1.   The length of identification time [6.28].
2.   The impulse response graphically obtained is of little use
      for <u>on-line</u> control of a system with unknown parameters.

6.3.2 Modified Discrete-Time Crosscorrelation

An improvement of performance, resulting from the necessity of
using a digital computer, is obtained through the discrete-time
version of the crosscorrelation-identification procedure. The
identification time is smaller than for other identification methods
and the recovery of the system's parameter input for the controller
is immediate.

The algorithm used here is the crosscorrelation scheme proposed
by Hill and McMurtry in [6.11] for discrete-time systems modified
for on-line identification [6.40]. The method [6.1], [6.11]
utilizes as a perturbation input $u(k)$ a *discrete interval binary
noise* (DIBN), and it identifies the pulse response $c(j)$ of the

discrete-time version of (6.1) for $x(0) = 0$, and for N number of information intervals. The pulse response $c(j)$ corresponds to the discrete-time system, assumed to be completely controllable and observable

$$x(k + 1) = Fx(k) + bu(k) + gw(k), F = \begin{bmatrix} 0 & | & I \\ \hline & f^T & \end{bmatrix} \quad b = \begin{bmatrix} b_1 \\ \vdots \\ b_n \end{bmatrix} \quad g = \begin{bmatrix} g_1 \\ \vdots \\ g_n \end{bmatrix}$$

$$h = \begin{bmatrix} 1 \\ 0 \\ \vdots \\ 0 \end{bmatrix} \tag{6.7}$$

$$z(k) = h^T x(k) + v(k)$$

where $x(k)$ is an n-dimensional vector, $u(k)$ and $z(k)$ are the scalar input and output, respectively, $w(k)$ and $v(k)$ are *independent and identically distributed* noise sequences, and F, b, g, and h are unknown matrices of appropriate dimensions. For zero-initial conditions and $b = g$

$$z(k) = \sum_{j=0}^{k-1} h^T F^j b \left[u(k - j - 1) + w(k - j - 1) \right] + v(k)$$

$$= \sum_{j=0}^{k-1} c(j) \left[u(k - j - 1) + w(k - j - 1) \right] + v(k) \tag{6.8}$$

where $c(j) \triangleq h^T F^j b$ is the pulse response of the system.

If a DIBN noise is applied at the input, then

$$u(k) \equiv \xi(k)$$

The crosscorrelation between input $u(k)$ and output $z(k)$ is:

$$R_{zu}(m) \triangleq E\{ z(N)u(N - m)\} = E\{ \sum_{j=0}^{N-1} c(j) \left[u(N - j - 1) \right.$$

$$\left. + w(N - j - 1) \right] u(N - m) \}$$

$$= \sum_{j=0}^{N-1} c(j) E\{ u(N - j - 1)u(N - m) \}$$

$$\triangleq \sum_{j=0}^{N-1} c(j) R_{uu}(m - j - 1) \tag{6.9}$$

By stacking equations (6.9) for $m = 1, 2, \ldots, N$

$$F_{zu}(N) = \begin{bmatrix} R_{zu}(1) \\ \vdots \\ R_{zu}(N) \end{bmatrix} = \begin{bmatrix} R_{uu}(o) \cdots R_{uu}(-N+1) \\ \cdots\cdots\cdots\cdots\cdots \\ R_{uu}(N-1) \cdots R_{uu}(o) \end{bmatrix} \begin{bmatrix} c(o) \\ \vdots \\ c(N-1) \end{bmatrix}$$

$$= F_{uu}(N)C(N-1) \tag{6.10}$$

For DIBN $\xi(k)$ with N information intervals and covariance a^2

$$F_{\xi\xi}(N) = \begin{bmatrix} a^2 & \dfrac{a^2}{N} & \cdots a^2 \\ \cdots\cdots\cdots\cdots\cdots \\ \dfrac{a^2}{N} & \dfrac{a^2}{N} & \cdots a^2 \end{bmatrix} \quad F_{\xi\xi}^{-1} = \dfrac{N}{a^2(N+1)} \begin{bmatrix} 2 & 1 & \cdots 1 \\ \vdots & \vdots & \cdots\cdots \\ 1 & 1 & \cdots 2 \end{bmatrix} \tag{6.11}$$

and

$$C(N-1) = \dfrac{N}{a^2(N+1)} \begin{bmatrix} 2 & 1 \cdots 1 \\ \cdots\cdots\cdots \\ 1 & 2 \cdots 2 \end{bmatrix} F_{z\xi}(N)$$

However, the crosscorrelation functions $R_{z\xi}(m)$, $m = 1,\ldots,N$ cannot be computed exactly, but only approximately as more input-output measurements are acquired.

$$\hat{R}_{z\xi}^{M}(m) = \dfrac{1}{M+1} \sum_{k=0}^{M} z(k)\, \xi(k-m) = \hat{R}_{z\xi}^{M-1}(m) + \dfrac{1}{m+1}[\, z(m)$$

$$\xi(M-m) - \hat{R}_{z\xi}^{M-1}(m)] \tag{6.12}$$

The algorithm updating for $R_{z\xi}^{M}(m)$ is a stochastic approximation algorithm. The approximate estimates of the pulse response $C(N-1)$ for N information intervals and M measurements $M \geq m$ is given by:

$$\hat{C}^{M}(N-1) = \dfrac{N}{a^2(N+1)} \begin{bmatrix} 2 & 1\cdots 1 \\ \vdots & \cdots\cdots \vdots \\ 1 & 1\cdots 2 \end{bmatrix} \begin{bmatrix} \hat{R}_{z\xi}^{M}(1) \\ \vdots \\ \hat{R}_{z\xi}^{M}(N) \end{bmatrix} = \dfrac{N}{a^2(N+1)}$$

$$\cdot \begin{bmatrix} 2 & 1\cdots 2 \\ \vdots & \vdots\cdots\vdots \\ 1 & 1\cdots 2 \end{bmatrix} \cdot$$

$$\begin{bmatrix} \begin{bmatrix} R_{z\xi}^{M-1}(1) \\ \vdots \\ R_{z\xi}^{M-1}(N) \end{bmatrix} + \dfrac{1}{M+1} \begin{bmatrix} z(M) \begin{bmatrix} \xi(M-1) \\ \vdots \\ \xi(M-N) \end{bmatrix} - \begin{bmatrix} \hat{R}_{z\xi}^{M-1}(1) \\ \vdots \\ \hat{R}_{z\xi}^{M-1}(N) \end{bmatrix} \end{bmatrix} \end{bmatrix} \tag{6.13}$$

$$= C^{M-1}(N-1) + \dfrac{1}{M+1} \begin{bmatrix} \dfrac{N}{(N+1)a^2} \begin{bmatrix} 2 & 1 & \cdots 1 \\ \vdots & \vdots & \cdots\cdots \\ 1 & 1 & \cdots 2 \end{bmatrix} z(M) \begin{bmatrix} \xi(M-1) \\ \vdots \\ \xi(M-N) \end{bmatrix}$$

$$- \hat{C}^{M-1}(N-1) \end{bmatrix}$$

It is obvious that algorithm (6.13) is updated sequentially, and the estimates $\hat{c}^M(N - 1)$ can be obtained on-line from the previous values and the new measurements.

However, the ultimate goal is to identify the parameters f and b of system (6.7). These are obtained from (6.10) for N = 2n, where n is the order of the system.

$$\hat{b}(M) = \begin{bmatrix} \hat{c}^M(o) \\ \vdots \\ \hat{c}^M(n-1) \end{bmatrix} \quad \hat{f}(M) = \begin{bmatrix} \hat{c}^M(o) & \cdots & \hat{c}^M(n-1) \\ \cdots & \cdots & \cdots \\ \hat{c}^M(n-1) & \cdots & \hat{c}^M(2n-1) \end{bmatrix}^{-1} \cdot \begin{bmatrix} \hat{c}^M(n) \\ \vdots \\ \hat{c}^M(2n - 1) \end{bmatrix}$$

(6.14)

The DIBN sequence must be at least N = 2n information intervals long, for $N \geq 2n$ and is repeated to recursively update the approximations (6.13) and (6.14). No assumptions have been made about the distributions of the noises involved or their moments. The major drawback of the algorithm is the forced perturbation applied to the system.

6.4   STOCHASTIC APPROXIMATION ALGORITHMS

A large category of identification algorithms has been developed, based on the sequential estimation procedure presented in Secs. 3.3.2 to 3.3.5. Stochastic approximation algorithms focus on the convergence to a consistent estimate without unnecessary claims of optimality of the estimator. Such algorithms, when used for parameter identification of dynamic systems, are extremely easy to implement, even though their convergence is relatively slow. Such algorithms have been proposed by Ho and Lee [6.13] and Saridis, Nikolic, and Fu [6.46] for discrete-time systems, and Sakrison [6.38] and Ricker and Saridis [6.35] for continuous-time systems. But it was not until Saridis and Stein [6.47] discovered and corrected biases on the estimators for dynamic systems that the complete identification algorithms were developed.

The consistent stochastic approximation identification algorithms for discrete-time open-loop systems will be outlined in this section. They are the most representative of the stochastic approximation algorithms when implemented on a digital computer for S.O.C. purposes. It will be shown in Chap. 7 that most of the open-loop algorithms cannot be used for closed-loop identification, i.e., when the identification information is used for on-line control purposes of the system. The only exception is a modification of the perturbation input stochastic approximation of Sec. 6.4.3, discussed in detail in Chap. 7.

Where no perturbation input is used, the least squares algorithm of Sec. 3.3.1 may be used for identification only if the biases of estimation are eliminated.

Since the validity of the algorithm is based now on the convergence properties only, it is presented as a second-order stochastic approximation, the name being derived from the one order of magnitude higher rate of convergence than the others.

The system, whose parameters f and b are identified, is given by (6.7) and repeated here:

$$x(k + 1) = Fx(k) + bu(k) + gw(k) \qquad\qquad (6.15)$$

$$z(k) = h^T x(k) + v(k)$$

where $x(k)$ is the n-dimensional state vector, $u(k)$ the scalar input, $z(k)$ the scalar output, and $w(k)$ and $v(k)$ are iid zero mean with bounded variances q^2 and σ^2, respectively, and

$$F = \left[\begin{array}{c|c} 0 & I \\ \hline f^T \end{array}\right] \quad b = \begin{bmatrix} b_1 \\ \vdots \\ b_n \end{bmatrix} \quad g = \begin{bmatrix} g_1 \\ \vdots \\ g_n \end{bmatrix} \quad h = \begin{bmatrix} 1 \\ 0 \\ \vdots \\ 0 \end{bmatrix} \qquad (6.16)$$

The system Eqs. (6.15) can be written as an input-output relation provided that the system is completely controllable and observable

$$z(k + n) = \sum_{i=1}^{n} f_i\, z(k + i - 1) + \sum_{i=1}^{n} b_i u(k + i - 1)$$

$$+ \sum_{i=1}^{n+1} d_i \xi(k + i - 1) \qquad\qquad (6.17)$$

where

$$d_i \xi(k + i - 1) = g_i w(k - i - 1) - f_i v(k + i - 1)$$

$$d_{n+1} \triangleq 1 \quad g_{n+1} \triangleq 0 \quad f_{n+1} \triangleq -1$$

This system realizations of (6.15) and (6.17) contain the minimum
number of parameters to be identified and are general enough; and
other realization can be obtained by an appropriate similarity
transformation [6.6]. The response equation for system (6.15) is
given by

$$z(k) = \sum_{j=0}^{\infty} h^T F^j bu(k - j - 1) + \sum_{j=0}^{\infty} h^T F^j gw(k - j - 1) \qquad (6.18)$$

6.4.1 A First-Order Stochastic Approximation Algorithm

The stochastic approximation method proposed by Saridis and
Stein in [6.47] is the scheme used here to identify the parameters
f and b of the system. The algorithm is of the error-correction
type and also accounts for a bias generated by the error-corrective
term. The algorithm reads

$$\begin{bmatrix} \hat{f}(k + n) \\ - - - - \\ \hat{b}(k + n) \end{bmatrix} = \begin{bmatrix} \hat{f}(k - 1) \\ - - - - \\ \hat{b}(k - 1) \end{bmatrix} + \gamma \left(\frac{k - 1}{n + 1}\right) S(k + n + 1)$$

$$\left[z(k + n) - s^T(k + n - 1)\right]$$

$$\begin{bmatrix} \hat{f}(k - 1) \\ - - - - \\ \hat{b}(k - 1) \end{bmatrix} + \begin{bmatrix} \sigma^2 I + q^2 GG^T & | & 0 \\ - - - - - - & | & - \\ 0 & | & 0 \end{bmatrix} \begin{bmatrix} \hat{f}(k - 1) \\ - - - - \\ \hat{b}(k - 1) \end{bmatrix} - \begin{bmatrix} q^2 g* \\ - - \\ 0 \end{bmatrix}$$

$$k = 1, n + 2, 2n + 3 \qquad (6.19)$$

where $s^T(k + n - 1) \triangleq [z(k),\ldots,z(k + n - 1), u(k),\ldots,u(k + n - 1)]$,
and q^2 and σ^2 are the variances of the noises w and v, respectively,
with unknown distributions otherwise, and d is assumed known, so
that

$$G \triangleq \begin{bmatrix} 0 & 0 & \cdots & 0 \\ g_1 & 0 & \cdots & 0 \\ \cdots & \cdots & & \cdots \\ g_{n-1} & g_{n-2} & \cdots & 0 \end{bmatrix} \qquad g* \triangleq G \begin{bmatrix} g_n \\ g_{n-1} \\ \vdots \\ g_1 \end{bmatrix} \qquad (6.20)$$

Under the assumptions

$$||\hat{f}(o)|| + ||\hat{b}(o)|| < \infty \ , \ \sum_{j=1}^{\infty} \gamma(j) = \infty, \ \sum_{j=1}^{\infty} \gamma^2(j) < \infty \qquad (6.21)$$

the algorithm (6.19) converges with probability one and in the mean square.

The algorithm, without any attempt for acceleration, demonstrates rather slow convergence and is simple and easy to implement; its only limitation is the need to know the noise variances and d.   The iteration of the algorithm is every $(2n + 1)$ measurement of the output to avoid additional biases in the estimation.

The proof of convergence and detailed discussion of the algorithm are given in [6.47].   Extension to a multi-input/multi-output system is straightforward and will be demonstrated in Chap. 7.

6.4.2   A Second-Order Stochastic Approximation Algorithm

The error-correction term of this algorithm is generated by minimizing a least-square-fit criterion,

$$I = \sum_{i=1}^{k} \left[z(i + n) - S(i + n - 1) \begin{bmatrix} \hat{f} \\ - \\ \hat{b} \end{bmatrix} \right]^2 \qquad (6.22)$$

However, a purely least-square-fit algorithm is biased, as with the first-order stochastic approximation scheme [6.47].   Therefore, additional terms to account for the bias are incorporated, and the second-order algorithm is now only a stochastic approximation identifying the parameters f and b of system (6.15) as follows:

$$\begin{bmatrix} \hat{f}(k + n) \\ \hline \hat{b}(k + n) \end{bmatrix} = \begin{bmatrix} \hat{f}(k - 1) \\ \hline \hat{b}(k - 1) \end{bmatrix} + P(k + n)S(k + n - 1)$$

$$\left[z(k + n) - S^T(k + n - 1) \begin{bmatrix} \hat{f}(k - 1) \\ \hline \hat{b}(k - 1) \end{bmatrix} \right]$$

$$+ \begin{bmatrix} \sigma^2 I + q^2 GG^T & | & 0 \\ \hline 0 & | & 0 \end{bmatrix} \begin{bmatrix} \hat{f}(k - 1) \\ \hline \hat{b}(k - 1) \end{bmatrix} - \begin{bmatrix} q^2 g* \\ \hline 0 \end{bmatrix}$$

$$P(k + n) = P(k - 1) - P(k - 1)S(k + n - 1)[S^T(k + n - 1)P(k - 1)$$

$$S(k + n - 1) + 1]^{-1}S(k + n - 1)P(k - 1) \quad P(o) = P_o$$

$$k=1, \ n+2, \ 2n+3 \qquad (6.23)$$

where

$$s^T(k + n - 1) \triangleq [z(k),\ldots,z(k + n - 1); u(k),\ldots,u(k + n - 1)]$$

This method represents a definite improvement over the previous algorithm regarding the rate of convergence. It utilizes the error-covariance matrix, this has a deflection effect on the error-correction vector which speeds up the convergence at the first iteration of the algorithm. For large k, the matrix $P(k)$ behaves approximately as $(1/k)C_k$, which justifies its name as a stochastic approximation

$$0 < E\{C_k^{-1}\} = \left[\begin{array}{c|c} W_k + r^2 BB^T + a^2 GG^T + \sigma^2 I & Br^2 \\ \hline r^2 B^T & r^2 I \end{array} \right] < \infty \qquad (6.24)$$

where $0 < \bar{W}_k = \dfrac{1}{k} \sum_{i=1}^{k} E\{x(i)x^T(i)\} \to \bar{W} < \infty$ and r^2 is the variance of

the input $u(k)$, and B is generated in the same way as G in (6.20).

$$B = \begin{bmatrix} 0 & 0 & \cdots & 0 \\ b_1 & 0 & \cdots & 0 \\ \cdot & \cdot & \cdots & \cdot \\ b_{n-1} & b_{n-2} & \cdots & 0 \end{bmatrix}$$

The trade-off for the improved convergence rate is the time-consuming recursive computation of the deflection matrix $P(k + n)$. The algorithm iterates every $(2n + 1)$ measurements to avoid biases caused by correlation of output data. No other assumptions have been made about the probability distribution of the noises involved except for the knowledge of their variances.

The proof of convergence and detailed discussions on the algorithms are given in Ref. [6.47].

6.4.3 A Perturbation Input Stochastic Approximation Scheme

The on-line identification problem for linear discrete-time systems considered in the previous section is now approached from a different point of view. The resulting stochastic approximation algorithm does not require the knowledge of the noise statistics and converges to the true value of the parameters in the mean-square sense.

It was developed by Saridis and Stein in [6.48], and extended by Saridis and Lobbia for the feedback control system [6.45].

In the present identification scheme, it is assumed that the initial conditions represent the past history of the system excited by the same sequences of random variables defined over $-\infty \le i \le 0$. The output of the system at time k was described by Eq. (6.18) and is partitioned here.

$$z(k) = \sum_{i=0}^{\infty} h^T F^i [bu(k - i - 1) + gw(k - i - 1)] + v(k) \qquad (6.25)$$

$$= \sum_{i=0}^{2n-1} h^T F^i [bu(k - i - 1) + gw(k - i - 1)] +$$

$$+ \sum_{i=2n}^{\infty} h^T F^i [bu(k - i - 1) + gw(k - i - 1)] + v(k)$$

Define the following auxiliary 2n-dimensional vectors:

$$\theta^T = [h^T b, h^T Fb, \ldots, h^T F^{2n-1} b]$$
$$\beta^T = [h^T g, h^T Fg, \ldots, h^T F^{2n-1} g]$$
$$U^{k^T} = [u(k), u(k - 1), \ldots, u(k - 2n + 1)]$$
$$W^{k^T} = [w(k), w(k - 1), \ldots, w(k - 2n + 1)] \qquad (6.26)$$

Then

$$z(k) = U^{k-1^T} \theta + W^{k-1^T} \beta + \varepsilon(k - 2n - 1) + v(k) \qquad (6.27)$$

where

$$\varepsilon(k - 2n - 1) = h^T \sum_{i=2n}^{\infty} F^i \left[bu(k - i - 1) + gw(k - i - 1) \right] \qquad (6.28)$$

Define the estimate $\hat{\theta}$ of the parameter vector θ as the vector which minimizes the mean-square error criterion

$$J(\hat{\theta}) = E\{z - U^T \hat{\theta}\}^2 \qquad (6.29)$$

This estimate $\hat{\theta}$ may be computed recursively by means of the following stochastic approximation procedure

$$\hat{\theta}(k + 2n) = \hat{\theta}(k - 1) + \gamma(\frac{k - 1}{2n + 1}) U(k+2n-1)[z(k + 2n) - U^T(k+2n-1)$$

$$\hat{\theta}(k - 1)] \quad k = 1, 2n + 2, 4n + 3, \ldots \qquad (6.30)$$

This stochastic approximation algorithm uses only measurements of the input and output, and does not require knowledge of the noise statistics or the noise coefficient vector g. It is shown that the resulting estimate $\hat{\theta}(i(2n + 1))$, i=1, 2,... converges in the mean-square sense to the true value of the parameter vector θ under the following conditions:

$$\sum_{i=1}^{\infty} \gamma(i) = \infty \quad \sum_{i=1}^{\infty} \gamma^2(i) < \infty \quad E||\hat{\theta}(0)||^2 < \infty \tag{6.31}$$

The convergence of this algorithm is given in Ref. [6.48], as well as further discussion concerning it. Estimates $\hat{f}$, $\hat{b}$ of the desired parameter vectors f and b can be recovered from the estimate $\hat{\theta}$ by using the first equation in (6.26) repetitively.

$$\hat{b} = \begin{bmatrix} \hat{\theta}_1 \\ \vdots \\ \hat{\theta}_n \end{bmatrix} \quad \hat{f} = \begin{bmatrix} \hat{\theta}_1 & \hat{\theta}_2, & \cdots, & \hat{\theta}_n \\ \cdot & \cdots & \cdots & \cdot \\ \hat{\theta}_n, & \cdots, & & \hat{\theta}_{2n-1} \end{bmatrix}^{-1} \cdot \begin{bmatrix} \hat{\theta}_{n+1} \\ \vdots \\ \hat{\theta}_{2n} \end{bmatrix} \tag{6.32}$$

The inverse of the matrix required to obtain f exists with probability 1, if the order n of the model system (6.15) is the same as the order of the real system. If, however, n is greater than the order of the real system, the matrix becomes singular as the estimation process converges. This property may be used to find the order of the system when it is not known [6.26], [6.30].

6.5   AN ON-LINE MAXIMUM-LIKELIHOOD ALGORITHM

The algorithm discussed here was originally presented by Åström [6.2] and Kashyap [6.21], [6.22] for off-line parameter identification. A recursive on-line version has been developed and is given in the sequel. The linear system treated here is in the form of Eq. (6.17), repeated here for completeness

$$z(k) = \sum_{i=0}^{n-1} f_i z(k - n + 1) + \sum_{i=0}^{n-1} b_i u(k - n + 1) + \zeta(k) \tag{6.33}$$

where

$$\zeta(k) = \sum_{i=0}^{n} d_i \xi(k - n + i) = \sum_{i=0}^{n} [g_i w(k - n + i) - f_i v(k - n + i)]$$

$$d_o \triangleq 1 \quad g_o = 0, \quad f_o = -1$$

If $w(k) \sim N[0,W^2]$ and $v(k) \sim N[0,V^2]$, $\xi(j)$ is also Gaussian as well as the correlated noise $\xi(j)$ with zero mean and covariance

$$E\{\zeta(i)\zeta(i - j)\} = \sum_{\ell=0}^{n} [f_{\ell+j} f_{\ell} V^2 + g_{\ell} g_{\ell+j} W^2] = \sum_{\ell=0}^{n} d_{\ell+j} d_{\ell} \sigma_x^2$$

$$\tag{6.34}$$

Define the unknown parameters to be identified as

$$\theta^T = [f_1,\ldots,f_n, b_1,\ldots,b_n, d_1,\ldots,d_n] \tag{6.35}$$

Let the best estimate of z, given θ, be

$$\hat{z}(i/\theta) = E\{z(i)/z^j, u^j; \theta\} \tag{6.36}$$

$$\sigma_e^2(i) = E\{[z(i) - \hat{z}(i/\theta)]^2\}$$

Then the probability density of the estimates of z(i) is

$$p(z(i)/Z^j, U^j, \theta) \sim N[\hat{z}(i/\theta), \sigma_e^2(i)] \tag{6.37}$$

To simplify, the following approximation [6.21] may be used for $\sigma_e^2(i)$;

$$\sigma_e^2 = \lim_{i \to \infty} \sigma_e^2(i) < \infty \tag{6.38}$$

Define the error

$$e(k/\theta) \triangleq z(k) - \hat{z}(k/\theta) = \xi(k) \tag{6.39}$$

$$\sigma_e^2 = E\{e^2(k/\theta)\}$$

satisfying

$$e(k/\theta) = -\sum_{j=0}^{n-1} d_j e(k - n + j/\theta) + z(k) - \sum_{j=0}^{n-1} [f_j z(k - n + j)$$

$$+ b_j u(k - n + j)] \tag{6.40}$$

Maximization of the *a posteriori* estimation probability

$$p(z(k),\ldots,z(1)/u(k-1),\ldots,u(o);\theta) = \prod_{j=1}^{k} p(z(j)/z(j-1);\theta)p(z_o/\theta)$$

for the linear system (6.33), is equivalent to maximizing the likelihood function $L_k(\theta,\sigma_e)$ with respect to the unknown parameters θ and σ_e. Assuming that the *a priori* probability $p(z_o/\theta)$ is given, then

$$L_k(\theta,\sigma_e) = \ln[p(z^k/u^{k-1};\theta)] = c - \frac{1}{2}\ln[\sigma_e^2] - \frac{1}{2}\sum_{j=1}^{k} e^2(j/\theta)/\sigma_e^2$$

$$c = \text{constant} \tag{6.41}$$

Maximization of $L_k(\theta,\sigma_e)$ implies

$$\frac{\partial L_k(\theta,\sigma_e)}{\partial \sigma_e^2} = 0 = \hat{\sigma}_e^2 = \frac{1}{k}\sum_{j=1}^{k} e^2(j/\theta) \tag{6.42}$$

and

$$\underset{\theta}{\text{Max}}\ L_k(\theta,\hat{\sigma}_e) = \underset{\theta}{\text{Min}}\ \frac{1}{k}\sum_{j=1}^{k} \hat{e}^2(j/\theta) = \underset{\theta}{\text{Min}}\ J(\theta) \tag{6.43}$$

subject to

$$\hat{e}(k/\theta) = -\sum_{j=0}^{n-1} \hat{d}_j\hat{e}(k - n + j/\theta) + z(k) - \sum_{j=0}^{n-1} [\hat{f}_j z(k - n + j)$$
$$+ \hat{b}_j u(k - n + j)] \tag{6.44}$$

Under certain controllability conditions [6.31] on the input sequence $u(i)$ $i = 0,\ldots,k - 1$, this Two Point Boundary Value (TPBV) problem can be solved recursively for <u>on-line</u> identification of the parameter θ. The optimization algorithm evolves as follows:

1. Assume initial conditions $\hat{e}(i/\theta)$, $\dfrac{\partial \hat{e}(i/\theta)}{\partial \theta}$ $i = 1,\ldots,n$; guess $\hat{\theta}(o)$; $N = 0$.

2. Compute $e(k/\theta)$ from (6.40) for $k = N + 1, N + 2,\ldots, N + M$.

3. Evaluate the sensitivity functions sequentially after M measurements

$$\frac{\partial \hat{e}(k/\theta)}{\partial f_j} = -z(k - n + j) - \sum_{i=0}^{n-1} d_i\ \frac{\partial \hat{e}(k - n + i/\theta)}{\partial f_j}$$

$$\frac{\partial \hat{e}(k/\theta)}{\partial b_j} = -u(k - n + j) - \sum_{i=0}^{n-1} d_i\ \frac{\partial \hat{e}(k - n + i/\theta)}{\partial b_j} \tag{6.45}$$

$$\frac{\partial \hat{e}(k/\theta)}{\partial d_j} = -\hat{e}(k - n + j) - \sum_{i=0}^{n-1} d_i\ \frac{\partial \hat{e}(k - n + i/\theta)}{\partial d_j}$$

k = n + 1,...,N + M

4.   Compute the gradient of J(θ)

$$\frac{\partial J}{\partial \theta} = -2 \sum_{j=N+1}^{N+M} \hat{e}(j/\theta) \frac{\partial \hat{e}(j/\theta)}{\partial \theta}$$ (6.46)

5.   Compute the deflection matrix approximately

$$S = \sum_{j=N+1}^{N+M} \left[\frac{\partial \hat{e}(j/\theta)}{\partial \theta} \frac{\partial \hat{e}^T(j/\theta)}{\partial \theta} \right] \simeq \frac{\partial^2 J}{\partial \theta^2}$$ (6.47)

6.   Revise estimates of parameters at every M measurements by the following stochastic approximation:

$$\hat{\theta}(N + M) = \hat{\theta}(N) - \frac{1}{(N/M+1)} S^{-1} \frac{\partial J}{\partial \theta}$$ (6.48)

7.   Make M measurements of input and output and set N = N + M. Use the last n values of $\hat{e}(i/\theta)$ and $\frac{\partial \hat{e}(i/\theta)}{\partial \theta}$ as initial guesses for the next iteration. Repeat 2. Stop when little improvement is accomplished.

This algorithm was originally designed to use a fixed set of data which was repeated again and again for the evaluation of the gradients without the stochastic approximation gains $\frac{1}{(N/M + 1)}$ This off-line procedure is extremely slow and time-comsuming. The on-line algorithm, on the other hand, for a reasonable number of measurements converge to steady-state values after the transient period, a proof of convergence is possible. The great disadvantage of the method is still the assumption that all the random variables involved are Gaussian. It also happens that for most cases the initial guesses of the parameters must be very close to the true values in order to guarantee convergence.

6.6   IDENTIFICATION AS AN EXTENSION OF FILTERING

The idea of using the filtering procedures developed in Sec. 3.4 of Chap. 3 was first suggested by Kalman and discussed by Kopp and Orford [6.25], Cox [6.8], and others. It assumes that parameters are time-invariant states and require Gaussian additive random

processes for the noises involved. The method is mostly suitable when
state estimation is required along with parameter identification, as
in space technology problems [6.8], [6.18], [6.19], [6.36], [6.37].
The single-input/single-output discrete-time system under consideration
(6.15) is repeated here for convenience.

$$x(k + 1) = Fx(k) + bu(k) + gw(k) \quad x(0) = x_0 \qquad (6.49)$$

$$z(k) = h^T x(k) + v(k)$$

The set of state variables is chosen so that

$$F = \left[\begin{array}{c|c} 0 & I \\ \hline & f^T \end{array} \right], h^T = (1, 0, \ldots, 0)$$

g' is an arbitrary n-dimensional vector, and the parameters $f^T = [f_1, \ldots, f_n]$ and $b^T = [b_1, \ldots, b_n]$ are unknown and must be identified
from input and output records. The matrix F is assumed stable and
the pair F,b completely controllable. In the system (6.49) the
following random processes are involved:

1.   {w(k)} is a sequence of Gaussian independent not directly
       measured random variables with $E\{w(k)\} = 0$, $|E\{w(k)w(j)\}| \leq q^2 \delta_{kj}$ for all k,j.
2.   {v(k)} is a sequence of Gaussian independent not directly
       measured random variables with $E\{v(k)\} = 0$, $|E\{v(k)v(j)\}| \leq \sigma^2 \delta_{kj}$ for all k,j.
3. x_0 is a Gaussian random variable with $E\{x_0\} = \bar{x}_0$, $E\{(x_0 - \bar{x}_0)(x_0 - x_0)^T\} = P_0$.

       In order to apply this method, the state variables of the
systems are augmented by adjoining to them the unknown parameter
vectors and treating them as part of the new state variable X(k),

$$X^T(k) = [x^T(k), f^T, b^T] \qquad (6.50)$$

The difference equations governing the unknown parameters are given
by

$$f(k + 1) = f(k) \quad f(o) = \bar{f} \qquad (6.51)$$
$$b(k + 1) = b(k) \quad b(o) = \bar{b}$$

By combining (6.49), (6.50), and (6.51) one may write the difference
equations for the augmented systems, which for obvious reasons are
nonlinear

$$X(k + 1) = F(X(k), u(k)) + g'w(k) \quad X(o) = \bar{X}_0$$

$$z(k) = h'^T X(k) + v(k) \tag{6.52}$$

where $X(k)$ is now a 3n-dimensional vector, and g and h are the 3n-dimensional augmentations of g' and h'

$$g'^T \triangleq [\ g^T, 0, \dots, 0] \quad h'^T \triangleq [h^T_{.}, 0, \dots, 0]$$

The estimation algorithm can be generated either by maximizing the *a posteriori* probability of the states (as in Sec. 3.4.4) or by considering the minimum mean-square estimator for nonlinear systems as in Sec. 3.5. Detailed discussion of these algorithms are given by Sage and Melsa in Ref. [6.36].

6.6.1 Maximum A Posteriori Probability Algorithm, The Linearized Kalman-Bucy Filter

The following *a posteriori* conditional probability density functions are associated with the augmented state vector $X(k)$:

$$p(X^k/z^k) \triangleq p(X(o), X(1), \dots, X(k)/z(1), z(2), \dots, z(k)) \tag{6.53}$$

Due to the way that the Gaussian noises enter Eq. (6.52), these conditional densities are also Gaussian:

$$p(X^k/z^k) = \frac{p(z^k/X^k)p(X^k)}{p(z^k)} = c' \prod_{i=1}^{k} \exp\left\{ -\frac{1}{2\sigma^2} [z(i) - h'^T X(i)]^2 \right\}$$

$$\prod_{j=1}^{k} \exp\left\{ -\frac{1}{2q^2 g'^T g'} \| \hat{x}(j) - F(\hat{x}(j - 1), u(j - 1)) \|^2 \right\}$$

$$\exp\left\{ -\frac{1}{2} \| X(o) - X_0 \|^2_{P_0^{-1}} \right\}$$

where c' is an appropriate constant.

The most probable estimate can be produced by maximizing the *a posteriori* probability $p(X^k/z^k)$ as in Sec. 3.4.4. This is equivalent to minimizing the following quadratic performance criterion

$$J = \sum_{j=1}^{k} \left\{ \frac{1}{2\sigma^2} [\ z(j) - h^T \hat{x}(j)]^2 + \frac{1}{2q^2} \hat{w}^2(j) \right\} + \| \hat{x}(0) - x_0 \|^2_{P_0^{-1}} \tag{6.55}$$

subject to the constraints

$$\hat{X}(j + 1) = F(\hat{X}(j), u(j)) + g\hat{w}(j) \quad j = 1, 2, \ldots, k \qquad (6.56)$$

for every incoming measurement $z(k)$, $k = 1, 2, \ldots, N$.

This constitutes a nonlinear optimization problem which does not possess a straightforward analytic solution since no reproducible form can be obtained.

To overcome this difficulty, Eq. (6.56) is linearized about a nominal trajectory predetermined from

$$\hat{X}_0(j + 1) = F(\hat{X}_0(j), u(j)), \hat{X}_0(0) = \bar{X}_0 \qquad (6.57)$$

$$\hat{X}(j) = \hat{X}_0(j) + \delta\hat{X}(j)$$

Linearization about (6.57) produces the linear incremental Eq. for (6.56)

$$\delta\hat{X}(j + 1) = \Phi(j)\delta\hat{X}(j) + g\hat{w}(j), \delta\hat{X}(0) = 0 \qquad (6.58)$$

while the performance criterion (6.55) is valid, if one defines

$$z(j) = z_0(j) + \delta z(j) = h^T X_0(j) + \delta z(j) \qquad (6.59)$$

where

$$\Phi(j) = \frac{\partial F}{\partial \hat{X}_0(j)} (\hat{X}_0(j), u(j)) = \begin{bmatrix} 0 & I & 0 & \vert & u_0(j)I \\ \hline f^T & X_0(j) & \\ \hline 0 & 1 & 0 \\ \hline 0 & 0 & I \end{bmatrix}$$

$$(6.60)$$

The minimization of (6.55), subject to (6.58) and (6.54), requires the solution of the following two-point boundary-value problem:

$$\delta\hat{X}(j+1/k) = \Phi(j)\hat{X}(j/k) + g\hat{w}(j/k)$$

$$\hat{w}(j) = q^2\lambda(j)$$

$$\lambda(j) = \Phi(j + 1)\lambda(j + 1) + \frac{1}{\sigma^2} h[\delta z(j + 1) - h^T\delta\hat{X}(j + 1/k)]$$

$$\hat{X}(0/k) = P_0\phi(0)\lambda(0) + \bar{X}_0, \lambda(k) = 0 \qquad (6.61)$$

By solving the problem recursively as in [6.26], the following equations of the linearized Kalman filter equation result, similar to (3.180)-(3.182):

$$\hat{X}(k/k) = \hat{X}_0(k) + \delta\hat{X}(k/k)$$

$$\delta\hat{X}(k/k) = \Phi(k - 1)\delta\hat{X}(k - 1/k - 1) + P(k/k)h[z(k) - h^T\Phi(k - 1)$$
$$\delta\hat{X}(k - 1/k - 1)], \quad \delta\hat{X}(0/0) = 0 \qquad\qquad (6.62)$$

$$P(k/k - 1) = \Phi(k - 1)P(k - 1/k - 1) \Phi^T(k - 1) + q^2..^T$$

$$P(k/k) = P(k/k - 1) - P(k/k - 1)h[h^TP(k/k - 1)h + \sigma^2]^{-1}$$
$$h^TP(k/k - 1), \quad P(0/0) = P_0$$

This algorithm has several disadvantages. It requires Gaussian
noise sequences with known statistics and known gains g'. It is
usually a poor approximation of the exact estimation scheme and may
be improved only by using second-order expansion terms [6.36].
Finally, it usually consumes computer time if the state estimation
is not required.

6.6.2 Nonlinear M.M.S. Estimator, the Extended Kalman Filter

A different identification algorithm is obtained if the
nonlinear estimator presented in Sec. 3.4.4 is used [6.37]. An
approximation is obtained by expanding the nonlinear augmented
Eq. (6.52) in Taylor series about the augmented state estimate $\hat{X}(k)$,
and retaining only the first-order terms.

$$\hat{X}(k + 1) \simeq F(\hat{X}(k),u(k)) + \frac{\partial F}{\partial X} (k)[X(k) - \hat{X}(k)] + g'w(k) \quad (6.63)$$

$$z(k) = h'^T\hat{X}(k) + h'^T[X(k) - \hat{X}(k)] + v(k)$$

where all the variables and parameters have been defined in Sec. 6.6.

Such a model for the augmented system is linear in $\hat{X}(k)$ and
time varying. Therefore, a discrete Kalman filter can be implemented
to yield estimates of the states and the unknown parameters [6.8],
[6.12]. The filter equations are as follows:

$$\hat{X}(k + 1) = \hat{X}(k + 1|k) + K(k + 1)\{z(k + 1) - h'^T\hat{X}(k + 1|k)\}$$
$$\hat{X}(k + 1|k) = F(\hat{X}(k),u(k)) \quad \hat{X}(0) = \bar{X}_0 \qquad\qquad (6.64)$$

$$K(k + 1) = P(k + 1|k + 1)h'\sigma^{-2}$$

$$P(k + 1|k + 1) = P(k + 1|k) - P(k + 1|k)h[h^TP(k + 1|k)h + \sigma^2]^{-1}$$
$$\cdot h^TP(k + 1|k)$$

$$P(0/0) = \begin{bmatrix} P_0 & \vdots & 0 \\ - & - & - & - \\ 0 & \vdots & 10I \end{bmatrix}$$

$$P(k+1/k) = F(\hat{X}(k),u(k))P(k/k)F^T(\hat{X}(k),u(k)) + q^2g'g'^T$$

The same result may be obtained as an approximate discrete conditional-mean filter for nonlinear systems obtained in Sec. 3.4.4. Improvements of this algorithm have been generated by considering higher-order terms in the Taylor series expansion or even by adaptive filtering [6.18].

This method suffers from a large computational programming problem, especially when only the unknown parameter and not the estimator of the states are required.   The Gaussian assumption of the random variable involved is essential for the implementation of the method, and a study of the error is necessary for the evaluation of possible biases in the estimates of the parameters.   However, the method has been very popular among researchers in the field of space technology.

6.7   THE RANDOM SEARCH IDENTIFICATION ALGORITHMS

The estimation method described in Sec. 3.3.6 as random search can be adapted for the parameter identification of dynamic or nondynamic linear or nonlinear systems as well as S.O.C., a subject discussed in Chap. 8.   The algorithm developed [6.39], [6.42] was given the name *expanding subinterval algorithms* because of the properties of the performance evaluation of the system and its relation to the overall performance criterion.

The advantages of the expanding subinterval algorithm are the convergence to a global minimum, the consistency of estimation for dynamic as well as nondynamic systems, its applicability to nonlinear and linear systems, and its simplicity of implementation, which considerably reduces the total search time on a digital computer for large-scale systems.

However, its most important property is the relation of the convergence properties of the instantaneous performance cost $J(c_i)$

to the overall accrued performance cost $V(n)$, for an appropriate nondecreasing time interval T_i. The meaning of this property is that such a search minimizes in probability not only the instantaneous but also the accrued cost of the performance of this system, and therefore it can be applied for <u>on-line</u> optimization of a performance of a system defined over the duration of the probess. This is obviously useful for learning control systems and application of identification algorithms with cumulative cost criteria, such as least squares. For the purpose of using the *expanding subinterval random search algorithm* for parameter identification purposes this method will be presented here and the appropriate convergence theorems stated, adapted from Ref. [6.39] where they were developed for the S.O.C. problem.

The problem of parameter identification may be posed as follows:

Suppose that a nonlinear asymptotically ergodic stochastic system depends on a fixed but unknown parameter vector c belonging to a closed and bounded set Ω_c of parameters, $\Omega_c = \{c\}$. Define a cost function $J(c)$ of the identification defined over a time interval $T_i \in [0,\infty)$, and depending on the available input-output noisy measurements of the system. Assume that a minimum value $J(c*) = J^*_{min}$ of $J(c)$ exists and that it satisfies the following conditions:

$$0 \leq J^*_{min} \leq J(c) < \infty \quad \forall c \in \Omega_c \qquad (6.65)$$

Define the accrued cost function

$$V(n) = \frac{\sum\limits_{i=0}^{n} T_i E_{\xi\eta}\{J(c_i)\}}{\sum\limits_{i=0}^{n} T_i} \quad n = 1,], 2,\ldots \qquad (6.66)$$

where $J(c_i)$ is evaluated on T_i for $c = c_i$ and $E_{\xi\eta}(\cdot)$ is the expectation with respect to the system random noise vectors ξ and η.

Then if T_i satisfies the conditions

$$T_{i-1} \leq T_i, \quad \lim_{n \to \infty} T_n = \infty, \quad \forall i \qquad (6.67)$$

one may define the following random search algorithm for an appropriate μ

$$c_{i+1} = \begin{cases} c_i & \text{if } J(\rho_i) - J(c_i) > 2\mu \\[2em] \rho_i & \text{if } J(\rho_i) - J(c_i) \leq 2\mu \end{cases} \tag{6.68}$$

where ρ_i is a random number generated by a prespecified independent and identically distributed density function $p(\rho) \neq 0$, $\rho \in \Omega_c$. It may be shown then that for some positive δ, not only

$$\lim_{n \to \infty} \text{Prob}[J(c_n) - J^*_{\min} < \delta] = 1 \tag{6.69}$$

but

$$\lim_{n \to \infty} \text{Prob}\left[\frac{\sum_{i=1}^{n} T_i E_{\xi_n} J(c_i)}{\sum_{i=1}^{n} T_i} - J^*_{\min} < 3\delta\right] = 1 \tag{6.70}$$

For identification purposes define the per-interval cost for a continuous-time system

$$J(c) = (t_i - t_{i-1})^{-1} \int_{t_{i-1}}^{t_i} L(u(t), z(t,c)) \, dt \quad i = 1, 2, \ldots \tag{6.71}$$

or equivalently for a discrete-time system

$$J(c) = (t_i - t_{i-1})^{-1} \sum_{k=t_{i-1}}^{t_i} L(u(k), z(c,k)) \quad i = 1, 2, \ldots \tag{6.72}$$

where

$$T_i = t_i - t_{i-1}$$

satisfies conditions (6.67), $L(\cdot \cdot)$ is an appropriate cost function, $u(t)$ or $u(k)$ are the inputs to the continuous- and discrete-time systems, and $z(c,t)$ or $z(c,k)$ are the measured respective outputs depending on the parameter vector c. It is obvious from (6.66) that the accrued cost is given by

$$V(n) = \frac{1}{t_n - t_0} E_{\xi_n}\left\{\int_{t_0}^{t_n} L(u(t), z(c,t)) \, dt\right\} \tag{6.73}$$

for the continuous-time case, or

$$V(n) = \frac{1}{t_n - t_0} \, E_{\xi n} \left\{ \sum_{k=t_0}^{t_n} L(u(k), z(c,k)) \right\} \tag{6.74}$$

for the discrete-time case. The expressions of $V(n)$ in (6.73) and (6.74) have the more familiar form of an overall performance criterion defined over the duration of the process. It is obvious that if the process is asymptotically ergodic and

$$c^* = \text{Arg}\left[\underset{c}{\text{Min}} \, \underset{T_i \to \infty}{\lim} \, J(c) = J_{min}^* \right] \tag{6.75}$$

in the limit

$$\underset{c}{\text{Min}} \left[\underset{n \to \infty}{\lim} \, V(n) = \underset{n \to \infty}{\lim} \, V(n) \bigg|_{c_i = c^*} \right] = J_{min}^* \quad i = 1, 2, \dots \tag{6.76}$$

Then the following theorem establishes the asymptotic optimality of the expanding subinterval algorithm for either continuous- or discrete-time systems identification.

Theorem 6.1

Given an arbitrary number $\delta > 0$ and the random search algorithm defined in (6.68) with the following properties:

P1.   The function $J(c)$ is defined by (6.71) or (6.72) so that the set $\Omega_c^* \triangleq \{c/J(c) - J_{min}^* < \delta, \forall \delta, c \in \Omega_c\}$ has a positive measure.

P2.   Conditions (6.67) are satisfied for all T_i.

P3.   For any n_1, n_2, n_3

$$\underset{\frac{n_3}{n_2} \to 0}{\lim} \frac{\sum_i^{n_3} T_i}{\sum_{i=n_1+1}^{n_1 + n_2} T_i} = 0$$

where $\sum_i^{n_3} T_i$ denotes a sum of n_3 terms of the denominator.

Then for $\delta > 0$ there exists a number $\mu > 0$ such that

$$\underset{n \to \infty}{\lim} \, \text{Prob}[V(n) - J_{min}^* < 3\delta] = 1 \tag{6.77}$$

Corollary 6.1

Assume that conditions of Theorem 6.1 are satisfied. Then $V(n)$ converges to $J_{min}^* + 3\delta$ in the rth-mean for an arbitrary $\delta > 0$

$$\lim_{n \to \infty} E\{ |V(n) - [J_{min}^{*}+3\delta]|^{+}|^{r}\} = 0 \quad 0 < r < n \tag{6.78}$$

where $[x]^{+} \triangleq \frac{1}{2}[x + |x|]$

It was found necessary to evaluate $J(c_i)$ at every iteration of algorithm (6.68) in order to use revised values of the performance criterion for comparison during the transient period of a dynamic system.   Since this procedure is performed <u>on-line,</u> the additional cost, due to the reevaluation of $J(c_i)$ after a successful iteration, may be included in the accrued cost function $V(n)$ by redefining it as

$$\bar{V}(n) = \frac{\sum\limits_{i=1}^{n} E_{\xi\eta}\{ T_i J(c_i) + \bar{T}_i J(\rho_i)\}}{\sum\limits_{i=1}^{n} [T_i + \bar{T}_i]} \qquad \bar{T}_i = \begin{cases} T_i, & c_{i+1} = \rho_i \\ 0, & c_{i+1} = c_i \end{cases} \tag{6.79}$$

For the new accrued cost Theorem 6.2 may be obtained.

Theorem 6.2

Assume that for a given $\delta > 0$, the random search (6.68) satisfies properties P1 to P3 of Theorem 6.1.   Let the additional
property be satisfied by

P4.
$$\lim_{n \to \infty} \frac{\sum\limits_{i=1}^{n} T_i}{\sum\limits_{i=1}^{n} [T_i + \bar{T}_i]} = 0$$

Then there exists a number $\mu > 0$ so that
$$\lim_{n \to \infty} Prob[\bar{V}(n) - J_{min}^{*} \le 4\delta] = 1 \tag{6.80}$$

and
$$\lim_{n \to \infty} E\{ |\bar{V}(n) - [J_{min}^{*} + 4\delta]|^{+}\} = 0 \quad 0 < r < n \tag{6.81}$$

Theorems 6.1 and 6.2 and Corollary 1 have been proved in
Ref. [6.39] for the similar S.O.C. problem.   Their proofs are directly
applicable to the present theorems used for identification.

This expanding subinterval random search algorithm has been
tested for nondynamic as well as linear and nonlinear dynamic system
identification in Ref [6.42] and has proved superior to the

stochastic approximation algorithm because of the speed of its
iteration.   It can even be implemented for closed-loop control systems.
Acceleration schemes similar to the ones in Sec. 3.3.6 are discussed
in Ref. [6.42].

6.8   EXAMPLES AND COMPARISONS

Because of preferential treatment of the various identification
algorithms by different authors, there are only a few comparative
studies [6.40], [6.42], [6.17] of the methods presented in this chapter.
Therefore some examples are included and some comparisons among the
algorithms worked out by the author and his colleagues[6.40], [6.42]
are presented.   Another important comparative study covering the
algorithms not presented in this chapter is given by   Isermann et al.
in Ref [6.17].

The impulse response of a continuous-time system with transfer
function

$$G(s) = \frac{200}{s^2 + 10s + 200}$$

has been identified on the Purdue crosscorrelator, a machine working
on the principle described in Sec. 6.3.1 with m = 35.   A periodic
sequence of N = 59 information interval of DIBN is used.   The results
of identification are presented in Fig. 6.6.

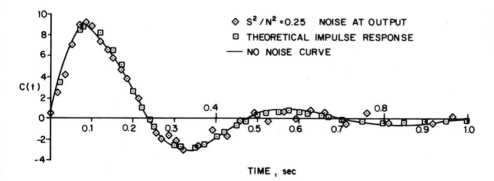

FIG. 6.6.   Impulse Response Actual and Through Crosscorrelations
for a Second-Order System.

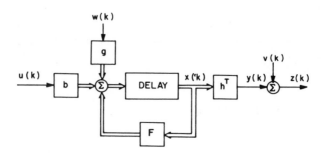

FIG. 6.7.   Linear Discrete-Time System.

A comparative study of open-loop identification has been
performed on a linear discrete-time dynamic system depicted in
Fig. 6.7 and described by the following set of difference equations:

$$x(k + 1) = F x(k) + g[w(k) + u(k)]$$

$$z(k) = h^T x(k) + v(k)$$

The crosscorrelation method, the three stochastic approximations
(S.A.), the maximum likelihood, and the two Kalman filtering methods
have been implemented on a specific fourth-order system where

$$f = \begin{bmatrix} 0 & 1 & 0 & 0 \\ 0 & 0 & 1 & 0 \\ 0 & 0 & 0 & 1 \\ f_1 & f_2 & f_3 & f_4 \end{bmatrix} \quad g = \begin{bmatrix} 1 \\ 1 \\ 2 \\ 1 \end{bmatrix} \quad u(k) = \begin{cases} \text{DIBN; cross-} \\ \text{correlation,} \\ \text{white noise;} \\ \text{perturb, S.A.} \\ 0, \text{ other methods} \end{cases}$$

and the noises are Gaussian,

$$w(k) \sim N[0,1.00] \quad v(k) \sim N[0,0.25] \quad x_o \sim N[5,10]$$

The problem is to identify the parameters f_1, f_2, f_3, and f_4 which,
for the real system, have the true values of -0.66, 0.78, -0.18, and
1.00, respectively.

The crosscorrelation method used as input $u(k)$ a DIBN noise
sequence of 67 bands of magnitude 1.0. The sequence was repeated
100 times. The stochastic approximation methods were used without
any acceleration devices. The initial guess for the second-order
method was $P_o = 10I$.

The maximum likelihood method had to be initiated at points

Table 6.1. Comparison of Identification Methods; Fourth-Order Linear System.

Algorithm	Computer Time		Core,k Octal	Avg. Sq. Error		Initial Guess	Implementation
	600 IT.	2000 IT.		2000 IT.	5000 IT.		
Crosscorrelation	3	5	22	0.45×10^{-2}	0.2×10^{-2}	anywhere	easy
Perturbation input stochastic approx.	3.5	6	20	0.9×10^{-2}	0.7×10^{-3}	anywhere	easy
1st-order S. A. (unb. lst. sq.)	1.8	5.4	6.2	0.5	0.45	anywhere	easy
2nd-order s. A. (unb. lst. sq.)	2.4	16.5	8	0.11	0.11	anywhere	less easy
Extended Kalman filter	19.8	66	10.6	0.46	0.44	close	less easy
Off-line maximum likelihood 100 pts.	98		46	0.02	---	close	difficult
On-line, mod. max. likelihood 100 pts. per step	9.5	21	16.5	0.02	0.02	close	more difficult
Max. a-posteriori prob. filter	50	204	12	0.68	0.68	close	most difficult

close to the true parameter vector. The measurements M were set equal
to 100. The initial conditions for the error had to be chosen only
for the first iteration since the last values of the previous iteration
were used to initiate the new one. An _off-line_ run with M = 10600
and N = 0 has been given for relative comparison.

The maximum _a-posteriori_ probability filter was used for both
noise gains but it was impossible to get even one converging run
with the second-gain vector. The initial conditions of the parameter
vector had to be chosen close to the true values to get any results.
Again $P(o) = 10I$ was chosen.

Finally, the extended Kalman filter was implemented for initial
guesses $F(o) = 2f$, $P_o = 10I$. No acceleration schemes were employed
here either. It should be noted that all the algorithms are equally
applicable for the case that some plant parameters are known.

The results are presented in Fig. 6.8 and Table 6.1.

It is interesting to note that the modified crosscorrelation
method and the perturbation input stochastic approximation gave
spectacular results for both noise-gain vector cases. However, it
should be remembered that one algorithm iteration corresponds to
one plant iteration which corresponds to Four plant iteration for the
stochastic approximation methods, 100 iterations for the maximum
likelihood, and one iteration for the maximum _a-posteriori_ probability
filter. It should also be remembered that the additional noise input
used contributes favorably to the advantage of the first method.
The stochastic approximation method behaves as expected; the first-
order method is slower, yielding convergence of at least one order
or magnitude less than the second-order method. The modified on-line
maximum likelihood gave surprisingly good results compared to the
other methods and also the _off-line_ method with ~ 10000 points of
output measurements. The real disappointment was the maximum
a posteriori probability filter (linearized Kalman filter) which
yielded only two out of eight converging runs. The maximum
likelihood method had also many divergent runs while the other three
did not have any.

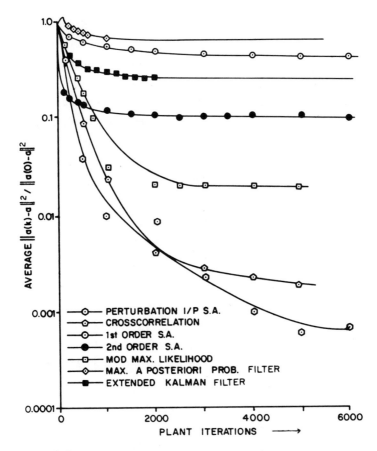

FIG. 6.8. Fourth-Order System Identification Case I.

The extended Kalman filter algorithms converged, when the
initial guess was higher in magnitude than zero, with the same rate
as the first-order stochastic approximation algorithm. This was
expected, since this filter is a first-order approximation. Higher
convergence speed is expected for higher-order approximation.
Slower convergence was observed for initial guesses below the true
parameter values.

Table 6.1 gives additional information about algorithms, such as
the computer time after 600 and 2000 plant iterations and the
average normalized squared error for each noise gain vector after

3000 and 5000 plant iterations, and core requirements and implementation difficulties. An off-line run of the maximum likelihood method
with 1000 points repeated six times is used for further comparisons
with the on-line methods.

Finally, the random search technique was compared to the
perturbation input stochastic approximation, modified as in Chap. 7
in the case of a feedback control system. These two methods
guarantee unbiased estimates of the parameters of the system if the
following identification criterion, satisfying (6.75), is used:

$$J(f) = \lim_{n \to \infty} V(n) = \lim_{n \to \infty} \frac{1}{t_n - t_0} E\{ \sum_{k=1}^{n} e^2(k+n|\theta) \}$$

$$e(k/\theta) = z(k) - E\{z(k)/Z^{(k-1)}, U^{k-1}; \theta\} = - \sum_{i=k-2}^{k-1} e(i/\theta)d_i$$

$$-Z^{T}{}_{k-4}^{k-1} f - U^{T}{}_{k-4}^{k-1} b + z(k)$$

The unknown parameter vector $f^T = [f_1, f_2]$ and the gain vectors
$b = [0, 1]$ and $d_i = [0, 4]$ correspond to the following plant:

An "optimal control" has been defined that asymptotically
minimizes the performance criterion

$$J(u) = E\{\frac{1}{N} \sum_{i=0}^{N} [x_1^2(i + 1) + 5 u^2(i)]\}$$

from approximations of the appropriate Kalman-Bucy filter cascaded
to an approximate optimal controller, depending on the estimates of
the parameters $\hat{f}$ and having the form

$$y(k + 1) = A(\hat{f})y(k) + d(\hat{f})z(k)$$
$$u(k) = (1,0)y(k) \quad y \in R^2$$

The results of a computer simulation are plotted in Fig. 6.9,
and a comparison with the stochastic approximation is given. A
special simulation of the random search with a perturbation input
which was used for the stochastic approximation method was given
for comparison on even terms. As a matter of fact, this remark
indicates another advantage of the random search over the stochastic
approximation method. There is no need to apply a perturbation
input at the plant's input in order to accomplish an unbiased

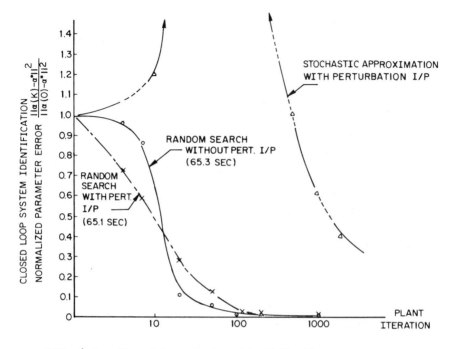

FIG. 6.9.   Closed-Loop System Identification.

identification. The random input is applied internally in the
identifier, and therefore does not accrue additional cost on the
performance criterion for the identification. Stability considerations
for the overall system were used to define the region for the search
of the unknown parameters $\hat{f}$ [6.42], [6.45]. The superiority of the
random search is due to the fact that the number of scheme iteration
that can be performed for every plant iteration is 150 more than
those of the stochastic approximation because of the simplicity and
speed of the algorithm.

6.9 DISCUSSION

Several on-line identification methods have been selected for
presentation in this chapter based on their suitability for parameter
adaptive S.O.C. Algorithms using crosscorrelation, stochastic

approximation, maximum likelihood, nonlinear filtering, and random
search were presented and their properties were evaluated and compared
for speed of convergence, computer specification, and ease of
implementation. However, of all these algorithms only the perturbation
input stochastic approximation, the crosscorrelation, and the random
search can be used for identification of feedback control system
where the information from the identifier is used in the feedback
loop, to produce unbiases estimates. The other algorithms may
introduce biases in the estimators [6.29]. Therefore, unless
modifications are available, the consistency of the open-loop
identifiers is questionable.

If there is a method favored by the author, if is the *expanding
subinterval random search* which has proved to yield the best results
in spite of its unappealing format. It has been used successfully
for the identification of linear and nonlinear dynamic systems with
results far superior to any other method [6.35]. The method has
been used for performance adaptive S.O.C. and will be further
discussed in Chap. 8. The properties of the method are summarized
here for reference.

1.   It is a global search.

2.   It is applicable in the same fashion to continuous- as well as
      discrete-time systems.

3.   It is independent of local convergence properties such as the
      gradient type of search techniques and therefore it retains a
      consistent rate of convergence.

4.   Due to its speed of iteration it may be performed 100 times
      or more faster than the plant iterations, and therefore
      converges faster than the stochastic approximation methods.

5.   As it has been demonstrated [6.42], its iteration time grows
      almost linearly with the system's dimension while the stochastic
      approximations and the "optimal" search methods grow
      exponentially.

6.   It identifies without disparity parameters appearing nonlinearly
      in the system, e.g., exponential coefficients.

7. When an appropriate identification criterion is used, it gives unbiased estimates to dynamic systems and especially to feedback control systems, where the information from identification is used to update the controller.

8. In most cases, it may be accelerated to yield a faster convergence.

9. It can identify systems with completely or partially unknown structures and without any characterization of the disturbances involved.

10. The search minimizes the instantaneous cost criterion as well as the accrued cost and therefore may be minimal for accumulative performance like least squares, etc.

11. It utilizes a random perturbation inside the identifier and therefore does not introduce additional cost in the performance of the system due to the identification.

12. It is simple to implement.

Finally, this chapter concludes all the background material needed to present the S.O.C. algorithms.

6.10 REFERENCES

[6.1]   Anderson, G., Booland, R., and Cooper, G., "Use of
          Crosscorrelation in an Adaptive System," *Proc. NEC*, 11,
          (1959).

[6.2]   Åström, K. J., and Bohlin, T., "Numerical Identification of
          Linear Dynamic Systems from Normal Operating Records," *Proc.
          IFAC Symposium on Self-Adaptive Control Systems*, Teddington,
          England, Sept. 1965.

[6.3]   Åström, K. J., and Eykoff, P., "System Identification, A
          Survey," *Automatica*, 7, (2) 123-162 (1971).

[6.4]   Badavas, P., and Saridis, G. N., "Response Identification
          of Distributed Systems with Noisy Measurements at Finite
          Points," *Intern. J. Inform. Sci.*, (2) 19-34 (1970).

[6.5]   Balakrishan, A. V., and Peterka, V., "Identification in
          Automatic Control Systems," *Automatica*, 5, (6), 607 (1969).

[6.6]   Box, G. E. P., and Jenkins, G-M., *Time Series Analysis
          Forecasting and Control*, Holden-Day, San Francisco, 1970.

[6.7]   Clarke, D. W., "Generalized Least Squares Estimates of the
          Parameters of a Dynamic Model," Preprints IFAC Symposium on
          Identification, Paper 3.17, Prague, 1967.

[6.8]   Cox, H., "On the Estimation of State Variables and Parameters
          for Noisy Dynamic Systems," *IEEE Trans. Automatic Control*,
          AC-9, (1), 5-12 (1964).

[6.9]   Graupe, D., *Identification of Systems*, Van Nostrand
          Reinhold, Princeton, N.J., 1972.

[6.10]   Gustavsson, I., "Comparison of Different Methods for
          Identification of Industrial Processes," *Automatica*, 8, (2)
          127-142 (1972).

[6.11]   Hill, J. D., and McMurtry, G., "An Application of Digital
          Computer to Linear System Identification," *IEEE Trans.
          Automatic Control*, AC-9, (4) 536-537 (1964).

[6.12]   Ho, Y. C., and Lee, R. C. K., "Identification of Linear
          Dynamic Systems," 3rd Symposium on Adaptive Systems, 1964.

[6.13]   Ho, Y. C., and Lee, R. C. K., "A Bayesian Approach to
          Problems in Stochastic Estimation and Control," *IEEE Trans.
          on Automatic Control*, AC-9 (4) 333-338 (1964).

[6.14] Ho, Y. C., and Whalen, B., "An Approach to Identification and Control of Linear Dynamics Systems with Unknown Parmaters," *IEEE Trans. Automatic Control*, AC-8 (3) 255-256 (1963).

[6.15] Hsia, T. C., and Vimolvanich, V., "An On-line Technique for System Identification," *IEEE Trans. Automatic Control*, AC-14, (1) 97-96 (1969).

[6.16] Isermann, R., *Identification*, Vols. I and II, Hochschultaschenbucher, Nos. 515, 764a, Bibliographisches Institut, Zurich, 1971.

[6.17] Isermann, R., et al, "Comparison of Six On-line Identification and Parmater Estimation Methods with Three Simulated Processes," *Automatica*, 10, (1) 81-104 (1974).

[6.18] Jazwinski, A. H., *Stochastic Processes and Filtering Theory*, Academic Press, New York, 1970.

[6.19] Jazwinski, A. H., "Nonlinear and Adaptive Estimation in Reentry," AIAA Paper No. 72-874, Guidance and Control Conference, Stanford, Calif., August 1972.

[6.20] Kailath, T., "An Innovations Approach to Least Squares Estimation," Part 1; P. Frost, Part 11; *IEEE Trans. Automatic Control*, AC-13, (6) 646-660 (1968) - Geesy, R. A., Part IV, AC-16, (6) 720-727, (1971).

[6.21] Kashyap, R. L., "Maximum Likelihood Identification of Stochastic Linear Systems," *IEEE Trans. Automatic Control*, AC-15, (1) 25-34 1970.

[6.22] Kashyap, R. L., *Estimation of Parameters in a Partially Specific Stochastic System*, TR-EE 72-11, Technical Report, School of Electrical Engineering, Purdue University, Lafayette, Ind., Feb. 1972.

[6.23] Kashyap, R. L., and Blaydon, C. C., "Recovery of Functions From Noisy Measurements Taking Randomly Selected Points," *Proc. IEEE*, 54, 1127-1128 (1966).

[6.24] Kirvaitis, K., and Fu, K. S., "Identification of Nonlinear Systems by Stochastic Approximation," Preprints JACC 1966, p. 255.

[6.25] Kopp, R. E., and Orford, R. J., "Linear Regression to System Identification for Adaptive Systems Identification," *J. AIAA*, 1, (40) 2302-2306 (1962).

[6.26] Lee, R. C. K., *Optimal Estimation Identification and Control*, MIT Press, Cambridge, Mass., 1965.

[6.27]   Levin, M., "Estimation of a System Pulse Transfer Function
          in the Presence of Noise," IEEE *Automatic Control*, AC-9
          (3) 229 (1964).

[6.28]   Lindenlaub, J., and Cooper, G. R., "Noise Limitations of
          System Identification Techniques," IEEE *Trans. Automatic
          Control*, AC-8, (1) 43-48 (1963).

[6.29]   Ljung, L., "Convergence of Recursive Stochastic Algorithms,"
          Preprints IFAC Symposium on Stochastic Control, Budapest,
          1974.

[6.30]   Lobbia, R. N., and Saridis, G. N., "On-line Identification
          and Control of Multi-variable Stochastic Feedback Systems,"
          J. *Cybern.*, 3 (1) 40-59 (1973).

[6.31]   Mendel, J. M., *Discrete Techniques of Parametric Estimation*,
          M. Dekker, New York, 1973.

[6.32]   Mehra, R., "Identification of Linear Dynamic Systems with
          Applications to Kalman Filtering," IEEE *Trans. Automatic
          Control*, AC-16, (1) 12-21 (1971).

[6.33]   Mishkin, E., and Braun, Jr., L., *Adaptive Control System*,
          McGraw-Hill, New York, 1961.

[6.34]   Perlis, H., "The Minimization of Measurement Error in a
          General Perturbation Correlation Process Identification
          System," IEEE *Trans. Automatic Control*, AC-9 (4) 339-345
          (1964).

[6.35]   Ricker, D., and Saridis, G. N., "On-line Identification of
          a Class of Non-linear Systems from Noisy Measurements,"
          Automatica, 7, (4) 517-522 (1971).

[6.36]   Sage, A. P., and Melsa, J. L., *System Identification*,
          Academic Press, New York, 1972.

[6.37]   Sage, A. P., and Melsa, J. L., *Estimation Theory with
          Applications to Communications and Control*, McGraw-Hill,
          New York, 1971.

[6.38]   Sakrison, D., "A Continuous Kiefer-Walfowitz Procedure for
          Random Processes," *Ann. Math. Stat.*, 31, (2) 591 (1964).

[6.39]   Saridis, G. N., "On a Class of Performance-Adaptive Self-
          Organizing Control Systems," In *Pattern Recognition and
          Machine Learning*, (K. S. Fu, ed.), Plenum Press, New York,
          1971.

[6.40]   Saridis, G. N., "Comparison of Six On-Line Identification
          Algorithms," *Automatica*, 10, (1) 69-80 (1974).

[6.41] Saridis, G. N., "Learning Applied to Successive Approximation
        Algorithms," *IEEE Trans. Systems Sci. Cybern.*, <u>SSC-6</u>, (2)
        97-103 (1970).

[6.42] Saridis, G. N., "Expanding Subinterval Random Search for
        System Identification and Control," *Proc. IV IFAC Symp.
        Identification, etc.* Tbilisi GSSR, Sept 1976.

[6.43] Saridis, G. N., and Badavas, P., "Identifying Solutions of
        Distributed Systems by Stochastic Approximation," *IEEE Trans.
        Automatic Control*, <u>AC-15</u>, (4), 393-5 (1970).

[6.44] Saridis, G. N., and Hofstadter, R. N., "A Pattern Recognition
        Approach to the Classification of Nonlinear Systems," *IEEE
        Trans. Systems, Man, Cybern.*, <u>SMC-4</u>, (4), 362-371 (1974).

[6.45] Saridis, G. N., and Lobbia, R. N., "Parameter Identification
        and Control of Linear Discrete-Time Systems," *IEEE Trans.
        Automatic Control*, <u>AC-17</u>, (1) 52-61 (1972).

[6.46] Saridis, G. N., Nikolic, Z. J., and Fu, K. S., "Stochastic
        Approximation Algorithms for System Identification,
        Estimation and Decomposition of Mixtures", *IEEE Trans.
        Systems Sci. Cybern.*, <u>5</u>, (1) 8-15 (1969).

[6.47] Saridis, G. N., and Stein, G., "Stochastic Approximation
        Algorithms for Discrete-Time System Identification,"
        IEEE Trans. Automatic Control, <u>AC-13</u>, (5) 515-523 (1968).

[6.48] Saridis, G. N., and Stein, G., "A New Algorithm for Linear
        System Identification," *IEEE Trans. Automatic Control*,
        (5), 592-594 (1968).

[6.49] Saridis, G. N., and Okita, S., "The On-Line Identification
        of Chemical Kinetic Parameters," Purdue University Technical
        Report, PLAIC No. 44, Lafayette, Ind., Nov. 1971.

[6.50] Tou, J., Joseph, P., and Lewis, J. B., "Plant Identification
        in Presence of Disturbances and Applications to Digital
        Systems," *AIEE Appl. Ind.*, <u>2</u>, March 1961.

[6.51] Van Trees, H. L., *Synthesis of Nonlinear Control Systems*,
        MIT Press, Cambridge, Mass., 1962.

[6.52] Wiener, N., "Interpolation Extrapolation and Time Series,"
        MIT Press, Cambridge, Mass., 1967.

[6.53] Wong, K. Y., and Polak, E., "Identification of Linear
        Discrete-Time Systems Using the Instrumental Variable
        Method," *IEEE Trans. Automatic Control*, <u>AC-12</u>, (6) 707-718
        (1967).

[6.54]   Young, P. C., "An Instrumental Variable Method for Real-Time
         Identification of a Noisy Process," *Automatica*, <u>6</u>, (2)
         271-287 (1970).

[6.55]   Zaborsky, J., and Flake, R. H., "Outfitting with Functional
         Polynomials," Preprints JACC 1966, p. 554.

PART III

SELF-ORGANIZING CONTROL

Chapter 7

PARAMETER-ADAPTIVE SELF-ORGANIZING CONTROL

7.1   INTRODUCTION

After a thorough discussion on the stochastic optimal and dual
control formulations, their potential and limitations should be
obvious. They definitely do not present universal solutions to the
control problem. The stochastic optimal control may produce
solutions to problems where the plant is of relatively low
dimension and can be accurately approximated by a linear dynamic
model while the process and measurement noises involved can be
approximated by white Gaussian processes as a result of the *central
limit theorem*   The dual control is applicable to problems of
higher sophistication where uncertainties inherent in the model
are reduced by some "learning" procedure in the controller, but
must be limited by the dimensionality of the plant due to the
complexity of the dual control formulation. Any further questions
regarding the validity of these formulations should arise from the
applications point of view. It should be remembered, however, that
"one should use the right size screwdriver for the right size
screw", in order to have effective control. Therefore the need for
a dual control solution should be dictated by processes requiring
the learning property of the dual formulation, the same way that
remote controlled systems require the inclusion of random noises in
their model, since deterministic sensitivity methods could not
eliminate their effect.

As stated in Chap. 1, modern general system theory, with its
analytic formulations of problems in areas such as bioengineering,
robotics, socioeconomics, environmental interactions with technology,
health care, etc., has produced the incentive and the need for

233

various levels of *intelligent control* solutions as in dual control.
Such a demand may arise from uncertain models of a societal system
to the highly intelligent decision making of a prosthetic device
or a robot.   However, dual control, on one hand, has been too
complex and too rigid to be used even with one of the most
sophisticated modern computers; on the other hand, it does not
always possess the generality in formulation to tackle all these
problems.

Self-organizing control has been defined to account for both
these problems. By relaxing the optimality requirement,
parameter-adaptive S.O.C., simplifies the dual control formulation
to yield implementable solutions to modern general systems that do
not require transient optimality. Such solutions may still yield
asymptotic optimality or other approximations of an optimal
solution, while they reduce the uncertainties of the plant as the
process evolves.

Performance-adaptive S.O.C. utilizes the "learning" property
of dual control in a more generic form to produce asymptotically
optimal solutions to systems with more general uncertainties which
will be discussed in the next chapter.

7.2   HISTORICAL REVIEW

The parameter-adaptive S.O.C. systems to be treated in this
chapter may be represented in general by the block diagram of
Fig. 7.1. In such cases the structure of the plant is assumed
known *a priori* in most cases linear, and the reducible
uncertainties compose a vector θ of parameters and noise statistics.
An identification procedure is used to "learn" sequentially the
unknown vector $\theta \triangleq [\theta_1,\ldots,\theta_s]^T$, through its estimate $\hat{\theta}(k)$ at time
k or its whole information state [7.3]:

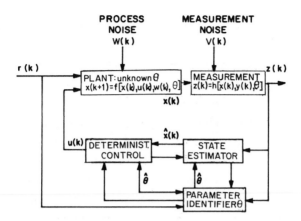

FIG. 7.1.   Block Diagram of a General Parameter Adaptive S.O.C.
System.

$$q_i(k) \triangleq \text{Prob}[\theta = \theta_i; k] \quad k = 1, \ldots, s \quad \sum_{i=1}^{s} q_i(k) = 1 \qquad (7.1)$$

Such identification algorithms may be borrowed from Chap. 5 and
must normally satisfy a Markov-Gauss type relation:

$$\hat{\theta}(k + 1) = A_1(k)\,\hat{\theta}(k) + \xi(k) \qquad (7.2)$$

or for the information states, Bayes' rule

$$q_i(k + 1) = \frac{p(z(k)/\theta_i)q_i(k)}{\sum_{i=1}^{s} p(z(k)/\theta_i)q_i(k)} \quad i = 1, \ldots, s \qquad (7.3)$$

In general $A_1(k)$, a matrix and $\xi(k)$ a noise vector of
appropriate dimensions are random variables depending on
$z^k = \{z(0), \ldots, z(k)\}^T$ and $U^k = \{u(0), \ldots, u(k)\}^T$, the output and
input past sequences, establishing the correlation of $\hat{\theta}$ as seen
in Chap. 5.   In certain instances, however, the system parameters
are assumed to be random as a result of the interpretation of
uncertainty and are defined to obey a Markov-Gauss relationship

$$\theta(k + 1) = A_2(k)\theta(k) + \eta(k) \quad \theta(0) = \theta_0 \qquad (7.4)$$

where $A_2(k)$ is a known matrix, $A_2(k) = I$ for time invariant
parameters, and $\eta(k)$ and θ_0 are white Gaussian random variables,
with $\eta(k) \equiv 0$ representing the fixed parameter case. In that case,
the parameter identifier is defined by the appropriate estimator θ.

  The information produced by the identifier is used to update
a deterministic controller, which may or may or may not be
asymptotically $(\theta(k) \rightarrow \theta)$ optimal and a state estimator, possibly
asymptotically optimal, when the states of the plant are not
available for direct measurement. The various approximations to
the optimal solution presented here are obtained through relaxation
of certain couplings among the various functions of the controller
which could constitute the dual control, and therefore simplify
its derivation.

  Parameter-adaptive control for systems operating in a
stochastic environment have been developed under various forms
since about 1960. Bellman [7.3] was one of the first to present
the basic ideas, while Mishkin and Braun [7.21] also discussed it
along with the deterministic adaptive algorithms of the first
stochastic equivalent. Some of the first parameter-adaptive
S.O.C.'s were produced as an extension of the formulation by
Kopp and Orford [7.16] and Cox's algorithms [7.6]. Jenkins and Roy's
algorithm [7.14], which will be discussed here, represents the
attempt to reduce the S.O.C. problem to an approximate stochastic
optimal control problem. Farison, Graham, and Shelton [7.10],
extending the work of Florentin [7.12], used an open-loop feedback
optimal (OLFO) approach as a parameter-adaptive S.O.C. solution.
Tse and Athans [7.15] extended this work, while Lainiotis and
while Lainiotis and his colleagues [7.8], [7.17], Spang [7.29], Gorman
and Zaborsky [7.13], and Bar-Shalom and Sivan [7.4] have used the
information states approach for parameter identification resulting
also in an OLFO algorithm. Aoki [7.1] and Sworder [7.32] follow a
similar procedure for parameter-adaptive S.O.C. algorithms that
effectively utilize the information state in an approximation of
the dual control algorithm. Stein and Saridis [7.30] and Saridis

and Dao [7.23] developed algorithms which minimize bounds on the
optimal solution producing also a measure of the approximation
performed. Saridis and Lobbia [7.19], [7.24] , developed algorithms
which utilize stochastic approximations algorithms for parameter
identification in parallel with state estimations and deterministic
control, after an idea suggested by Lee [7.18]. Finally, Tse,
Bar-Shalom, and Meier [7.34], [7.35] developed a parameter-adaptive
S.O.C. algorithm as a direct approximation of the dual control
formulation.

A selection of algorithms which are believed to be the most
representative from the point of view of the definitions given
in Chap. 1 are discussed in the sequel. Other algorithms are not
included in the detailed presentation or are not even mentioned
here because of space limitations. No deterministic parameter-
adaptive algorithms are discussed here because they were already
covered in Chap. 2.

The presentations are given here for discrete-time systems
only. Even though continuous-time parameter-adaptive S.O.C.
algorithm can be immediately produced, due to the complexity of the
algorithm, a digital computer is absolutely necessary for its implementation, rendering a piece-wise constant input and therefore a
discrete-time system representation as the most important one.

7.3 LINEARIZED STOCHASTIC OPTIMAL ALGORITHMS

The philosophy of this approach is to reduce the optimal
stochastic control problem with uncertain parameters to an
approximate stochastic optimal control problem to which the
separation theorem can be applied. It is based on an idea
originated by Kalman and developed by Kopp and Orford [7.16]
Cox [7.6], according to which the unknown parameters are adjoined to
the state vector as additional states augmenting the plant by the
appropriate difference equations. The proper linearization of the
augmented system will render it amenable to the separate optimal
control and state estimation methods described in Chap. 4.

Specifically, consider the nonlinear discrete-time system

$$x(k + 1) = f(x(k),u(k),\theta,k) + g(\theta,x(k))w(k) \quad x(0) = x_0$$
$$z(k) = h(\theta)x(k) + v(k) \tag{7.5}$$

where $x(k)$ is the n-dimensional state vector, $u(k)$ the m-dimensional control input, $z(k)$ the r-dimensional output, θ is an s-dimensional unknown parameter vector, f, g, and h are appropriate functions of their arguments, and $w(k)$ and $v(k)$ appropriate zero-mean i.i.d. Gaussian random variables with covariances

$$E\{w(k)w^T(j)\} = Q(k)\,\delta_{kj} \quad E\{v(k)v^T(j)\} = R(k)\,\delta_{kj}$$

The initial state x_0 is also a random variable with mean $\bar{x}_0$ and covariance P_0. Define the parameters θ as additional state variables which, in case they are constant, satisfy the difference equations, representing somehow the parameter estimator algorithms

$$\theta(k+1) = \theta(k) \quad \theta(0) = \theta_0 \tag{7.6a}$$

or in case they are time varying and their change with time is known

$$\theta(k + 1) = K(k)\,\theta(k) + c(k) \quad \theta(0) = \theta_0 \tag{7.6b}$$

where $K(k)$ and $c(k)$ are known, and θ_0 is a Gaussian random variable with some $p(\theta_0)$, representing the a priori knowledge of θ. Equation (7.6a) or (7.6b) combined with (7.4) is a function of a new (n + s)-dimensional state vector $X(k)$

$$X^T(k) \underset{=}{\Delta} \{x^T(k),\theta^T(k)\} \quad X^T(0) = \{x_0^T,\theta_0^T\} \tag{7.7}$$

The resulting state equations are

$$X(k + 1) = f'(X(k),u(k),k) + g'(X(k))w(k) \quad X(0) = X_0$$
$$z(k) = h'(X(k)) + v(k) \tag{7.8}$$

Note that in case that (7.6b) is used and $c(k)$ is defined as a

Gaussian random variable representing the uncertainty of the
parameter estimation algorithm, $w(k)$ should be replaced in (7.8) by
$w'^T(k) = \{w^T(k), c^T(k)\}$ with proper adjustments in g'. A
performance criterion is defined as a criterion for minimization

$$J(u) = E \frac{1}{2} \sum_{i=1}^{N+1} x^T(i + 1)M(i)x(i + 1) + u^T(i)N(i)u(i)$$

$$(7.9)$$

Linearization of (7.8) in order to apply the separation
theorem results in awkward expressions very difficult to evaluate.
The reason is that the optimal linear controller has to be
recomputed for every estimate of the states produced after a new
measurement. This was encountered in Sec. 3.5 as the
nonreproducibility property of the estimation algorithm.

In order to overcome this difficulty, a one-step updating of
the controller is performed as proposed by Jenkins and Roy [7.14].
A linearized Kalman filter (Sec. 3.5.1) and an extended Kalman
filter (Sec. 3.5.2) are used separately in the algorithm, and a
comparison of the results is presented in the sequel.

7.3.1 The Jenkins-Roy Linearization Algorithm

This version of the algorithm [7.14] assumes linearization of
the process (7.8) about a nominal trajectory defined in this case
by

$$X_0(k + 1) = f(X_0(k), u(X_0(k)), k) \quad X_0(0) = \bar{X}_0$$
$$z_0(k) = h'(X_0(k), k)$$

$$(7.10)$$

and $u(X_0(k))$ will be produced the same way as the $u^*(k)$ in the
sequel for $X(k)$ replaced by $X_0(k)$.

Assuming then that

$$\delta X(k) \triangleq X(k) - X_0(k)$$

$$(7.11)$$

$$\delta z(k) \triangleq z(k) - z_0(k)$$

the following linear incremental model for the system is possible:

$$\delta X(k + 1) = F(k)\delta X(k) + B(k)u(k) + \Gamma(k)w(k)$$

$$\delta z(k) = C(k)\delta X(k) + v(k) \tag{7.12}$$

where the matrices are precomputed off-line and stored from

$$F(k) = \frac{\partial f'(X_0(k),u(X_0),k)}{\partial X_0}$$

$$B(k) \triangleq \frac{\partial f'(X_0,u(X_0),k)}{\partial u(X_0)} \tag{7.13}$$

$$\Gamma(k) \triangleq g'(X_0(k),k)$$

$$C(k) \triangleq \frac{\partial h'(X_0(k),k)}{\partial X_0}$$

For (7.12) the linearized Kalman filter is given in Sec. (3.5.1) and modified, as in Sec. 4.5.4, to include the control function by

$$\delta\hat{X}(k + 1)/k + 1) = F(k)\delta\hat{X}(k/k) + P(k + 1/k + 1)$$
$$C^T(k+1)R^{-1}(k + 1)[\delta z(k + 1)$$
$$- C(k + 1)F(k)\delta X(k/k)]$$
$$+ P(k + 1/k + 1)P^{-1}(k + 1/k)B(k)u^*(k)$$
$$\delta X(0/0) = 0$$

$$P(k + 1/k + 1) = P(k + 1/k) - P(k + 1/k)C^T(k + 1)[C(k + 1)$$
$$P(k + 1/k)C^T(k + 1) + R(k + 1)]^{-1}C^T(k + 1)$$
$$P(k + 1/k) \quad P(0/0) = P_0 \tag{7.14}$$

$$P(k + 1/k) = F(k)P(k/k)F^T(k) + \Gamma(k)Q(k)\Gamma^T(k)$$

and the linearized estimate is given by

$$\hat{X}(k + 1/k + 1) = X_0(k) + \delta\hat{X}(k + 1/k + 1) \tag{7.15}$$

$$k = 0,\dots,N-1$$

The optimal control $u^*(k)$ for the linearized combined systems
(7.11) and (7.12) is assumed to be given by

$$u^*(k) = -A_{N-k}^T \hat{X}(k/k)$$

$$A_{N-k}^T \triangleq [N(k) + B^T(k)(M(k + 1) + S(k + 1))B(k)]^{-1}$$

$$B^T(k)(M(k + 1) + S(k + 1))F(k) \qquad (7.16)$$

as by eq. (4.46) where $S(k + 1)$ is obtained to satisfy the
following one-step equation:

$$S_0(k) = (\hat{F}(k) - \hat{B}(k)\hat{A}_{N-k}^T)^T [M(k + 1) + S(k + 1)]$$

$$(\hat{F}(k) - \hat{B}(k)\hat{A}_{N-k}^T) + \hat{A}_{N-k}^T N(k)\hat{A}_{N-k} \quad S(N) = 0 \qquad (7.17)$$

$$\hat{F}(k) = [F(k)]_{X_0 = \hat{X}(k/k)} \, , \, \hat{B}(k) = [B(k)]_{X_0 = \hat{X}(k/k)}$$

$$\hat{A}_{N-k} = [A_{N-k}]_{X_0 = \hat{X}(k/k)}$$

In other words, $S(k + 1)$ is computed by an equation for which
all the coefficient matrices $F(k)$, $B(k)$, and A_{N-k} have replaced the
nominal values X_0 by the best estimates $\hat{X}(k/k)$, and are recomputed
on-line. The quantity $S_0(k)$ is precomputed off-line and stored
from the equation

$$S_0(k) = (F(k) - B(k) A_{N-k}^T)^T [M(k + 1) + S(k + 1)] \qquad (7.18)$$

$$\cdot (F(k) - B(k)A_{N-k}^T) + A_{N-k}^T N(k)A_{N-k} \quad S_0(N) = 0$$

The above approximation of the optimal control law simplifies the
computation considerably. However, because of the inaccuracies of
estimation of the linearized Kalman filter and the poor
approximations of the controller $u^*(k)$, due to precomputed values,
the total approximation is rather poor and the overall system
presents stability problems.

7.3.2 Linearization with an Extended Kalman Filter

The disadvantages of the previous method may be partially corrected if the extended Kalman filter, presented in Sec. 3.5.2, is used for the augmented state estimation of (7.10). In such a case the linearization is performed about the previous estimate $\hat{X}(k/k)$ and the state estimation equations are

$$\hat{X}(k + 1/k + 1) = f'(\hat{X}(k/k),u^*(k),k) + \hat{P}(k + 1/k + 1)$$
$$\cdot \hat{c}^T(k + 1)R^{-1}(k + 1)[z(k + 1) - \hat{c}(k + 1)$$
$$\cdot f(\hat{X}(k/k)\hat{u}^*(k),k)]$$
$$\hat{P}(k + 1/k + 1) = R(k) + \hat{c}^T(k + 1)\hat{P}(k + 1/k)\hat{c}(k)^{-1}\hat{P}(k + 1/k)$$
$$\hat{P}(k + 1/k) = \hat{F}(k)\hat{P}(k/k)\hat{F}^T(k) + \hat{\Gamma}(k)Q(k)\hat{\Gamma}^T(k) \qquad (7.19)$$

where the matrices $\hat{F}(k)$, $\hat{B}(k)$, $\hat{T}(k)$, and $\hat{c}(k)$ are functions of the current estimates as in (7.17)

$$\hat{F}(k) \triangleq \frac{\partial f'(\hat{X}(k),\hat{u}^*(k),k)}{\partial \hat{X}}$$

$$\hat{B}(k) \triangleq \frac{\partial f'(\hat{X}(k),\hat{u}^*(k),k)}{\partial u^*}$$

$$\hat{\Gamma}(k) \triangleq g'(\hat{X}(k),k) \qquad (7.20)$$

$$\hat{c}(k) \triangleq \frac{\partial h'(\hat{X}(k),k)}{\partial \hat{X}}$$

and are estimated on-line.

The control law is computed in a similar way as in the Jenkins and Roy linearization but with the revised coefficients

$$\hat{u}^*(k) = -\hat{A}^T_{N-k}\hat{x}(k/k)$$

$$\hat{A}^T_{N-k} \triangleq [N(k) + \hat{B}^T(k)(M(k + 1) + S(k + 1))\hat{B}(k))^{-1}$$

$$\hat{B}^T(k)(M(k + 1) + S(k + 1)] \hat{F}(k)$$

$$S_0(k) = [\hat{F}(k) - \hat{B}(k)\hat{A}^T_{N-k}]^T(M(k + 1) + S(k + 1))$$

$$[\hat{F}(k) - \hat{B}(k)\hat{A}_{N-k}] + \hat{A}_{N-k}N(k)\hat{A}_{N-k} \quad S(N) = 0 \qquad (7.21)$$

where $S_0(k)$ has been computed off-line from (7.18) as in the previous section and $S(k + 1)$ presents only a one-step updated approximation.

This algorithm presents several improvements in convergence and stability over the preceding one. Further improvement may be obtained by using a several-step updating on $S(k + 1)$. However, such improvements are obtained at the sacrifice of computational complexity and time.

7.3.3 Comparison of the Two Linearization Algorithms

Jenkins and Roy developed their algorithm to obtain the pitch control of a large space vehicle.

The two algorithms presented here have been tested on the following second-order system with unknown coefficients

$$x(k + 1) = \begin{bmatrix} 0 & 1 \\ \theta_1 & \theta_2 \end{bmatrix} x(k) + \begin{bmatrix} 0 \\ \theta_3 \end{bmatrix} (u(k) + (w(k)) \qquad (7.22)$$

$$z(k) = [1 \ 0]x(k) + v(k) \qquad\qquad k = 0, 1,\dots$$

where

$$x_0 \sim N\left[\begin{pmatrix} 1 \\ 0 \end{pmatrix} ; \begin{pmatrix} 9 & 0 \\ 0 & 4 \end{pmatrix}\right]$$

$$w(k) \sim N[0,0.1] \qquad\qquad k = 0, 1,\dots$$

$$v(k) \sim N[0,0.1]$$

and the true parameter values, not known to the control, are

$$\theta_1 = -0.974 \quad \theta_2 = 1.874 \quad \theta_3 = 0.01$$

The control problem was considered for the infinite time case
$(N \to \infty)$, for which, as in Sec. 4.5.5 the performance index was
defined by

$$J(u) = \lim_{N \to \infty} \frac{0.1}{N} E \left\{ \sum_{k=1}^{N} [(z^d - z(k))^2 + 0.01u^2(k)] \right\} \qquad (7.23)$$

with z^d the desired output is assumed to be a step function at
$k = 0$.   Both algorithms presented in Secs. 7.3.1 and 7.3.2 were
run for several initial conditions X_o.   The extended Kalman filter
was superior to the linearized Kalman filter in all cases, the
latter demonstrating instability in many cases.

The normalized identification error averaged over ten runs
were plotted in Fig. 7.2 for both filters, with and without the
control, while the average performance criterion over ten runs
was plotted in Fig. 7.3.   The conclusions are obvious.

7.4   THE OPEN-LOOP FEEDBACK OPTIMAL ALGORITHMS (OLFO)

The next group of parameter-adaptive S.O.C. algorithms to be
examined are the ones obtained through the open-loop feedback
optimal concept which was discussed in Sec. 4.4 as one of the
passive stochastic control modes.   This principle can be described
as follows:

Considering a plant of the type of (7.5) with an unknown
parameter vector θ, and given at time k a measurement sequence
$z^{k^T} = \{z(0),\ldots,z(k)\}$ and an already applied control sequence
$u^{k^T} = \{u(0),\ldots,u(k)\}$, one is searching for a control sequence
$u_{k+1}^{N-1^T} = \{u(k + 1),\ldots,u(N - 1)\}$ that will minimize the average
performance to go

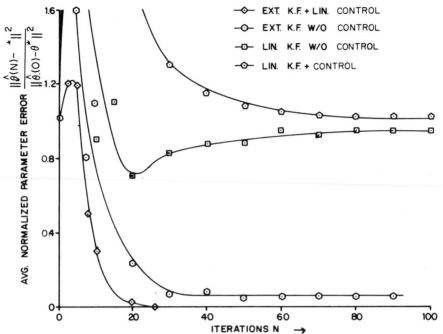

Fig. 7.2.   Comparison of Normalized Errors in Parameter Estimation
for Linearized Stochastic Optimal Algorithms.

$$J(U_{k+1}^{N-1}, \hat{\theta}(k)) \triangleq E\left\{\frac{1}{2} \sum_{j=k+1}^{N-1} x^T(j+1)M(j)x(j+1) + u^T(j)N(j)u(j) \;/U^k,\right.$$

$$\left. z^k, \hat{\theta}(k)\right\} \tag{7.24}$$

Out of the sequence U_{k+1}^{N-1} which is assumed deterministic and
yields open-loop controls, since the average performance to go
(7.24) is minimized, one selects the first one $u^*(k+1)$, measures
$z(k+1)$, updates $\hat{\theta}(k+1)$, and repeats the process using
$J(U_{k+2}^{N-1}, \hat{\theta}(k+1))$, etc.

Various algorithms have been developed that in some way
satisfy this principle as an approximation to the dual optimal
solution, even though originally they were designed on some
other principle. The three algorithms presented here are the most
typical OLFO-type parameter-adaptive algorithms from a large
collection available in the literature [7.1],[7.4],[7.7],[7.12],
[7.14],[7.29].

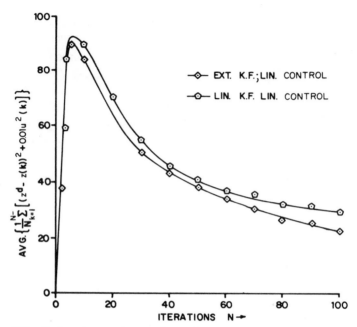

FIG. 7.3.   Comparison of Performance Index for the Linearized
Stochastic Optimal Algorithms.

7.4.1   Farison's Identification-Control Algorithm

Farison, Graham, and Shelton [7.10] considered an OLFO algorithm
for linear discrete-time systems with exactly measurable states.
Their method was an extension of Florentin's earlier pioneering
work [7.12] which is not presented here because of its rather
specialized nature.

Consider a plant described by the following difference
equation:,

$$x(k + 1) = F(\theta,k)x(k) + B(\theta,k)u(k) + G(k)w(k) \tag{7.25}$$

where $x(k)$ is the n-dimensional state vector available for direct
measurement, and $F(\theta,k)$ and $B(\theta,k)$ are linear functions of the
unknown parameter vector θ of dimension s. Assume that the
parameters are evolving with time as correlated random variables,

according to (7.4)

$$\theta(k + 1) = A(k)\theta(k) + \eta(k) \quad \theta(0) = \theta_0 \tag{7.26}$$

where the known matrix $A(k)$ reflects the correlation of the
parameter sequence, $\eta(k)$ is a random vector producing independent
variations in the parameter sequence, and θ_0 a random vector
representing the initial uncertainty regarding the knowledge
of the parameters. If, as a special case, the parameters are
deterministic but initially unknown $A(k) \equiv I$ and $\eta(k) \equiv 0$.

Assuming that $w(k)$ and $\eta(k)$ are zero-mean Gaussian uncorrelated
random variables with positive definite covariances $Q(k)$ and $R(k)$,
respectively, then Eq. (7.25) can be rewritten as a function of the
parameters

$$x(k + 1) = F(k,\theta)x(k) + B(k,\theta)u(k) + G(k)w(k) =$$

$$= D(k)\theta(k) + G(k)w(k) \tag{7.27}$$

where $D(k)$ is a function of $x(k)$ and $u(k)$. Assuming that $\hat{\theta}(k/k)$ is
the estimate of θ at k when X^k is given, one may obtain the
one-step estimates of x and θ and their covariances, which are
also Gaussian random variables, via normal estimation procedures.

$$\hat{x}(k + 1/k) = \hat{F}(k/k,\hat{\theta})x(k) + \hat{B}(k/k,\hat{\theta})u(k)$$

$$\hat{\theta}(k + 1/k) = A(k)\hat{\theta}(k/k)$$

$$P(k + 1/k) = D^T(k)\Gamma(k/k)D(k) + G(k)Q(k)G^T(k) \tag{7.28}$$

$$\Gamma(k + 1/k) = A(k)\Gamma(k/k)A^T(k) + R(k)$$

where $\hat{F}(k/k,\hat{\theta}) = F(k,\theta)\big|_{\theta=\hat{\theta}(k)}, \hat{B}(k/k,\hat{\theta}) = B(k,\theta)\big|_{\theta=\hat{\theta}(k)}$.

$$x(k + 1) - \hat{x}(k + 1/k) = D(k)[\theta(k) - \hat{\theta}(k)] + G(k)w(k) \tag{7.29}$$

The recursive state estimator for the parameters and the error
covariance $\Gamma(k + 1/k + 1)$ are defined by

$$\hat{\theta}(k + 1/k + 1) = \hat{\theta}(k + 1/k) + K(k + 1)P^{-1}(k + 1/k)[x(k + 1)$$
$$- \hat{x}(k + 1/k)]$$

$$K(k + 1) = A(k)\Gamma(k/k)D^T(k) \qquad\qquad\qquad (7.30)$$

$$\Gamma(k + 1/k + 1) = \Gamma(k + 1/k) - K(k + 1)P^{-1}(k + 1/k)K^T(k + 1)$$

Future values of $\hat{\theta}(j/k), \Gamma(j/k)$, $j = k + 1, \ldots, N$ are obtained by the $(j - k)$ predictor

$$\hat{\theta}(j/k) = \prod_{i=k}^{j} A(i)\hat{\theta}(k/k) \qquad\qquad\qquad (7.31)$$

$$\Gamma(j/k) = \left[\prod_{i=k}^{j} A(i) \right] \Gamma(k/k) \left[\prod_{i=k}^{j} A(i) \right]^T + \sum_{i=k}^{j-1} \left[\prod_{\ell=i+1}^{j} A(\ell) \right]$$

$$R(k) \left[\prod_{\ell=i+1}^{j} A(\ell) \right]^T$$

Since $\hat{\theta}(j/k)$ are Gaussian random variables, the conditional density functions of $\theta(j)$ are completely defined from (7.31).

$$p(\theta(j)/k) \sim N[\hat{\theta}(j/k), \Gamma(j/k)] \qquad\qquad\qquad (7.32)$$

The control algorithm may be defined now in a typical OLFO manner to minimize the average performance to go, as follows:

$$J(U_k^{N-1}) = E\left\{ \frac{1}{2} \sum_{i=k}^{N-1} x^T(i + 1)M(i)x(i + 1) \right.$$
$$\left. + u^T(i)N(i)u(i) \; |z^k, u^k \right\} \qquad\qquad (7.33)$$

where $M(i) = M^T(i) \geq 0$ and $N(i) = N^T(i) > 0$. Using the dynamic programming approach [7.3] to find the sequence U_{k+1}^{N-1} that minimizes (7.33), subject to (7.27), one obtains:

$$u^*(j/k) = -C(j/k)x(j) \quad j = k + 1, \ldots, N - 1 \tag{7.34}$$

$$C(j/k) = [E\{ B^T(j,\theta)S(j + 1/k)B(j,\theta) + R(j)\}]^{-1}$$

$$E\{ B^T(j)S(j + 1/k)F(j,\theta)\}$$

$$S(j/k) = E\{ [F(j,\theta) - B(j,\theta)C(j/k)]^T(M(j) + S(j + 1/k))$$

$$[F(j,k) - B(j,\theta) \ C(j/k)] + C(j/k)N(j)C^T(j/k)\}$$

where the expectations are taken with respect to θ to account for θ being a random variable. In view of (7.32) such an expectation is defined as

$$E_\theta\{y(k,\theta(j))\} \triangleq \int_{\Omega_\theta} y(j,\theta)p(\theta(j)/k) \ d\theta(j) \tag{7.35}$$

In the OLFO manner, from (7.34) one selects $u^*(k + 1)$, measures $x(k + 1)$, updates $p(\theta(j)/_X k + 1)$ and $J(U_{k+1}^{N-1})$, and repeats the process.

The solution considered here is limited to exactly measurable states and therefore represents only a special case of the problem. The next algorithm represents extensions of the work by Farison. An example of the estimation algorithm is given here for a third-order system with

$$F = \begin{bmatrix} 0 & 1 & 0 \\ 0 & 0 & 1 \\ \theta_1 & \theta_2 & \theta_3 \end{bmatrix} \quad B = \begin{bmatrix} 0 \\ 0 \\ 1 \end{bmatrix} \quad Q = 10^{-4} \begin{bmatrix} 3 & 1 & 0 & 0 \\ 0 & 1 & 2 & 0 \\ 0 & 0 & 2 & 3 \end{bmatrix}$$

and the true values $\theta_1 = 0.044$, $\theta_2 = 0.32$, and $\theta_3 = 0.55$. The control part, being straightforward, has been omitted for simplicity of the computer simulation, and the input was assumed $u(k) \triangleq (-0.40)^k$. Initial conditions were

$$x(0) = \begin{bmatrix} 0 \\ 1 \\ 0 \end{bmatrix} \quad \theta(0) = \begin{bmatrix} -0.06 \\ 0 \\ 0 \end{bmatrix} \quad \Gamma(0) = \begin{bmatrix} 0.013 & 0 & 0 \\ 0 & 0.33 & 0 \\ 0 & 0 & 0.33 \end{bmatrix}$$

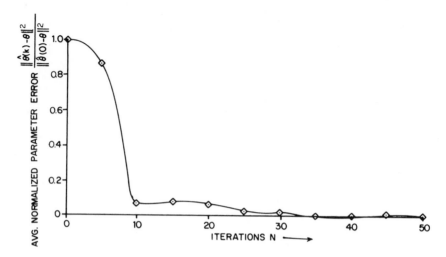

FIG. 7.4.   Normalized Parameter Identification Error For
             Farison's Method.

The normalized estimation error is plotted in Fig. 7.4.   The
results obviously show that the identification algorithm converges
extremely fast, mainly because of the lack of measurement error.

7.4.2   The Tse and Athans Method

Another application of the OLFO principle was made by Tse and
Athans [7.33].   This actually is an extension of Farison's work
discussed in the preceding chapter, covering plants with states
accessible for measurement through a measurement equation and
only their gain vector unknown.   The derivation of the OLFO control
in this case, is quite informative about the explanation of the
function of the controller and its influence to the overall
stability of the system.

Consider the discrete-time linear system with unknown gains
of a scalar controller

$$x(k + 1) = F(k)x(k) + b(k,\theta)u(k) + g(k)w(k) \quad x(0) = x_0$$

$$z(k) = H(k)x(k) + v(k) \tag{7.36}$$

where $x(k)$, $u(k)$, $w(k)$, and $v(k)$ are defined as previously, $F(k)$, $G(k)$, and $H(k)$ are known matrices, and the gain vector is defined as a function of the unknown parameters

$$b(k,\theta) = \begin{bmatrix} \theta_1(k) \\ \vdots \\ \theta_n(k) \end{bmatrix} \qquad (7.37)$$

which satisfy the evolution Eq. (7.4) with the coefficients of appropriate dimensions

$$\theta(k + 1) = A(k)\theta(k) + n(k) \qquad (7.38)$$

The identification algorithm results from the estimation (7.38). The control sequence U^{N-1} is sought to minimize

$$J(u) = \frac{1}{2} E \left\{ \sum_{i=0}^{N-1} x^T(i + 1)M(i)x(i + 1) + n(i)u^2(i) \right\} \qquad (7.39)$$

In order to use the OLFO principle, (7.38) is adjoined to (7.36), to yield the following linear system:

$$\begin{bmatrix} x(i + 1) \\ \hline \theta(i + 1) \end{bmatrix} = \tilde{F}(i,u^*(i)) \begin{bmatrix} x(i) \\ \hline \theta(i) \end{bmatrix} + \tilde{G}(i) \begin{bmatrix} w(i) \\ \hline n(i) \end{bmatrix} \qquad (7.40)$$

$$z(i) = \tilde{H}(i) \begin{bmatrix} x(i) \\ \hline \theta(i) \end{bmatrix} + v(i) \quad i = 0, 1, \ldots, k - 1$$

where

$$\tilde{F}(i,u^*(i)) \triangleq \begin{bmatrix} F(i) & u^*(i)I \\ \hline 0 & A(i) \end{bmatrix} \quad \tilde{G}(i) \triangleq \begin{bmatrix} G(i) & 0 \\ \hline 0 & I \end{bmatrix}$$

$$\tilde{H}(k) \triangleq \begin{bmatrix} H(i) & 0 \\ \hline 0 & I \end{bmatrix}$$

This system is linear with known coefficients, because the sequence U^{*k-1} of optimal controls is deterministic and known as required.

Then a linear optimal estimator may be used to estimate the
states and identify the parameters for the rest of the process
$k \leq j \leq N-1$

$$\hat{x}(j + 1/k) = F(j)\hat{x}(j/k) + \hat{\theta}(j/k)u(j); \; \hat{x}(k/k) \qquad (7.41)$$

$$\hat{\theta}(j + 1/k) = A(j)\hat{\theta}(j/k); \; \hat{\theta}(k/k)$$

$$P(j + 1/k) = \begin{bmatrix} P_{xx}(j + 1/\hat{k}) & P_{\theta x}(j + 1/k) \\ \hline P_{x\theta}(j + 1/k) & P_{\theta\theta}(j + 1/k) \end{bmatrix}$$

$$= \tilde{F}(j,u(j))P(j/k)\tilde{F}^T(j,u(j)) + \tilde{G}(j)\begin{bmatrix} Q(j) & 0 \\ \hline 0 & \Gamma(j) \end{bmatrix}\tilde{G}^T(j)$$

where $P(j + 1/k))$ is the error-covariance matrix for the augmented
system with components defined by

$$P_{xx}(j + 1/k) \triangleq E\{ [\hat{x}(j + 1/k) - x(j + 1)][\hat{x}(j + 1/k)$$
$$- x(j + 1)]^T /z^k, u^{*k-1} \}$$

$$P_{x\theta}(j + 1/k) \triangleq E\{ [\hat{x}(j + 1/k)] - x(j + 1)] [\hat{\theta}(j + 1/k)$$
$$- \theta(j + 1)]^T/z^k, u^{*k-1} \}$$

$$P_{\theta x}(j + 1/k) \triangleq E\{ [\hat{\theta}(j + 1/k) - \theta(j + 1)][\hat{x}(j + 1/k)$$
$$- x(j + 1)]^T/z^k, u^{*k-1} \}$$

$$P_{\theta\theta}(j + 1/k) \triangleq E\{ [\hat{\theta}(j + 1/k) - \theta(j + 1)][\hat{\theta}(j + 1/k)$$
$$- \theta(j + 1)]^T/z^k, u^{*k-1} \}$$

Once the optimal linear estimator is obtained for $k \leq j \leq N-1$
where k is the present instance of time, one may proceed to obtain
the open-loop control in the OLFO fashions by obtaining the
average performance to go:

$$J(U_k^{N-1};U^{*k-1}) \triangleq \frac{1}{2} E\left\{ \sum_{j=k}^{N-1} [x^T(j + 1)M(j)x(j + 1) \right.$$

$$\left. + n(j)u^2(j)]/z^k,U^{*k-1}\right\}$$

$$= \frac{1}{2} \sum_{j=k}^{N-1} [\hat{x}^T(j + 1/k)M(j)\hat{x}(j + 1/k) + tr[\tilde{Q}(j + 1)$$

$$P(j + 1/k)] + n(j)u^2(j)\} \tag{7.42}$$

where

$$\tilde{Q}(j + 1) = \begin{bmatrix} Q(j + 1) & | & 0 \\ \text{---------} & | & \text{---} \\ 0 & | & 0 \end{bmatrix}$$

The optimal open-loop control sequence

$$U_k^{*T_{N-1}} \triangleq \{ u_{OL}^*(k/k),\ldots,u_{OL}^*(N - 1/k)\}$$ which minimizes (7.42) is obtained in the usual way by

$$u_{OL}^*(j/k) = -[[\tilde{n}(j/k) + \tilde{\theta}(j^T/k)\tilde{K}(j + 1/k)\tilde{\theta}(j/k)]^{-1} \tag{7.43}$$
$$\hat{\theta}^T(j/k)\tilde{K}(j + 1/k)T(j/k) + \tilde{n}(j/k)d^T(j + 1)]\begin{bmatrix} \tilde{x}(j/k) \\ \text{----} \\ \sigma(j/k) \end{bmatrix}$$

where

$$\tilde{K}(j/k) = T^T(j/k) [\tilde{K}(j + 1/k) - \tilde{K}(j + 1/k)\tilde{\theta}(j/k) [\tilde{n}(j/k)$$
$$+ \tilde{\theta}^T(j/k)\tilde{K}(j + 1/k)\tilde{\theta}(j/k)]^{-1}\tilde{\theta}(j/k)K(j + 1/k)]$$
$$\cdot T(j/k) + \tilde{D}(j/k) \quad j = k+1,\ldots,N-1$$

$$K(N/k) = 0 \tag{7.44}$$

and

$$T(j/k) \triangleq \begin{bmatrix} F(j) & 0 & \cdots & 0 \\ 0 & F(j)a_{11} & \cdots & F(j)a_{n1} \\ \cdots & \cdots & \cdots & \cdots \\ 0 & F(j)a_{1n} & \cdots & F(j)a_{nn} \end{bmatrix} - \tilde{\theta}(j/k)\tilde{n}^{-1}(j/k)d^T(j + 1)$$

$$D(j/k) \triangleq \tilde{Q}(j) - d(j + 1)\tilde{n}^{-1}(j/k)d^T(j + 1)$$

$$d(j) \triangleq \begin{bmatrix} 0_n* \\ F(j - 1)S(j)\ell_1 \\ \vdots \\ F(j - 1)S(j)\ell_n \end{bmatrix}$$

$$\tilde{\theta}(j/k) \triangleq \begin{bmatrix} \hat{\theta}(j/k) \\ P_{\theta\theta}(j/k)A^T(j)\ell_1 \\ \vdots \\ P_{\theta\theta}(j/k)A^T(j)\ell_n \end{bmatrix}$$

$$\sigma(j/k) \triangleq \begin{bmatrix} P_{x\theta}(j/k)\ell_1 \\ \vdots \\ P_{x\theta}(j/k)\ell_n \end{bmatrix} \tag{7.45}$$

$$\tilde{n}(j/k) \triangleq n(j) + tr[P_{\theta\theta}(j/k)S(j + 1)]$$

$$S(j) = F^T(j)S(j + 1)F(j) + M(j) \quad S(N) = 0 \quad j = k,\ldots,N-1$$

The parameter predictor is given from the estimation Eqs. (7.41)

$$\hat{\theta}(j + 1/k) = A(j)\hat{\theta}(j/k) \quad \hat{\theta}(k/k) = \theta(k/z^k, u^{*k-1})$$

$$P_{\theta\theta}(j + 1/k) = A(j)P_{\theta\theta}(j/k)A^T(j) + \Gamma(j) \tag{7.46}$$

$$P_{\theta\theta}(k/k) = P_{\theta\theta}(k/z^k, u^{*k-1}) \quad j = k,\ldots,N - 1$$

where u^{*k-1} is the optimal sequence derived for $[0, k - 1]$. Then, according to the OLFO principle,

$$u^*(k) = u^*_{OL}(k/k) \tag{7.47}$$

is selected from (7.43) and the procedure is repeated after the
measurement of $z(k + 1)$.

Tse and Athans made some very interesting observations [7.33]
about the control $u^*(k)$:

1. The OLFO optimal control exists and is unique if the
    open-loop control sequences exist and are unique. This
    comes down to the satisfaction of condition
    $$B \geq A(j)BA^T(j) \text{ for all } B \geq 0 \text{ and } j = 0,...,N - 1. \qquad (7.48)$$

2. The OLFO optimal control (7.47) may be partitioned into
    two terms, the adaptive OLFO term and the
    control-correction terms

$$u^*(k) = \phi(k/k)\begin{bmatrix}\hat{x}(k/k) \\ ------ \\ 0\end{bmatrix} + \gamma(k/k)\begin{bmatrix} 0 \\ ------ \\ \sigma(k/k)\end{bmatrix} \qquad (7.49)$$

where

$$\phi(k/k) = -[\tilde{n}(k/k) + \tilde{b}\tilde{K}(k + 1/k)\tilde{b}(k/k)]^{-1}\tilde{b}^T(k/k)\tilde{K}(k + 1/k)$$
$$\cdot \Gamma(k/k)\begin{bmatrix}I & | & 0 \\ --+-- \\ 0 & | & 0\end{bmatrix} \qquad (7.50)$$

$$\gamma(k/k) = -[[\tilde{n}(k/k) + \tilde{b}^T(k/k)\tilde{K}(k + 1/k)\tilde{b}(k/k)]^{-1}\tilde{b}^T(k/k)$$
$$\cdot K(k + 1/k)\Gamma(k/k) + \tilde{n}(k/k)d^T(k + 1)]\begin{bmatrix}0 & | & 0 \\ --+-- \\ 0 & | & I\end{bmatrix}$$

The adaptive OLFO term depends on the estimates of the state with
gains parametrically adapted at every state. The control-correction
term depends on parameter state error crosscovariance $P_{x\theta}(k/k)$,
where the cross dependence between parameter identification and state
estimation enters the controller while the dependence between control and estimation has not been accounted for as the dual control
would require.

Finally, if the process is defined over an infinite interval,
$N \to \infty$, and the performance criterion is modified to account for a
finite value by considering its time average, as in (4.47), the
OLFO procedure has to be modified to account for the infinite
computation time required to obtain each of the open-loop optimal
controls. Tse and Athans [7.33] propose an N-stage shifting window

approximation where the open-loop optimal control has to be
computed for N stages ahead of time only when reducing the
computation time.

In general this method is very computer-time consuming due to
the repetitive evaluation of the open-loop controls. It is also
dependent on the systems stability.

Simulation results in a third-order dynamic system with

$$F(k) = \begin{bmatrix} 0 & 1 & 0 \\ 0 & 0 & 1 \\ -5 & -7 & -3 \end{bmatrix} \quad b(k) = \begin{bmatrix} \theta_1 \\ \theta_2 \\ \theta_3 \end{bmatrix} \quad x(0) = \begin{bmatrix} 6 \\ -3 \\ 12 \end{bmatrix}$$

where $\theta_1 = 1$, $\theta_2 = 2$, and $\theta_3 = -7$ have verified the observation of
Tse and Athans that the OLFO controller is not always a good
identifier but only a good controller. As can be observed from
Figs. 7.5 and 7.6 the OLFO controller converges rapidly to values
close to the optimal feedback with known parameters while the
parameter identification is very poor. The Tse and Athans method
was further extended by Ku and Athans [7.15]

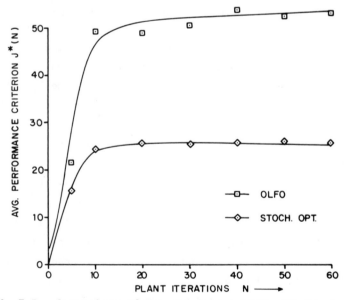

FIG. 7.5.   Comparison of Tse and Athans OLFO with Stochastic
            Optimal Algorithms.

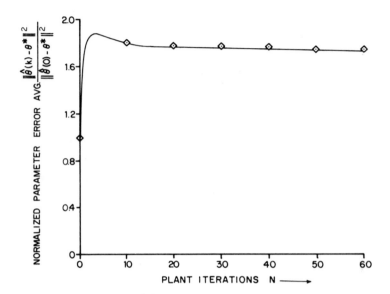

FIG. 7.6.   Normalized Parameter Estimation Error in Tse-Athans
Algorithms.

7.4.3 The Lainiotis Method

Lainiotis, Deshpande, and Upadhyay wrote a series of papers
[7.8], [7.17] on what they called "Adaptive Control" and "A Nonlinear
Separation Theorem." Basically, the method is another open-loop
feedback optimal (OLFO) approach to the stochastic control of
linear systems with unknown parameters. The control is designed
to minimize the average performance-to-go condition on the present
measurements and past control actions without any active
anticipation of new measurements. The result is a feedback control
law similar to the optimal LQG one of 4.5, but averaged over the
space of the space of the unknown parameters Ω_θ. The method,
utilizes the information states (7.1) for identification and
covers the general case of unknown parameters appearing in all the
coefficients of the plant, performance index, or noise statistics.
Algorithms have been developed for both the continuous [7.17] and
discrete-time systems [7.7] but only the latter case will be
presented here.

Consider the system

$$x(k + 1) = F(k,\theta)x(k) + B(k,\theta)u(k) + G(k,\theta)w(k) \quad x(0) = x_0$$

(7.51)

$$z(k) = H(k,\theta)x(k) + v(k)$$

where the variables $x(k)$, $u(k)$, $w(k)$, $z(k)$, and $v(k)$ as defined in (7.5) and the matrices $F(k,\theta)$, $B(k,\theta)$, $G(k,\theta)$, and $H(k,\theta)$ of appropriate dimensions, are functions of the unknown parameter vector defined as random variables with an *a-priori* density function

$$p(\theta) = q(0)$$

The performance index to be minimized

$$J(u) = E\left\{\sum_{j=0}^{N-1} [x^T(j + 1)M(j,\theta)x(j + 1) + u^T(j)N(j,\theta)u(j)]\right\}$$

(7.52)

where $M(j,\theta) = M^T(j,\theta) \geq 0$, $N(j,\theta) = N^T(j,\theta) > 0$, $j = 0,\ldots,N - 1$. If the parameter vector was known, the LQG optimal control as a result of the separation theorem would satisfy the estimation Eq. (4.40) and the control Eqs. (4.37), derived in Chap. 4.

Assuming now that the unknown parameters are random variables, the expectations in (7.53) and the subsequent Eqs. (4.37) and (4.40) are extended to average over the parameter space Ω_θ as well. The resulting parameter adaptive control is given by

$$u^*(k) = \int_{\Omega_\theta} \hat{u}(k/k,\theta)p(\theta/_z k) \; d\theta$$

(7.53)

where

$$\hat{u}(k/k,\theta) = -K(k,\theta)\hat{x}(k/k,\theta)$$

$$K(k,\theta) = [B^T(k,\theta)(M(k,\theta) + S(k + 1,\theta)_{N-k-1})B(k,\theta) + N(k,\theta)]^{-1}$$

$$\cdot B^T(k,\theta)(M(k,\theta) + S(k + 1,\theta)_{N-k-1})F(k,\theta)$$

(7.54)

$$S(k,\theta)_{N-k} = [F(k,\theta) - B(k,\theta)K(k,\theta)]^T(M(k,\theta) + S(k+1,\theta)_{N-k-1})$$

$$\cdot [F(k,\theta) - B(k,\theta)K(k,\theta)] + K^T(k,\theta)N(k,\theta)K(k,\theta)$$

$$S_N = 0$$

and

$$\hat{x}(k/k,\theta) \triangleq E\{x(k)/z^k,\theta\}$$

$$\hat{x}(k/k,\theta) = F(k - 1,\theta)\hat{x}(k - 1/k - 1,\theta) + P(k/k,\theta)H^T(k,\theta)R^{-1}(k,\theta)$$

$$\cdot[z(k) - H(k,\theta)F(k - 1,\theta)\hat{x}(k - 1/k - 1,\theta)]$$

$$+ P(k/k,\theta)P^{-1}(k/k - 1)B(k - 1,\theta)u(k - 1)$$

$$P(k/k,\theta) = P(k/k - 1,\theta) - P(k/k - 1,\theta)H^T(k,\theta)[H(k,\theta)P(k/k - 1,\theta)$$

$$\cdot H^T(k,\theta) + R(k,\theta)]^{-1}H^T(k,\theta)P(k/k - 1,\theta)$$

$$P(k/k - 1,\theta) = F(k - 1,\theta)P(k - 1/k - 1,\theta)F^T(k - 1,\theta)$$

$$+ G(k - 1,\theta)Q(k - 1,\theta)G^T(k - 1,\theta)$$

$$x_0 = \bar{x}_0 \quad P(0/0) = P_0 \tag{7.55}$$

Finally the a posteriori probability densities required in (7.53) are obtained from

$$p(\theta/_zk) = \frac{L(k/\theta)p(\theta/_zk-1)}{\displaystyle\int_{\Omega_\theta} L(k/\theta)p(\theta/_zk-1) \, d\theta} \tag{7.56}$$

$$L(k/\theta) = \det[P_z(k/k-1,\theta)]^{-1/2}\exp\left\{-\frac{1}{2}||z(k) - H(k,\theta)F(k - 1,\theta)\right.$$

$$\left.\cdot\hat{x}(k - 1/k - 1,\theta)||^2_{P_z^{-1}}\right\}$$

$$P_z(k/k - 1,\theta) \triangleq H(k,\theta)P(k/k - 1,\theta)H^T(k,\theta) + R(k,\theta)$$

$$L(0/\theta) = 1$$

The algorithm is straightforward and easier to implement than the equivalent algorithm of Tse and Athans or Farison. It may be generated by the average of the specific controllers $\hat{u}(k/k,\theta)$ for some value of the parameters weighted by the *a posteriori* probability $p(\theta/_zk)$, which are Gaussian [7.8].

As $N \to \infty$, and the matrices of coefficients are constant, the control equations tend to their respective steady-state equivalent of (4.48) and

$$p(\theta/_Zk) \rightarrow \delta(\theta - \theta^*)$$ (7.57)

where $\delta(\cdot)$ is the Dirac delta function and θ^* is the true value of
the parameters. Then the controller (7.53) converges to the L.Q.G.
optimal. This is another property which the previous algorithms do
not have.

The control (7.53) cannot be separated into two terms as in
previous cases, since the control correction depending on the error
covariance of the parameter vector enters the control through the
a posteriori probabilities.

One of the major drawbacks of the method, though, is that it
is almost impossible to obtain analytic expressions of $K(k,\theta)$ and
$\hat{x}(k/k,\theta)$ as a function of θ, and therefore the computation of
(7.53) for a continuously distributed parameter space Ω_θ is almost
impossible. As a reasonable approximation Lainiotis proposes
discretization of the parameter space with discrete parameter
vectors θ_i, $i = 1,\ldots,s$ replacing θ in (7.54)-(7.56). Then (7.53)
is replaced by

$$u^*(k) = \sum_{i=1}^{s} \hat{u}(k/k,\theta_i)p(\theta_i/_Zk)$$ (7.58)

for which $\hat{u}(k/k,\theta_i)$ can be numerically obtained for each θ_i. Also

$$p(\theta_i/_Zk) = \frac{L(k/\theta_i)\ p(\theta_i/_Z^{k-1})}{\sum_{i=1}^{s} L(k/\theta_i)p(\theta_i/_Z^{k-1})}$$ (7.59)

It will be shown later that under these conditions this
algorithm is one of a class with a learning property that selects
asymptotically the LQG optimal controller out of a finite
collection of controllers, to which the optimal belongs.

Simulation results have been obtained for a second-order
system with

$$F(k,\theta) = \begin{bmatrix} 0 & 1 \\ \theta_{11} & \theta_{12} \end{bmatrix} \quad B(k,\theta) = \begin{bmatrix} 0 \\ 1 \end{bmatrix} \quad G(k,\theta) = \begin{bmatrix} 1 & 0 \\ 0 & 1 \end{bmatrix}$$

$$H^T(k,\theta) = \begin{bmatrix} 2 & 1 \end{bmatrix}, x(0) \sim N\left[\begin{pmatrix} 1 \\ .5 \end{pmatrix} \begin{pmatrix} 2 & 0 \\ 0 & 4 \end{pmatrix} \right], \quad w(k) \sim N\left[\begin{pmatrix} 0 \\ 0 \end{pmatrix}, \begin{pmatrix} 3 & 0 \\ 0 & 2 \end{pmatrix} \right]$$

$v(k) \sim N(0,2),$

and θ belongs to the set of possible parameters $\begin{bmatrix} -0.29 \\ 0.4 \end{bmatrix}, \begin{bmatrix} 0.12 \\ 0.8 \end{bmatrix},$
$\begin{bmatrix} -0.63 \\ 0.2 \end{bmatrix}$ with $p_i(0) = 1/3$, $i = 1, 2, 3$.

The performance criterion used is

$$J(u) = E\{ 4x_1^2(50) + 3x_2^2(50) + \sum_{j=0}^{49} [3x_1^2(j) + 2x_2^2(j) + 5u^2(j)] \}$$

The results of this algorithm are quite impressive as seen from Figs. 7.7 and 7.8. The reason is that the a posteriori probabilities converge very fast (after 35 iteration) to (1,0,0), and therefore the control converges to the LQG optimal.

A stability investigation is needed. However, such an investigation is extremely difficult in view of the complexity of the algorithm.

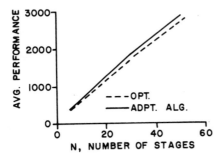

FIG. 7.7. Comparison of Average Performance of Lainiotis' Algorithms with LQG Optimal as a Third-Order System.

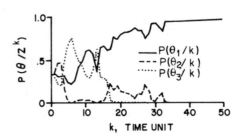

FIG. 7.8.   A Posteriori Probabilities for Second-Order System.

7.5 THE MINIMIZATION OF UPPER- AND LOWER-BOUND ALGORITHMS

In many instances when the exact solution of an optimization problem is not directly accessible due to computational difficulties, bounds on the optimal performance may be found useful for approximations. If upper and lower bounds can be computed, they may serve to define an acceptable region of approximation, and approximate solutions may be established by finding control sequences that minimize these bounds.

The incentive to generate such bounds is to obtain approximate solutions of the dual control problem, resulting from a LQG system with unknown parameters [7.30]. For the same realizability conditions considered in the previous sections, it is assumed that the unknown parameter vector θ belongs to a discrete finite set of parameter vectors Ω_θ with *a priori* probability densities $p_0(\theta)$

$$\theta \; \epsilon \; \Omega_\theta = \{\theta_1,\ldots,\theta_s\} \quad p_0(\theta_i) = q_i(0) \quad i = 1,\ldots,s \qquad (7.60)$$

Consider the same system used in Lainiotis algorithm (7.51) given here again for completeness

$$x(k + 1) = F(k,\theta)x(k) + B(k,\theta)u(k) + G(k,\theta)w(k) \quad x(0) = x_0$$
$$z(k) = H(k,\theta)x(k) + v(k) \qquad (7.61)$$

where the variables $x(k)$, $u(k)$, $w(k)$, and $v(k)$ have been defined in

(7.5), and the matrices $F(k,\theta)$, $B(k,\theta)$, $G(k,\theta)$, and $H(k,\theta)$ are of appropriate dimensions and are functions of the unknown parameter vector θ. It may also be assumed that the covariances of the random variables of the system x_0, $w(k)$, and $v(k)$ are also functions of θ, e.g., $P_0(\theta)$, $Q(k,\theta)$, or $R(k,\theta)$. The performance criterion to be minimized is defined as in (7.53)

$$J(u) = E\left\{\sum_{j=0}^{N-1} [x^T(j + 1)M(j,\theta)x(j + 1) + u^T(j)N(j,\theta)u(j)]\right\} \quad (7.62)$$

where $M(j,\theta) \geq M^T(j,\theta) \geq 0$, $N(j,\theta) = N^T(j,\theta) > 0$, $j = 0,\ldots,N-1$.

Following the minimization procedure of dynamic programming used in Chap. 5 for the dual control problem by defining the optimal return function $V_{N-k}(x(k)/_Zk)$ for an $(N - k)$-stage process with initial state $x(k)$,

$$V_{N-k}(x(k)/_Zk) \triangleq \min_{\substack{U_k^{N-1}}} E\left\{\sum_{j=k}^{N-1} [x^T(j + 1)M(j,\theta)x(j + 1) + u^T(j)N(j,\theta)\right.$$

$$\left. \cdot u(j)]/_Zk\right\} = \min_{\substack{U_k^{N-1}}} \sum_{i=1}^{s} p(\theta_i/_Zk) E\left\{\sum_{j=k}^{N-1} x^T(j + 1)\right.$$

$$\left. \cdot M(j,\theta)x(j + 1) + u^T(j)N(j,\theta)u(j)] /_Zk,_{\theta_i}\right\} \quad (7.63)$$

The last expression was obtained by considering θ as a random variable over Ω_θ and expanding the expectation used. The *a posteriori* probabilities being Gaussian can be calculated as in (7.59) by

$$p(\theta_i/_Zk) = \frac{L(k/\theta_i)p(\theta_i/_Z{k-1})}{\sum_{i=1}^{s} L(k/\theta_i)p(\theta_i/_Z{k-1})}, \; p(\theta_i/_Z0) \triangleq q_0(\theta_i)$$

$$L(k/\theta_i) = cdet|P_Z(k/k - 1,\theta_i)|^{-1/2}exp\left\{-\frac{1}{2}||z(k) - H(k,\theta_i)\right.$$

$$\left. F(k - 1,\theta_i)\hat{x}(k - 1/k-1,\theta_i)||^2_{P_Z^{-1}}\right\} \quad (7.64)$$

$$P_z(k/k - 1,\theta_i) \triangleq H(k,\theta_i)P(k/k - 1,\theta_i)H^T(k,\theta_i) + R(k,\theta_i)$$

where $\hat{x}(k/k,\theta_i)$ is the output of a Kalman-Bucy filter for the plant (7.61) with $\theta = \theta_i$.

The optimal return function must satisfy Bellman's equation [7.3]

$$V_{N-k}(x(k)/_zk) = \min_{u(k)} \sum_{i=1}^{s} p(\theta_i/_zk)E\{ [x^T(k + 1)M(k,\theta_i)x(k + 1)$$

$$+ u^T(k)N(k,\theta_i)u(k)] + V_{N-k-1}(x(k + 1)/_zk + 1)/$$

$$z^k,\theta_i\}$$

$$V_0(x(n)/_zN) = 0 \tag{7.65}$$

The solution of (7.65) and (7.64) yields a quadratic form [7.32] which is not reproducible

$$V_{N-k}(\hat{x}(k)/_zk) = \hat{x}^T(k)U(p(\theta/_zk),k)\hat{x}(k) + T(p(\theta/_zk),k) \tag{7.66}$$

where

$$\hat{x}^T(k) \triangleq [E\{ x^T(k)/_zk, \theta_1 \},\ldots,E\{x^T(k)/_zk, \theta_s \}] \tag{7.67}$$

$$\triangleq [\hat{x}^T(k/k,\theta_1),\ldots,\hat{x}^T(k/k,\theta_s)]$$

$\hat{x}(k/k,\theta_i)$ is the output of a Kalman estimator (7.55) for $\theta = \theta_i$, $U(p(\theta/_zk),k)$ is an $(s \cdot n) \times (s \cdot n)$ matrix and $T(p(\theta(_zk),k)$ a scalar nonlinear function of the *a posteriori* probabilities vector $p(\theta/_zk)$, and therefore inherently to the states to be defined later. There is no known way of obtaining recursive expressions for the computation of U or T. On the other hand, U and T must be computed on-line in order to incorporate the active control required through $p(\theta/_zk)$ for every k. Therefore, an attempt for obtaining the optimal dual solution is abandoned. Instead upper V^u and lower V^ℓ bounds of (7.66), which may be recursively computed, are possible so that

$$V_{N-k}^{\ell}(\hat{X}(k)/_Zk) \leq V_{N-k}(\hat{X}(k)/_Zk) \leq V_{N-k}^{u}(\hat{X}(k)/_Zk) \qquad (7.68)$$

The derivation of such bounds is due to Stein and is given in detail in Ref. [7.30]. A brief discussion of the motivation and development of these bounds without any proofs is presented here.

A *lower-bound* $V_{N-k}^{\ell}(\hat{X}(k)/_Zk)$ on the optimal return cost may be obtained by considering the following ideal problem:

Consider s plants (7.61) each one with θ_i, $i = 1,...,s$, running simultaneously with the same initial and noise conditions.

One may design LQG optimal controls for each one of them and obtain the respective return costs for an $(N - k)$-storage process starting at k. If the *a posteriori* probabilities $p(\theta_i/_Zj)$ for $k \leq j \leq N$ were known, then the optimal performance averaged over the s plants with weighting coefficients the *a posteriori* probabilities must be less than the average performance of one plant with unknown parameters given by (7.65); or

$$V_{N-k}^{\ell}(\hat{X}(k)/_Zk) = \sum_{i=1}^{s} p(\theta_i/_Zk) V_{N-k}^{\ell}(\hat{X}(k)/_Zk_{\theta_i}) \triangleq \sum_{i=1}^{s} p(\theta_i/_Zk)$$

$$\cdot [\hat{x}^T(k/k,\theta_i) U_{N-k}^{\ell}(\theta_i,k)\hat{x}(k/k,\theta_i) + T_{N-k}^{\ell}(\theta_i,k)]$$

$$\leq V_{N-k}(\hat{X}(k)/_Zk) \qquad (7.69)$$

where $\hat{x}(k/k,\theta_i)$ is given by (7.67) and can be produced by individual Kalman-Bucy filters for each of the linear systems (7.61) with $\theta = \theta_i$, $i = 1,...,s$, and error covariances $P(k/k,\theta_i)$:

$$U_{N-k}^{\ell}(\theta_i,k) = S_{N-k}^{\ell}(\theta_i,k) - M(k,\theta_i)$$
$$S_{N-k}^{\ell}(\theta_i,k) = M(k,\theta_i) + F^T(k,\theta_i)S_{N-k-1}^{\ell}(\theta_i,k + 1)$$
$$\cdot [I - B(k,\theta_i)[N(k,\theta_i) + B^T(k,\theta_i)S_{N-k-1}^{\ell}(\theta_i,k + 1)$$
$$\cdot B(k,\theta_i)]^{-1}B^T(k,\theta_i)S_{N-k-1}^{\ell}(\theta_i,k + 1)]F(k,\theta_i)$$
$$S_0^{\ell} = 0 \qquad (7.70)$$

$$T^{\ell}_{N-k}(\theta_i,k) = T^{\ell}_{N-k-1}(\theta_i,k+1) + \text{tr}[P(k/k,\theta_i)M(k,\theta_i)]$$

$$+ \text{tr}[S^{\ell}_{N-k-1}(\theta_i,k+1)P(k/k,\theta_i)H^T(k,\theta_i)$$

$$\cdot[H(k,\theta_i)L(k,\theta_i)H^T(k,\theta_i) + R(k)]^{-1}$$

$$\cdot H(k,\theta_i)P(k/k,\theta_i) \qquad\qquad T^{\ell}_0 = 0$$

$$L(k+1,\theta_i) = F(k,\theta_i)P(k/k,\theta_i)F^T(k,\theta_i) + G(k,\theta_i)Q(k,\theta_i)G^T(k,\theta_i)$$

A proof of the conjecture (7.69) by induction is given by Stein and Saridis [7.30] and is omitted here for space limitations. It is obvious that the combined controller which results from the generation of such a lower bound is fictitious and cannot be built, since only one of the s plants utilized in the derivation is present in reality.

An *upper-bound* $V^u_{N-k}(X(k)/_Zk)$ of the optimal return cost may be generated by linearization of the $((ns) \times (ns))$ matrix $U(p(\theta/_Zk),k)$ of Eq. (7.66) about some arbitrary fixed point $\bar{p}(\theta)$ of the probability space Ω_θ, and retention of the linear terms only

$$\tilde{U}(p(\theta/_Zk),k) = \sum_{i=1}^{s} p(\theta_i/_Zk)\ \tilde{U}^u_{N-k}(i) \qquad\qquad (7.71)$$

where

$$\tilde{U}^u_{N-k}(i) \triangleq U(\bar{p}(\theta),k) - \sum_{i=1}^{s-1} \frac{\partial U(\bar{p}(\theta),k)}{\partial \bar{p}(\theta_i)}\ \bar{p}(\theta_i)$$

$$\tilde{U}^u_{N-k}(s) \triangleq \tilde{U}^u_{N-k}(s) + \frac{\partial U(\bar{p}(\theta),k)}{\partial \bar{p}(\theta_i)}, \qquad\qquad i = 1,\ldots,s-1$$

For an appropriate $T^u_{N-k}(i)$, which does not affect the control, the above linearization produces a return function which is linear in

$p(\theta/Z^k)$:

$$V^u_{N-k}(\hat{X}(k)/_Zk) = \sum_{i=1}^{s} p(\theta_i/_Zk)[\hat{X}^T(k)\tilde{U}^u_{N-k}(i)\hat{X}(k) + T^u_{N-k}(i)]$$

$$(7.72)$$

It was shown in [7.29] that the quadratic form of (7.72) represents a supporting (tangent) hyperplane of the convex surface defined by the optimal return cost $V(\hat{X}(k)/_Zk)$ at the point $\bar{p}(\theta) \triangleq \bar{p}$ of the space probability function $\Omega_{p(\theta)}$.

According to the properties of the supporting hyperplanes, $V^u_{N-k}(\hat{X}(k)/_Zk)$ is larger than $V_{N-k}(\hat{X}(k)/_Zk)$ for all points except for $\bar{p}(\theta)$ where they are equal. Therefore $V^u_{N-k}(\hat{X}/_Zk)$ is an upper bound to the optimal cost.

$$V_{N-k}(\hat{X}(k)/_Zk) \leq V^u_{N-k}(\hat{X}(k)/_Zk) \quad k = 0,\ldots,N \tag{7.73}$$

In order to compute the upper bound from (7.72) the following recursive expressions are required, derived by Stein [7.30]:

$$\tilde{U}^u_{N-k}(i) = \overline{FWF}^u_{N-k-1}(i) - \overline{BWF}^{u\,T}_{N-k-1}(i)A_{N-k}(\bar{p}) - A^T_{N-k}(\bar{p})\overline{BWF}^u_{N-k-1}(i)$$

$$+ A^T_{N-k}(\bar{p})[\overline{BWB}^u_{N-k-1}(i) + N(k,\theta_i)]A_{N-k}(\bar{p}) \quad \tilde{U}^u_0(i) = 0 \tag{7.74}$$

$$T^u_{N-k}(i) = T^u_{N-k-1}(i) + tr[P(k/k,\theta_i)M(k,\theta_i)] + tr[W^u_{N-k-1}(i)$$
$$\cdot \bar{P}(k/k,\theta_i)]$$
$$T^u_0(i) = 0 \quad i = 1,\ldots,s$$

where the (sn) x (sn) matrices

$$U^u_{N-k}(i) \triangleq \{U^u_{N-k}(j,\ell;i)\}$$
$$W^u_{N-k}(i) \triangleq \{W^u_{N-k}(j,\ell;i)\} \tag{7.75}$$
$$\bar{P}(k/k,\theta_i) \triangleq \{\bar{P}_k(j,\ell;\ i)\}$$

are defined from their $j\ell$th partition

$$W^u_{N-k}(j,\ell,i) = \begin{cases} M(k-1,\theta_i) + U^u_{N-k}(j,\ell,i) & j=\ell=i \\ U^u_{N-k}(j,\ell;i) & \text{otherwise} \end{cases}$$

$$\bar{P}_k(j,\ell;i) \triangleq P(k/k,\theta_j)H^T(k,\theta_j)[H(k,\theta_i)(F(k/\theta_i)P(k/k,\theta_i)F^T(k/\theta_i)$$

$$+ G(k,\theta_i)Q(k,\theta_i)G^T(k,\theta_i))H^T(k,\theta_i) + I] H(k,\theta_\ell)$$

$$\cdot P(k/k,\theta_\ell) \quad i = 1,\ldots,s. \tag{7.76}$$

and

$$\overline{FWF}^u_{N-k}(i) \triangleq \bar{F}_{k-1}^{\ T}(i)W^u_{N-k}(i)\bar{F}_{k-1}(i)$$

$$\overline{BWF}^u_{N-k}(i) \triangleq \bar{B}_{k-1}^{\ T}(i)W^u_{N-k}(i)\bar{F}_{k-1}(i) \tag{7.77}$$

$$\overline{BWB}^u_{N-k}(i) \triangleq \bar{B}_{k-1}^{\ T}(i)W^u_{N-k}(i)\bar{B}_{k-1}(i)$$

$$A_{N-k}(\bar{p}) \triangleq \left[\sum_{i=1}^s \bar{p}(\theta_i)[\overline{BWB}^u_{N-k}(i) + N(k-1,\theta_i)] \right]^{-1}$$

$$\cdot \left[\sum_{i=1}^s \bar{p}(\theta_i)\overline{BWF}^u_{N-k}(i) \right]$$

and $\bar{F}$ and $\bar{B}$ are $(sn) \times (sn)$ matrices

$$\bar{F}_{k-1}(i) \triangleq \{\bar{F}(k-1,\theta_i;j,\ell)\} \tag{7.78}$$

$$\bar{B}_{k-1}(i) \triangleq \{\bar{B}(k-1,\theta_i;j)\}$$

defined by their partitions:

$$\bar{F}(k-1,\theta_i;j,\ell) = \begin{cases} [I - P(k/k,\theta_j)H^T(k,\theta_j)H(k,\theta_j)]F(k-1,\theta_j) & \\ \qquad\qquad\qquad\qquad\qquad j = \ell \neq i \\ P(k/k,\theta_j)H^T(k,\theta_j)H(k,\theta_i)F(k-1,\theta_i) & \\ \qquad\qquad\qquad\qquad\qquad j \neq i, \ \ell = i \\ F(k-1,\theta_i) & \\ \qquad\qquad\qquad\qquad\qquad j = \ell = i \\ 0 & \\ \qquad\qquad\qquad\qquad\qquad \text{elsewhere} \end{cases}$$

$$\tag{7.79}$$

$$\bar{B}(k,\theta_i;j) \triangleq [I - P(k/k,\theta_j)H^T(k,\theta_j)H(k,\theta_j)]B(k,\theta_i)$$
$$+ P(k/k,\theta_j)H^T(k,\theta_j)H(k,\theta_i)B(k,\theta_i)$$

The expressions of the upper bound are tedious and complex, in contrast to the expressions for the lower bound. The reason is that the lower bound is completely decoupled and linear in terms of $p(\theta/_zk)$. However, they can both be obtained <u>off-line</u> before the process starts and give a measure of the acceptable range of approximations α, by testing ΔV against it, e.g.,

$$\Delta V_N(\hat{X}(0)/_zo) \triangleq V_{N-k}^u(\hat{X}(0)/_zo) - V_N^\ell(\hat{X}(0)/_zo) \gtrless \alpha \qquad (7.80)$$

Two recursive algorithms giving return cost function within ΔV are obtained in the sequel, by minimizing either the upper or the lower bound. They demonstrate interesting properties and give new insight to the dual and OLFO solution.

7.5.1 The Stein and Saridis Upper-Bound Minimization Algorithm

Stein and Saridis presented an algorithm [7.31] that yields a control u giving a minimum of the upper bound of the optimal return cost $V_{N-k}^u(\hat{X}(k)/_zk)$ at every step, starting backwards at $k = N, N - 1, \ldots$ and assuming active estimation of the parameters. Therefore, this algorithm may be considered as an approximation to the dual optimal and not an OLFO solution of the upper bound in the recursive equation of the optimal return cost (7.65) and evaluation of the vectors $\hat{X}(k)$

$$\hat{X}^T(k) = [\hat{x}(k/k, \theta_1), \ldots, \hat{x}(k/k, \theta_s)]$$

by the Kalman-Bucy optimal linear estimators resulting by setting $\theta = \theta_i$

$$\hat{x}(k + 1/k + 1, \theta_i) = F(k, \theta_i)\hat{x}(k/k, \theta_i) + P(k + 1/k + 1, \theta_i)$$

$$H^T(k + 1, \theta_i)R^{-1}(k + 1, \theta_i)[\ z(k + 1)$$

$$- H(k + 1, \theta_i)F(k, \theta_i)\hat{x}(k/k, \theta_i)]$$

$$+ P(k + 1/k + 1, \theta_i)\ P^{-1}(k + 1/k, \theta_i)u(k)$$

$$\hat{x}(0/0) = \bar{x}_0 \qquad\qquad (7.81)$$

$$P(k + 1/k + 1, \theta_i) = P(k + 1/k, \theta_i) - P(k + 1/k, \theta_i)H^T(k + 1, \theta_i)$$

$$[H(k + 1, \theta_i)P(k + 1/k, \theta_i)\ H^T(k + 1, \theta_i)$$

$$+ R(k + 1, \theta_i)]^{-1}H^T(k + 1, \theta_i)P(k + 1/k, \theta_i)$$

$$P(0/0) = P_0$$

$$P(k + 1/k, \theta_i) = F(k, \theta_i)P(k/k, \theta_i)F^T(k, \theta_i) + G(k, \theta_i)Q(k, \theta_i)G^T(k, \theta_i)$$

$$i = 1, \ldots, s$$

Consider the recursive equation of optimal return cost for an (N - k)-stage process at time k given by (7.65), and replace $V_{N-k-1}(\hat{X}(k + 1)/_Z k+1)$ on the right hand side by its upper bound producing an inequality.

$$V_{N-k}(\hat{X}(k)/_Z k) = \underset{u(k)}{\text{Min}} \sum_{i=1}^{s} p(\theta_i/_Z k)E\{x^T(k + 1)M(k, \theta_i)x(k + 1)$$

$$+ u^T(\theta_i)N(k, \theta_i)\ u(k) + V_{N-k-1}(\hat{X}(k + 1)/_Z k+1)/$$

$$z^k, \theta_i\}$$

$$\leq \underset{u(k)}{\text{Min}} \sum_{i=1}^{s} p(\theta_i/_Z k)E\{x^T(k + 1)M(k, \theta_i)x(k + 1)$$

$$+ u^T(k)N(k, \theta_i)u(k) + V_{N-k-1}^u(\hat{X}(k + 1)/_Z k+1)/$$

$$z^k, \theta_i\}$$

$$= \underset{u(k)}{\text{Min}} \sum_{i=1}^{s} p(\theta_i/_Z k)[\hat{x}(k + 1/k + 1, \theta_i)M(k, \theta_i)$$

$$\hat{x}(k + 1/k + 1, \theta_i) + \text{tr}[P(k + 1/k+1, \theta_i)M(k, \theta_i)]$$

$$+ u^T(k)N(k,\theta_i)u(k) + \sum_{i=1}^{s} p(\theta_i/_zk+1)[\hat{x}(k+1) \; \tilde{U}^u_{N-k-1}\hat{x}(k+1)$$

$$+ T^u_{N-k-1}(i)] = \bar{V}_{N-k}(\hat{X}(k)/_zk) \qquad\qquad (7.82)$$

In (7.82), $p(\theta_i/_zk+1)$ may be computed from (7.64) as a nonlinear function of $p(\theta_i/_zk)$ and $L(k + 1/\theta_i)$, but $\bar{V}_{N-k}(\hat{X}(k)/_zk)$ still represents a convex function in the probability density space $\Omega_{p(\theta)}$. Linearization about some $\bar{p}(\theta)$, in this particular case $\bar{p}(\theta) = q(0)$ the *a priori* densities, would yield the upper-bound $V^u_{N-k}(\hat{X}(k)/_zk)$ obtained in the preceding section.

$$V_{N-k}(\hat{X}(k)/_zk) \le \bar{V}_{N-k}(\hat{X}(k)/_zk) \le V^u_{N-k}(\hat{X}(k)/_zk) \quad k = 0,\ldots,N$$
$$(7.83)$$

$\bar{V}_{N-k}(\hat{X}(k)/_zk)$ does not possess a self-reproducing form and cannot be used for a recursive computation of the upper bound. However, it can be used to obtain an approximately optimal control $\bar{u}^*(k)$ which yields a performance function smaller than the upper bound as from (7.83)

$$\bar{u}^*(k) = -A_{N-k}(p(\theta/_zk))\hat{X}(k) \qquad\qquad (7.84)$$

$$A_{N-k}(p(\theta/_zk)) = \left[\sum_{i=1}^{s} p(\theta_i/_zk)[\overline{BWB}^u_{N-k-1}(i) + N(k,\theta_i)] \right]^{-1}$$

$$\left[\sum_{i=1}^{s} p(\theta_i/_zk)\overline{BWF}^u_{N-k-1}(i) \right]$$

The control gains $A_{N-k}(p(\theta/_zk))$ were first defined for $\bar{p}(\theta)$ in (7.77), where the rest of the quantities required are also defined. The recursive expressions for the upper-bound matrices, etc., are given by (7.74), (7.75), (7.76), (7.78), and (7.79). The state estimation is produced by s parallel Kalman-Bucy filters of (7.81) and the *a posteriori* probability densities through the identification

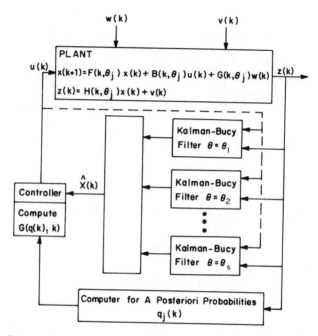

FIG. 7.9.   The Stein and Saridis Parameter-Adaptive S.O.C.
Algorithm.

algorithm (7.64). The whole system is depicted in Fig. 7.9. It
represents a rather complicated solution to the problem. As was
mentioned before, linearization of the matrices $\tilde{U}_{N-k}^{u}$ is performed
about the *a priori* probabilities.

$$\bar{p}(\theta) = q(0)$$

However, this is not necessary, and Stein [7.30] proposed a
search for an optimal $\bar{p}^{*}(\theta)$ that further minimizes $\bar{V}_{N-k}(\hat{x}(k)/_{z}k)$.
This, however, further complicates the algorithm for very little
gain.

The algorithm was further extended [7.30] to cover the
infinite-time case, e.g., $N \to \infty$. In that case, all the coefficients
in (7.84) and (7.74) to (7.79) are constants having attained their
steady-state value. Since the solution is obtained through several

Kalman filters and controls which may diverge, the system may be
unstable. There is another disadvantage to the method: while the
a posteriori probabilities learn the true parameter θ^*, e.g.,

$$\lim_{k \to \infty} p(\theta_i/z^k) = \begin{cases} 1 & \theta_i = \theta^* \\ 0 & \theta_i \neq \theta^* \end{cases} \tag{7.85}$$

and the respective ith-state estimator converges to the optimal,
the control gains do not converge to their optimal deterministic
values. Therefore the algorithm *cannot* give an asymptotic optimal
solution. This additional implication makes the study of the
stability of the algorithm almost impossible.

Simulation results of this algorithm will be discussed in
Sec. 7.5.3 where it will be compared with the lower-bound
algorithm.

7.5.2 The Saridis and Dao Lower-Bound Minimization Algorithm

It was mentioned in the previous sections that the lower bound
can be easily obtained through the recursive relations (7.70)
because the quadratic forms in (7.69) are linear in $p(\theta_i/z^k)$ and
therefore self-reproducible. However, the control that minimizes
the lower bound is not realizable by definition of the fictitious
problem that generated it. After all, if it were realizable one
would get a controller "more optimal" than the "LQG optimal"
which is a contradiction.

Instead, a realizable controller can be designed at k that
depends on the old values of measurements z^k and controls U^{k-1}
but will minimize the future lower bound of the performance
functional

$$\underline{V}_{N-k}(\hat{X}(k)/z^k) = \underset{u(k)}{\text{Min}} \, [\sum_{i=1}^{s} p(\theta_i/z^k) E\{x(k+1)^T M(k,\theta_i) x(k+1)$$

$$+ \, u^T(k) \quad N(k,\theta_i) u(k) + V_{N-k-1}^{\ell}(\hat{X}(k+1/z^k,\theta_i))]$$

$$\tag{7.86}$$

However, this minimization is not the same as in the derivation
of the lower bound, since the outputs of the s estimators are not
all optimal except for the one which corresponds to the true
parameter vector; they all depend on the same z(k). Therefore,
as expected,

$$V_{N-k}(\hat{x}(k)/_zk) \leq \underline{V}_{N-k}(\hat{x}(k)/_zk) \qquad (7.87)$$

The resulting control may assume an OLFO practice and use only the
first minimization step in the procedure (7.86) and then, when
z(k + 1) is updated, repeat. The equation of such a control is

$$\underline{u}^*(k) = - \sum_{i=1}^{s} p(\theta_i/_zk) K(k,\theta_i) \hat{x}(k/k,\theta_i) \qquad (7.88)$$

$$K(k,\theta_i) = \left[\sum_{i=1}^{s} p(\theta_i/_zk) [N(k,\theta_i) + B^T(k,\theta_i) S^{\ell}_{N-k-1}(i) B(k,\theta_i)] \right]^{-1}$$
$$\cdot B^T(k,\theta_i) S^{\ell}_{N-k-1}(i)$$

This control was derived by Saridis and Dao [7.23], and involves the
separation of the identification performed by $p(\theta_i/_zk)$, $i = 1,\ldots,s$,
as defined by Eq. (7.64), the state estimation performed by
$\hat{x}(k/k,\theta_i)$, $i = 1,\ldots,s$, as defined by (7.81), and the deterministic
control performed by (7.88) with matrices $U^{\ell}_{N-k}(i)$, $i = 1,\ldots,s$
obtained in (7.70) for the lower bound.

This algorithm is asymptotically optimal since $p(\theta_i/_zk) \to 1$,
when $\theta_i = \theta^*$ and $p(\theta_i/_zk) \to 0$, $\theta \neq \theta^*$, and therefore converges to
the LQG optimal system. As $N \to \infty$, all coefficients assume their
steady-state values, e.g., $A(k,\theta_i) = A(\theta_i)$ for all $A = F, G, H, B$,
Q, or R in (7.81), (7.80), and (7.70). Then the solution of the
infinite time problem exists for appropriate time-averaging of the
performance criterion as in Eq. (4.47)

$$J(u(k)) = \lim_{N \to \infty} \frac{1}{N} E \left\{ \sum_{k=0}^{N-1} ||x(k + 1)||^2_{M(\theta)} + ||u(k)||^2_{N(\theta)} \right\} \qquad (7.89)$$

if the matrix $F(\theta_i)$ is stable for all θ_i. However, the last
condition is only sufficient and may be somehow relaxed, since the
stochastic optimal solution with the true parameter θ^* is
asymptotically stable.

The infinite time control equations are given by

$$u(k) = -\sum_{j=1}^{s} q_j(k) K_\ell(k,\theta_j) \hat{x}(k,\theta_j)$$

where

$$K_\ell(k,\theta_j) = \left[\sum_{j=1}^{s} q_j(k) [N(\theta_j) + B^T(\theta_j)S(\theta_j)B(\theta_j)] \right]^{-1} B^T(\theta_j)$$

$$\cdot S(\theta_j)F(\theta_j) \tag{7.90}$$

$$U(\theta_j) = S(\theta_j) - M(\theta_j)$$

$$S(\theta_j) = M(\theta_j) + F^T(\theta_j)S(\theta_j)[I-B(\theta_j)[N(\theta_j) + B^T(\theta_j)S(\theta_j)$$

$$B(\theta_j)]^{-1} \cdot B^T(\theta_j)S(\theta_j)]F(\theta_j) \qquad j = 1,\ldots,s$$

The state estimation equations are

$$\hat{x}(k + 1/k + 1,\theta_j) = [I - P(\theta_j)H^T(\theta_j)R(\theta_j,k + 1)H(\theta_j)][A(\theta_j)$$

$$\hat{x}(k/k,\theta_j) + B(\theta_j)u(k)] + P(\theta_j)H^T(\theta_j)z(k + 1)$$

$$\hat{x}(0,\theta_j) = x_0 \tag{7.91}$$

$$P(\theta_j) = M(\theta_j) + M(\theta_j)H^T(\theta_j)[H(\theta_j)L(\theta_j)H^T(\theta_j) + R(\theta_j)]^{-1}$$

$$H(\theta_j)L(\theta_j)$$

$$L(\theta_j) = F(\theta_j)P(\theta_j)F^T(\theta_j) + G(\theta_j)Q(\theta_j)G^T(\theta_j)$$

while the identification algorithm is given by

$$p(\theta_j/z^k) = \frac{p(\theta_j/z^{k-1})det|P_z(\theta_j)|^{-1/2}\exp\left\{-\frac{1}{2}||z(k) - \hat{z}(\theta_j,k)||^2_{P_z^{-1}(\theta_j)}\right\}}{\sum_{\ell=1}^{s} p(\theta_\ell/z^{k-1})det|P_z((\theta_\ell)|^{-1/2}\exp\left\{-\frac{1}{2}||z(k)-\hat{z}(\theta_\ell,k)||^2_{P_z^{-1}(\theta_\ell)}\right\}}$$

$$\tag{7.92}$$

where

$$\hat{z}(\theta_j,k) = H(\theta_j)\hat{x}(k/k,\theta_j) \text{ and } P_z(\theta_j) = H(\theta_j)P(\theta_j)H^T(\theta_j) + R(\theta_j)$$

The Saridis and Dao algorithm is depicted in a block diagram in Fig. 7.10. Further discussion and comparison with the Stein and Saridis algorithm are given in the next section.

7.5.3 Comparisons and Variation

When the two algorithms were developed it was clearly stated that, while the Stein and Saridis algorithm is an approximation to the optimal dual control problem with active information processing, the Saridis and Dao algorithm is an open-loop feedback optimal solution. It will be shown in the sequel that, in spite of intuitive opinions about the opposite, the OLFO solution yields a better performance value than the approximate dual optimal. By

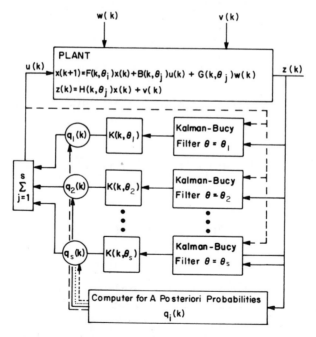

FIG. 7.10.   The Parameter-Adaptive S.O.C. System of Saridis and Dao.

successive substitutions of the upper-bound $V^u_{N-k-1}(\hat{X}(k + 1)/_Z k+1)$
and the lower-bound $V^\ell_{N-k-1}(\hat{X}(k + 1/_Z k+1)$ in (7.86) one obtains

$$\underline{V}_{N-k}(\hat{X}(k)/_Z k) = \underset{U(k)}{\text{Min}} \sum_{i=1}^{s} p(\theta_i/_Z k)E\{x^T(k + 1)M(k,\theta_i)x(k + 1)$$

$$+ u^T(k)N(k,\theta_i)u(k) + V^\ell_{N-k-1}(\hat{X}(k + 1/_Z k+1,\theta_i)/$$

$$z^k,\theta_i) \} \leq \sum_{i=1}^{s} p(\theta_i/_Z k)E\{x^{uT}(k + 1)$$

$$M(k,\theta_i)x^u(k + 1) + u^T(k)N(k,\theta_i)u(k)$$

$$+ V^\ell_{N-k-1}(\hat{x}^u(k + 1)/_Z k+1, \theta_i)/_Z k,\theta_i)\} \leq \underset{u(k)=u^u(k)}{\text{Min}}$$

$$\sum_{i=1}^{s} p(\theta_i/_Z k)E\{x^{uT}(k + 1)M(k,\theta_i)x^u(k + 1)$$

$$+ u^T(k)N(k,\theta_i)u(k) + V^u_{N-k-1}(\hat{x}^u(k + 1/_Z k+1, \theta_i)\}$$

$$= \bar{V}_{N-k}(\hat{X}(k)/_Z k) \leq V^u_{N-k}(\hat{X}(k)/_Z k) \quad k = 0,...,N$$
$$(7.93)$$

In the above chain of inequalities, the symbol $x^u(k)$ indicates the
system trajectory generated from the minimum upper-bound
approximation. This indicates clearly that the performance criterion
of the lower-bound minimization algorithm is superior to the one
obtained from the upper-bound minimization algorithm. In general,
the chain of inequalities generated originally by (7.68) can be
rearranged in view of the above results (7.87) as follows:

$$V^\ell_{N-k}(\hat{X}(k)/_Z k) \leq V_{N-k}(\hat{X}(k)/_Z k) \leq \underline{V}_{N-k}(\hat{X}(k)/_Z k) \leq \bar{V}_{N-k}(\hat{X}(k)/_Z k)$$
$$\leq V^u_{N-k}(\hat{X}(k)/_Z k) \quad k = 0,...,N \qquad (7.94)$$

The conclusion from this comparison is that the Saridis and
Dao algorithm is not only superior because of simplicity of
implementation, but also because of superiority of performance,
as shown by (7.93).

The Saridis and Dao algorithm bears a strong resemblance with the OLFO algorithm by Lainiotis et al. given in Sec. 7.4.3. As a matter of fact, they may both be produced as special cases of a more general algorithm which is based on the learning property of the *a posteriori* probabilities.

$$\lim_{k \to \infty} p(\theta_i / z^k) = \begin{cases} 1 & \text{if } \theta_i = \theta^* \\ 0 & \text{if } \theta_i \neq \theta^* \end{cases} \qquad (7.95)$$

This property which has been proved by Saridis and Dao [7.23] can be easily verified by solving recursively (7.64) starting at k = 0. Therefore, the *a posteriori* algorithms identify the unknown parameter vector θ, and they do it very fast as demonstrated in Fig. 7.11.

Using the *a posteriori* probabilities learning property, one may generate the following general algorithm:

Define a finite class of feedback control algorithms corresponding to each of the parameter θ_i, i = 1,...,s, among which the combination of optimal LQG control-state estimation with the

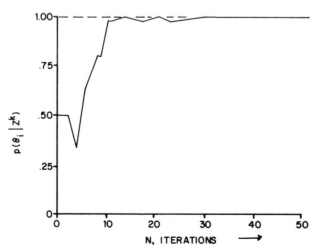

FIG. 7.11. Convergence of A Posteriori Probabilities for $\theta_i = \theta^*$.

true parameters is a number but not known to the designer. Then
weigh them with the respective *a posteriori* probabilities and use
their average as a feedback control. It will converge asymptotically
to the LQG optimal control because of the learning property of
$p(\theta/_Zk)$, Eq. (7.95). It is obvious that the Lainiotis and the
Saridis and Dao algorithms belong to this class of asymptotically
optimal algorithms. Actually, they only differ in the derivation
of the selection of the control gains. The Lainiotis algorithm
selects the average of the control that minimizes the conditional
lower-bounds $V_{N-k}^{\ell}(\hat{X}(k)/_Zk,\theta_i)$ individually, e.g.,

$$u_L^*(k) = \sum_{i=1}^{s} p(\theta_i/_Zk)u_L^*(k,\theta_i); \quad u_L^*(k,\theta_i): \quad V_{N-k}^{\ell}(\hat{X}(k)/_Zk,\theta_i) = \text{Min}$$

while the Saridis and Dao algorithm is selected to minimize the
average lower-bound $V_{N-k}^{\ell}(\hat{X}(k)/_Zk)$

$$u_{SD}^*(k): \quad V_{N-k}^{\ell}(\hat{X}(k)/_Zk) = \text{Min}$$

From the above consideration and in view of the definition of the
performance criterion and cost function involved representing
average performance, the Saridis and Dao algorithm has a lower
cost function $\underline{V}_{N-k}(\hat{X}(k)/_Zk)$ than the Lainiotis algorithm $V_L(\hat{X}(k)/_Zk)$

$$\underline{V}_{-N-k}(\hat{X}(k)/_Zk) \leq V_L(\hat{X}(k)/_Zk) \tag{7.96}$$

Another learning algorithm may be generated by weighting the control
gain by

$$\delta_i = 1 \quad p_i(\theta_i/_Zk) = \text{Max} , \quad \delta_j = 0 \quad j \neq i$$

Finally, simulation results have been obtained on a
second-order system representing a simplified version of the S.O.C.
of an aircraft longitudinal autopilot problem with unknown
lifting-time constant and process noise variance [7.23],[7.30]. The

dynamic equations are discretized and given in the state variable
form by

$$x(k+1) = \begin{bmatrix} 1 & 0.1 \\ \alpha_1(\theta) & \alpha_2(\theta) \end{bmatrix} x(k) + \begin{bmatrix} 0 \\ 0.1 \end{bmatrix} u(k) + \begin{bmatrix} 0 \\ 1 \end{bmatrix} w(k)$$

$$z(k) = \begin{bmatrix} 1 & 0 \end{bmatrix} x(k) + v(k)$$

$$x(0) \sim N[x_o, P_o], w(k) \sim N[0, 0.1\sigma^2], v(k) \sim N[0, 0.1]$$

where the unknown parameters, the damping ratio ξ, and the process
noise covariance σ^2, belong to the two-member parameter space

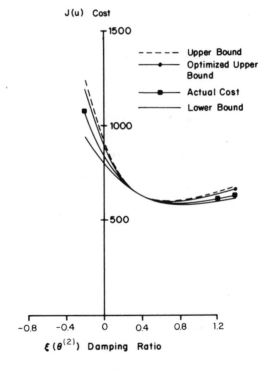

FIG. 7.12.   Comparison of Various Stein and Saridis Solutions
             with the Actual Cost and Upper and Lower Bounds.

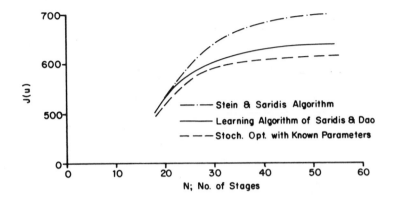

FIG. 7.13.   Comparison of Various Parameter-Adaptive S.O.C.
Techniques with the Stochastic Optimal Solution.

$$\theta \in \Omega_\theta = \left\{ \begin{bmatrix} \sigma_1 = 1.0 \\ \\ \xi_1 = 0.1 \end{bmatrix}, \begin{bmatrix} \sigma_2 = 2.0 \\ \\ \xi_2 = 0.0 \end{bmatrix} \right\}$$

and enter the system parameters as, $\alpha(\theta) = 0.1$, $\alpha_2(\theta) = 1-\xi$.   The
performance criterion is given by

$$J(u) = E\left\{ \sum_{k=0}^{N-1} [x_1^2(k + 1) + x_2^2(k + 1) + u^2(k)] \right\}$$

The problem has been solved by the Stein and Saridis algorithm for
N = 25 and the plots of the performance criterion and the upper and
lower bounds for various true parameter value ξ.   Both algorithms
also were compared with each other and the LQG optimal in Figs. 7.12
and 7.13 and Table 7.1.   The results verify the claims made.

Table 7.1. Cost Functions and Computer Time for a 50-Stage Process

Algorithm	Performance cost	On-line computer time, 20 Samples (CDC-6500)
Stein and Saridis	693.14	241 SEC
Saridis and Dao	636.05	202 SEC
Stochastic optimum	625.00	196 SEC

7.6 PARALLEL IDENTIFICATION AND CONTROL OF SARIDIS AND LOBBIA

One of the most intuitive ideas for a parameter-adaptive S.O.C. system is to separate the state estimation, parameter identification, and control functions from each other and use their outputs to generate the parameter-adaptive controller. Such an idea was suggested among others by Lee [7.18], and was implemented by Saridis and Lobbia for single-input/single-output plants [7.24] and Lobbia and Saridis for multi-input/multi-output plants [7.19]. The main difficulty encountered in solving this problem was that most of the identification algorithms described in Chap. 6 develop biases when used in a feedback control, because of unavoidable correlations between the output of the identifier and the output of the system. The perturbation input stochastic approximation method and the random search algorithm are the only ones known that may eliminate these correlations and provide consistent parameter estimator in a feedback system.

Saridis and Lobbia [7.19],[7.24],[7.25] used a modification of the perturbation input stochastic approximation identification algorithm in a parallel identification estimation and control scheme, which is depicted in Fig. 7.14. Two different types of deterministic controllers have been used and their performance compared. In order to obtain an asymptotically optimal solution, the controller consists of a Kalman-Bucy filter cascaded with a

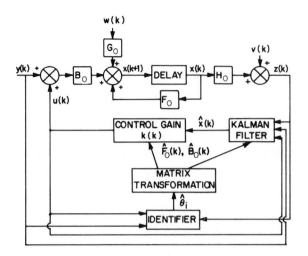

FIG. 7.14. Block Diagram of the Saridis and Lobbia Algorithms.

certain deterministic optimal controller as if the system
parameters were known. However, instead of the true values,
estimates of the parameters are used to implement the elements of
the feedback loop, generated by a parameter identifier. A
stochastic approximation algorithm is used to implement the
identifier for the feedback control system.

7.6.1 Single-Input/Single-Output Feedback System Identification

For a single-input/single-output process the plant equations
are given by

$$x(k + 1) = Fx(k) + b[u(k) + y(k)] + gw(k)$$
$$z(k) = h^T x(k) + v(k)$$

$$(7.97)$$

where

$$F = \begin{bmatrix} 0 & \vdots & I \\ \text{-} & \text{-} & \text{-} \\ f^T & & \end{bmatrix}, \quad b = \begin{bmatrix} b_1 \\ \vdots \\ b_n \end{bmatrix} \quad h^T = [1,0,\ldots,0] \quad g = \begin{bmatrix} g_1 \\ \vdots \\ g_n \end{bmatrix}$$

and $x(k)$ is the n-dimensional state vector, $u(k)$ is the scalar feedback control input, $z(k)$ is the scalar output. $w(k)$ and $v(k)$ are the scalar sample functions of independent identically distributed stationary random processes with zero-mean and variances q^2 and σ^2, respectively, and finite moments up to fourth order, and $\{y(k)\}$ is a random input sequence with known density function used for identification purposes.

The unknown parameter vector $f^T = [f^T, b^T]$ is identified via a first-order stochastic approximation algorithm, appropriate for systems which use the parameter estimates in the feedback loop. The response of the plant (7.97) is

$$z(k + 2n) = [Y_k^{T\,k+2n-1} + U_k^{T\,k+2n-1}]\theta + W_k^{T\,k+2n-1} \cdot \beta$$
$$+ \epsilon(k-1) + v(k + 2n) \tag{7.98}$$

where

$$\theta^T = [h^T b, h^T F b, \ldots, h^T F^{2n-1} b]$$
$$\beta^T = [h^T g, h^T F g, \ldots, h^T F^{2n-1} g]$$
$$Y_k^{T\,k+2n-1} = [y(k + 2n - 1), \ldots, y(k)]$$
$$U_k^{T\,k+2n-1} = [u(k + 2n - 1), \ldots, u(k)]$$

The parameter identification algorithm is

$$\hat{\theta}(k + 2n) = \theta(k - 1) + \gamma\left(\frac{k - 1}{2n + 1}\right) [Y_k^{k+2n-1}[z(k + 2n) -$$
$$- [Y_k^{k+2n-1} + U_k^{k+2n-1}]^T \hat{\theta}(k-1)] \tag{7.99}$$
$$k = 1, \ 2n + 2, \ 4n + 3, \ldots,$$

If $\gamma(\frac{k - 1}{2n + 1})$ and $\theta(0)$ satisfy Dvoretzky's conditions for stochastic approximation (6.31), the algorithm converges to the true parameter vector with probability one and in the mean-square sense. The original parameters may be recovered by the transformation

$$\hat{b}(k) = \begin{bmatrix} \hat{\theta}_1(k) \\ \vdots \\ \hat{\theta}_n(k) \end{bmatrix}, \quad \hat{f}(k) = \begin{bmatrix} \hat{\theta}_1(k), \dots, \hat{\theta}_n(k) \\ \cdot \quad \dots \quad \cdot \\ \hat{\theta}_n(k), \dots, \hat{\theta}_{2n-1}(k) \end{bmatrix}^{-1} \cdot \begin{bmatrix} \hat{\theta}_{n+1}(k) \\ \vdots \\ \hat{\theta}_{2n}(k) \end{bmatrix}$$

$$(7.100)$$

The parameter values $\hat{F}$ and $\hat{b}$ are replacing the true parameters f and b in the Kalman-Bucy state estimator

$$\hat{x}(k + 1/k + 1) = \hat{F}(k)x(k/k) + \hat{b}(k)[u(k) + y(k)] + K(k + 1)$$
$$\left[z(k + 1) - h^T[\hat{F}(k)\hat{x}(k/k) + b(k)(u(k)\right.$$
$$\left. + y(k))]\right]$$

$$(7.101)$$

$$K(k + 1) = P(k + 1/k)h[h^T P(k + 1/k)h + \sigma^2]^{-1}$$

$$P(k + 1/k) = \hat{F}(k)P(k/k)\hat{F}^T(k) + q^2 gg^T$$

$$P(k + 1/k + 1) = [I - K(k + 1)h^T]P(k + 1/k); \qquad P(0/0) = P_0$$

7.6.2   Multiinput/Multioutput Feedback System Identification

A generalization of this algorithm is straightforward for the multi-input/multi output case [7.19]

$$x'(k + 1) = Fx'(k) + B[u(k) + y(k)] + Gw(k) \tag{7.102}$$

$$z(k) = Hx'(k) + v(k)$$

where $u(k)$ is an m-dimensional input vector, $z(k)$ is an r-dimensional output vector, and $H = [h_1 h_2, \dots, h_r]^T$ is an n x r measurement matrix. By using the observability matrix P as a similarly transformation $x(k) = Px'(k)$

$$P^T = [h_1, F^T h_1, h_2, \dots, F^{P_r - 1} h]$$

where p_i, called the *observability subindices* satisfy

$$\sum_{i=1}^{r} p_i = n \tag{7.103}$$

system (7.101) can be transformed into a canonical structure
suitable for identification purposes.

$$x(k + 1) = F_0 x(k) + B_0 [u(k) + y(k)] + G_0 w(k)$$

$$z(k) = H_0 x(k) + v(k)$$

(7.104)

where

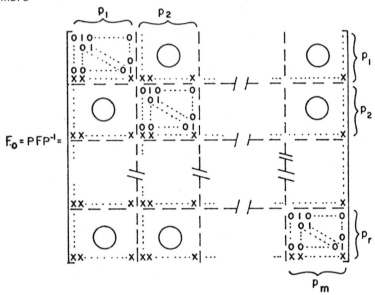

$$B_0 = PB$$
$$D_0 = PD$$

$$H_0 = HP^{-1} = \begin{bmatrix} 1 & 0 & \cdots & & & & & & & & \cdots & 0 \\ 0 & 0 & \cdots & & & 0 & 1 & 0 & & & \cdots & 0 \\ 0 & 0 & \cdots & & & & & 0 & 1 & 0 & \cdots & 0 \\ \cdot & \cdot & & & & & & & & & & \cdot \\ \cdot & \cdot & & & & & & & & & & \cdot \\ \cdot & \cdot & & & & & & & & & & \cdot \\ 0 & 0 & \cdots & & & & & & & 0 & 1 & 0 \cdot 0 \end{bmatrix}$$

The matrix in (7.104) is composed of blocks of phase-variable submatrices situated along the main diagonal. The r rows of x's in F_0 correspond to possible nonzero elements and occur in rows p_1, $p_1 + p_2, \ldots, p_1 + p_2 + \cdots + p_r$. The matrices B_0 and D_0, in general, assume no special form. With the above system structure, it becomes necessary to identify only r x n elements of F_0 and n x m elements of B_0, a total of n x (m+r) unknown parameter coefficients. The r subsystems thus generated have a response

$$z_i(k + g) = [Y_k^{k+g-1} + U_k^{k+g-1}]^T \hat{\theta}_i + W_k^{T k+g-1} \beta_i + \varepsilon_i + v(k + g)$$

$$i = 1, 2, \ldots, r$$

where

$$\theta_i^T = [h_i^T F_0, \ h_i^T F_0 B_0, \ldots, h_i^T F_0^{s-1} B_0)]$$

$$\beta_i^T = [h_i G_0, \ h_i^T F_0 G_0, \ldots, h_i^T F_0^{s-1} G_0)] \tag{7.105}$$

$$\text{and } s = \max [p_i] + \max [q_i] \quad i = 1, \ldots, m$$

p_i and q_i are the observability and controllability subindices and

$$\varepsilon_i = h_i^T \sum_{i=0}^{\infty} F_0^{i+s} [B_0 y(k - i - 1) + B_0 u(k - i - 1)] + G_0 w(k - i - 1)$$

The identification algorithm becomes

$$\hat{\theta}_i(k + g) = \hat{\theta}_i(k - 1) + \gamma\left(\frac{k - 1}{g + 1}\right) Y_k^{k+g-1} [z_i(k+g)$$

$$- [Y_k^{k+g-1} + U_k^{k+g-1}]^T \hat{\theta}(k-1)] \quad k = 1, g+2, 2g+3, \ldots$$

$$i = 1, 2, \ldots, r \tag{7.106}$$

It will be argued later that the convergence of the identification algorithm (7.106) depends strongly on the stability of the closed-loop system. Such a proof of convergence is given by Lobbia and Saridis [7.19]. The estimates of $F_0(k)$ and $B_0(k)$ of the unknown system matrices are given by

$$F_0 = \begin{bmatrix} \theta_1^{m+1} & \cdots & \theta_1^{q_1 m+1} & \theta_1^{m+2} & \cdots & \theta_1^{q_m m+m} \\ \theta_1^{2m+1} & \cdots & \theta_1^{(q_1+1)m+1} & \theta_1^{2m+2} & & \theta_1^{q_m m+2m} \\ \vdots & & \vdots & \vdots & & \\ \theta_1^{p_1 m+1} & \cdots & \theta_1^{(q_1+p_1-1)m+1} & \theta_1^{p_1 m+2} & \cdots & \theta_1^{(q_m+p_1-1)m+m} \\ \theta_2^{m+1} & \cdots & \theta_2^{q_1 m+1} & \theta_2^{m+2} & \cdots & \theta_2^{q_m m+m} \\ \vdots & & \vdots & \vdots & & \vdots \\ \theta_r^{p_r m+1} & \cdots & \theta_r^{(q_1+p_r-1)m+m} & \theta_r^{p_r m+2} & \cdots & \theta_r^{(q_m+p_r-1)m+m} \end{bmatrix}$$

$$\begin{bmatrix} \theta_1^{1} & \cdots & \theta_1^{(q_1-1)m+1} & \theta_1^{2} & \cdots & \theta_1^{q_r m} \\ \theta_1^{m+1} & \cdots & \theta_1^{q_1 m+1} & \theta_1^{m+2} & \cdots & \theta_1^{q_m m+m} \\ \vdots & & \vdots & \vdots & & \vdots \\ \theta_1^{(p_1-1)m+1} & \cdots & \theta_1^{(q_1+p_1-2)m+1} & \theta_1^{(p_1-1)m+2} & \cdots & \theta_1^{(q_m+p_1)m} \\ \theta_2^{1} & \cdots & \theta_2^{(q_1-1)m+1} & \theta_2^{2} & \cdots & \theta_2^{q_m m} \\ \vdots & & \vdots & \vdots & & \vdots \\ \theta_r^{(p_r-1)m+1} & \cdots & \theta_r^{(q_1+p_r-2)m+1} & \theta_r^{(p_m-1)m+2} & \cdots & \theta_r^{(q_m+p_r-1)m} \end{bmatrix}^{-1}$$

$$(7.107)$$

$$B_0 = \left[\begin{array}{ccccc}
\theta_1^1 & & \theta_1^2 & \cdots & \theta_1^m \\
\vdots & & \vdots & & \vdots \\
\theta_1^{(P_1-1)m+1} & & & & \theta_1^{P_1 m} \\
\hline
\theta_2^1 & & & \cdots & \theta_2^m \\
\vdots & & & & \vdots \\
\theta_2^{(P_2-1)m+1} & & & \cdots & \theta_2^{P_2 m} \\
\hline
& & & \vdots & \\
\hline
\theta_r^1 & & \theta_r^2 & \cdots & \theta_r^m \\
\vdots & & & & \\
\theta_r^{(P_r-1)m+1} & & & \cdots & \theta_m^{P_r m}
\end{array}\right] \begin{array}{l} \Big\}\, P_1 \\[1em] \Big\}\, P_2 \\[1em] \\ \Big\}\, P_r \end{array}$$

The observability matrix P in (7.103) and the similarly obtained controllability matrix Q, may serve to identify the order of the systems n and the proper partitions.

$$Q = [b_1, Fb_1, \ldots, F^{q_1-1} b_1, b_2, Fb_2, \ldots, F^{q_2-1} b_2, b_3, \ldots, F^{q_m-1} b_m] \qquad (7.108)$$

where B has been partitioned as

$$B = [b_1, b_2, \ldots, b_m]$$

and the q_i's in (7.106), referred to as *controllability subindices* must satisfy,

$$\sum_{i=1}^{m} q_i = n$$

Assuming that the system is completely observable and completely controllable, the matrix PQ is nonsingular only if the number of elements and therefore the order of the system n and the **observability and controllability subindices** have been selected correctly.

$$PQ = \begin{bmatrix}
\theta_1^1 & \theta_1^{m+1} & \cdots & \theta_1^{(q_1-1)m+1} & \theta_1^2 & \cdots & \theta_1^{q_2-m} \\
\theta_1^{m+1} & \theta_1^{2m+1} & \cdots & \theta_1^{q_i m+1} & \theta_1^{r+2} & \cdots & \vdots \\
\vdots & \vdots & & \vdots & & & \vdots \\
\theta_1^{(p_1-1)m+1} & \theta_1^{p_1 m+1} & \cdots & \theta_1^{(q_1+p_1-2)m+i(p_1-1)m+2} & \theta_1^{(q_m+p_1-1)m} & \cdots & \theta_1 \\
\theta_2^1 & \theta_2^{m+1} & \cdots & \theta_2^{(q_1-1)m+1} & \theta_2^2 & \cdots & \theta_2^{q_m-m} \\
\vdots & \vdots & & \vdots & & & \vdots \\
\theta_m^{(p_r-1)r+1} & \theta_r^{p_r r+1} & \cdots & \theta_r^{(q_1+p_r-2)r+1(p_r-1)m+2} & \theta_r^{(q_m+p_r-1)r} & \cdots & \theta_r
\end{bmatrix}$$

$$\text{(7.109)}$$

$$= \begin{bmatrix}
h_1^T b_1 & h_1^T F b_1 & \cdots & h_1^T F^{q_1-1} b_1 & h_1^T b_2 & \cdots & h_1^T F^{q_m-1} b_m \\
h_1^T F b_1 & h_1^T F^2 b_1 & \cdots & h_1^T F^{q_1} b_1 & h_1^T F b_2 & \cdots & h_1^T F^{q_m} b_m \\
\vdots & \vdots & & \vdots & \vdots & & \vdots \\
h_1^T F^{p_1-1} b_1 & \cdots & & h_1^T F^{p_1+q_1-2} b_1 & h_1^T F_c^{p_1-1} b_2 & \cdots & h^T F^{q_m-1} b_m \\
h_2^T b_1 & \cdots & & h_2^T F^{q_1-1} b_1 & \cdots & & h_2^T F^{q_m-1} b_m \\
\vdots & & & \vdots & & & \vdots \\
h_r^T F^{p_r-1} b_1 & \cdots & & h_r^T F^{p_r+q_1-2} b_1 & \cdots & & h_r^T F^{p_r+q_m-2} b_m
\end{bmatrix}$$

This may generate a test for the proper selection of n, p_i, and q_j, if they are not known *a priori* [7.19].

The multiinput/multioutput case will be used for the development of the control gains in the next section and an appropriate multivariable version of the Kalman-Bucy filter of (7.101), also given in the next sections, will be used for the state estimation.

7.6.3 Minimization of an Overall Performance Criterion

The control problem is defined as the search for a control sequence $U^{N-1} \triangleq \{u(0),...,u(n-1)\}$ that minimizes the performance criterion

$$J(u) = \lim_{N \to \infty} E \frac{1}{N} \left\{ \sum_{i=0}^{N} [x^T(i+1)Mx(i+1) + u^T(i)Nu(i)] \right\} \quad (7.110)$$

for the multivariable case where $M = M^T \geq 0$ and $N = N^T > 0$. The process is defined over an infinite time interval, since the asymptotic behavior of this algorithm is in focus.

A controller which will give asymptotically optimal results may be designed as follows:

$$u^*(k) = -A(k)\hat{x}(k/k)$$

$$A(k) = [N + \hat{B}_0^T(k)S^\infty B_0(k)]^{-1}\hat{B}_0^T(k)S^\infty F_0(k) \quad (7.111)$$

$$S^\infty = M + \hat{F}_0^T(k)S^\infty [I - \hat{B}_0(k)[N - \hat{B}_0^T(k)S^\infty \hat{B}_0(k)]^{-1}S^\infty \hat{B}_0^T(k)$$

$$- S^\infty]\hat{F}_0(k) \quad k = 1, 2,...$$

where S^∞ and $A(k)$ must be updated whenever new estimates $\hat{F}_0$ and $\hat{B}_0$ are produced as well as $\hat{x}(k/k)$ from the Kalman-Bucy estimator.

$$\hat{x}(k+1/k+1) = \hat{F}_0(k)\hat{x}(k/k) + \hat{B}_0(k)[u(k) + y(k)] + K(k+1)$$

$$\cdot \left[z(k+1) - H[\hat{F}_0(k)\hat{x}(k/k) + \hat{B}_0(k)(u(k)+y(k))] \right]$$

$$K(k+1) = P(k+1/k)H_0^T[H_0P(k+1/k)H_0^T + R]^{-1}$$

$$P(k+1/k) = \hat{F}_0(k) P(k/k)\hat{F}_0^T(k) + G_0QG_0^T \quad (7.112)$$

$$P(k+1/k+1) = [I - K(k+1)H]P(k+1/k)$$

This particular controller has given interesting results for both the single/input-output and multi-variable cases, especially in view of the asymptotic optimality of the algorithm. The stability of this solution will be discussed in Sec. 7.6.5.

7.6.4   Minimization of a Per-Interval Performance Criterion

According to the Saridis and Lobbia algorithm, the parameter
identification and the state estimation and control are performed
independently of one another and one may select a control to require
a simpler design and implementation while it still yields a good
approximation to the overall minimization problem.   Therefore a
large collection of potential controllers is available to the
designer.

The subgoal approach is a method whereby the approximate
controller's form can be appreciably simplified by minimizing,
instead of the overall performance criterion of (7.110), a sequence
of single-stage processes of the same form as (7.104) with a
performance criterion defined per stage

$$J_I(u) = E\{x^T(k + 1)Mx(k + 1) + u^T(k)Nu(k)\} \qquad k = 0, 1, 2,\ldots$$

$$\tag{7.113}$$

A control law may be formulated to minimize (7.113), subject to
(7.104) at each step k.   The result is

$$u_I^*(k) = -K_I(k)\hat{x}(k/k)$$

$$K_I(k) = [\hat{B}_0(k)^T M\hat{B}_0(k) + N]^{-1}\hat{B}_0^T(k)M\hat{F}_0(k) \tag{7.114}$$

In (7.112) $\hat{x}(k/k)$ is produced by the Kalman-Bucy filter of
(7.112) and the matrices $\hat{B}_0(k)$ and $\hat{F}_0(k)$ are the ones produced by
(7.107) when the new estimates $\hat{\theta}(k)$ are generated by (7.105).

This approach, however, does not yield the asymptotically
optimal control obtained in the preceding section, but it represents
a far simpler design requiring simpler computation, since (7.114)
do not have to be resolved for every k.   It is very simple also to
switch to the asymptotically optimal control (7.111) when the
identification is complete and the parameters θ are recovered with
satisfactory accuracy.   The transient as well as the asymptotic
stability investigation of this algorithm is similar to the

previous one and will be discussed in the next section.

The interesting fact is that experimental results indicated that this algorithm yielded in all the cases tested an "accrued" cost smaller than the overall performance case. This also will be discussed and explained in the next section.

7.6.5 Stability, Convergence and Comparisons

The first concern in an asymptotically behaving algorithm is the overall stability of the system. However, the stability investigation in a stochastic system is highly complicated by the various notions and definitions of stochastic stability [7.1], [7.3]. Such an investigation has been explored in the sequel and has found rather limited applicability due to its complexity.

A simpler procedure may be developed to guarantee bounded input bounded output (BIBO) stability, because the plant (7.104) and the feedback estimator control (7.112) with (7.111) or (7.114) are linear in $x(k)$ and $\hat{x}(k/k)$. A further assumption must be made about the process and measurement noises $w(k)$ and $v(k)$ and the perturbation input $y(k)$, that they are bounded random variables, in which case only truncated Gaussian random variables are acceptable. For a very large truncation value this is a reasonable approximation to Gaussian random variables, and therefore would insignificantly affect the formulation of the optimal LQG solution. After all, there are no sources in nature that produce unbounded signals even with infinitesimal probability. Combining (7.104) and (7.112) one obtains the 2n-dimensional overall closed-loop system with inputs the noises $w(k)$, $v(k)$ and the perturbation input $y(k)$

$$
\begin{bmatrix} x(k+1) \\ \hline \hat{x}(k+1/k+1) \end{bmatrix} = \begin{bmatrix} F_0 & -B_0 K(k) \\ \hline K(k+1)H_0 F_0 & L(k) \end{bmatrix} \begin{bmatrix} x(k) \\ \hline \hat{x}(k/k) \end{bmatrix}
$$

$$
+ \begin{bmatrix} G_0 & 0 \\ \hline K(k+1)H_0 G_0 & K(k+1) \end{bmatrix} \begin{bmatrix} w(k) \\ \hline v(k) \end{bmatrix} + \begin{bmatrix} B_0 \\ \hline \hat{B}_0(k) \end{bmatrix} y(k) \quad k = 0, 1, \ldots
$$

$$(7.115)$$

where F_0, B_0, and G_0 are defined in (7.104) and $K(k)$ in either (7.111) or (7.114) and

$$L(k) \triangleq \hat{F}_0(k) - \hat{B}_0(k)K(k) - K(k + 1)H_0[\hat{F}_0(k) - \hat{B}_0(k)K(k)]$$

For bounded inputs $w(k)$, $v(k)$, and $y(k)$, $k = 0, 1, 2, \ldots,$ the above system is BIBO stable if and only if the eigenvalues

$$\lambda_i \begin{bmatrix} F_0 & -B_0 K(k) \\ \hline K(k + 1)H_0 F_0 & L(k) \end{bmatrix} \quad i = 1, 2, \ldots, 2n$$

lie inside the unit circle

$$|\lambda_i(k)| < 1 \qquad i = 1, \ldots, 2n \quad k = 0, 1, \ldots \qquad (7.116)$$

The 2n-inequalities (7.116) or the equivalent ones that may be produced by the Schur-Cohn criterion or any other stability criteria [7.25] are functions of both true parameters θ and their estimates $\hat{\theta}(k)$. If bounds on θ are known, e.g.,

$$\theta \in \Omega_\theta \triangleq \{\alpha_i \leq \theta_i \leq \beta_i; \ i = 1, 2, \ldots, 2n\} \qquad (7.117)$$

then (7.117) can be adjoined to (7.116) or any other criterion establishing the BIBO stability of the system (7.115) can be solved [7.25] to yield the range Ω_θ of the estimates $\hat{\theta}$, which would guarantee (7.117) or the equivalent for all k.

$$\hat{\theta}(k) \in \Omega_{\hat{\theta}} = \{\gamma_i \leq \hat{\theta}_i(k) \leq \zeta_i; \ i = 1, \ldots, 2n : |\lambda_i(k)| \leq 1\}$$
$$k = 0, 1, \ldots \qquad (7.118)$$

It was shown by Saridis and Lobbia [7.25] that the convergence of the identification algorithms (7.99) is closely related to the system BIBO stability. As a matter of fact, the condition

$$1 > \sum_{i=1}^{n} \sum_{j=1}^{n} |\hat{\theta}_i \hat{n}_j| \qquad\qquad (7.119)$$

where $\hat{n}_j = h^T L^{j-1}(\hat{\theta})K(k+1)h$, is sufficient for both the
convergence of (7.99) and the BIBO stability of the system (7.115).
Extension to cover algorithm (7.106) is straightforward.

The comparison of the two algorithms presented in Secs. 7.6.3
and 7.6.4 was experimentally tested by the author and his colleagues
on many examples [7.19],[7.24],[7.26]. The general conclusion was
that the per-interval control usually yielded a lower value of performance and identification error than the overall control. A con-
jectural explanation of this phenomenon is that even though only pas -
sive information is transmitted by the overall control to the end of
the process and back, the parameter estimation errors, at time k,
are multiplied thus increasing the cost of performance. On the
other hand, the per-interval algorithm localizes the identification
error and does not accumulate additional cost. Variations of the
per-interval algorithm by considering a j-interval criterion,
j = 2, 3,..., are possible compromises between the two presented
algorithms with possible improvement of performance.

A second-order single-input/single-output system has been
selected as an example for the Saridis and Lobbia algorithms

$$x(k+1) = \begin{bmatrix} 0 & 1 \\ f_1 & f_2 \end{bmatrix} x(k) + \begin{bmatrix} 0 \\ 1 \end{bmatrix} (u(k) + y(k)) + \begin{bmatrix} 0 \\ 0.4 \end{bmatrix} w(k)$$

$$z(k) = [1 \ 0]x(k) + v(k)$$

with

$$J(u) = E\left\{ \frac{1}{3000} \sum_{i=1}^{3000} [x_1^2(i+1) + 5u^2(i)] \right\}$$

and $f_1 = -0.8$, $f_2 = 0.1$, and $\beta = 1$,

$$w(k) \sim N[0,0.04], \quad v(k) \sim N[0,0.06], \quad y(k) \sim N[0,2]$$

The computational results and comparisons are depicted in Figs. 7.15
and 7.16. Comparisons with other algorithms are given in Chap. 9.

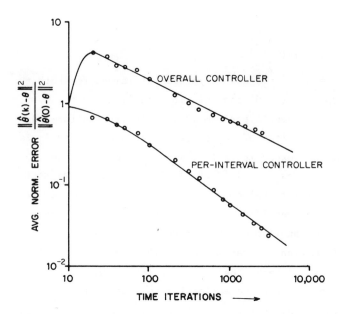

FIG. 7.15.   Identifier Time-History Comparison Second-Order
Example for Saridis-Lobbia Algorithm.

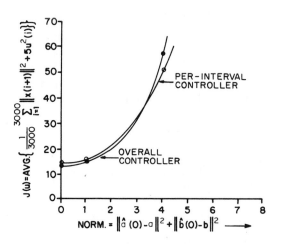

FIG. 7.16.   Performance Comparison Second-Order Example of
Overall and Per-Interval Control for the Saridis
and Lobbia Algorithm.

7.7 THE ACTIVELY ADAPTIVE ALGORITHM OF TSE AND BAR-SHALOM

An approximately optimal algorithm that takes into account the active use of information or, in other words, takes into account the future measurements when it compiles the present control, has been developed by Tse and Bar-Shalom and has been presented in its various stages of development in Refs. [7.34],[7.35],[7.36]. It considers a nonlinear system of the form

$$X(k + 1) = F(X(k),u(k),k) + w(k) \quad k = 0, 1,...,N - 1 \qquad (7.120)$$

$$z(k) = h(X(k),k) + v(k) \quad k = 1,...,N$$

where $u(k)$ is the m-dimensional control input, $f(...)$ and $h(..)$ are bounded functions of this argument, and X is the augmented state vector of $(n + s)$ dimension in which the last s components compose the unknown parameter vector θ, obeying the following general difference equation discussed previously as (7.2):

$$\begin{bmatrix} X_{n+1}(k + 1) \\ \vdots \\ X_{n+s+1}(k + 1) \end{bmatrix} = \hat{\theta}(k + 1) = A(k)\hat{\theta}(k) + \begin{bmatrix} w_{n+1}(k) \\ \vdots \\ w_{n+s}(k) \end{bmatrix} \qquad (7.121)$$

In the above equations the initial state $x(0)$ is a random variable with mean $\hat{x}(0/0)$ and covariance P_0 known, and the process noise $w(k)$ and measurement noise $v(k)$ are, respectively, independent identically distributed random variables, mutually independent with statistics up to second order known, e.g., zero mean and covariances $Q(k)$ and $R(k)$, respectively. The performance criterion or cost function for the system is defined by

$$J(u) = E\left\{ \sum_{k=0}^{N-1} L(X(k + 1),k) + N(u(k),k) \right\} \qquad (7.122)$$

It is assumed that, at time k, all the information accumulated from the past measurements Z^k has been used in obtaining the best

estimate of the state $\hat{x}(k/k)$. This is obtained through the one of
the nonlinear filtering techniques described in Sec. 3.5. The
optimal control must satisfy the Bellman equation of dynamic
programming on the optimal cost $V_{N-k}(X(k)/_Z k)$ given by (4.35)

$$V_{N-k}(X(k)/_Z k) = \min_{u(k)} E\{L(X(k + 1),k) + N(u(k),k) + V_{N-k-1}(X(k+1)/_Z k+1)/$$

$$_Z k \} \qquad\qquad\qquad (7.123)$$

It has been demonstrated before that the dual optimal solution of
this equation is practically impossible to obtain. An approximation
that will preserve the active use of information is sought. To do
that a linearization of the system is performed about a nominal
trajectory $X_0(j)$, $k + 1 \leq j \leq N$. The nominal trajectory is defined
to satisfy

$$X_0(j + 1) = f(X_0(j)u_0(j),j) \quad j = k + 1,\ldots,N-1$$
$$X_0(k + 1) = \hat{X}(k + 1/k,u(k)) \qquad\qquad (7.124)$$

$U_{k+1}^{N-1} = \{u_0(j),j = k+1,\ldots,N - 1\}$ is the control sequence to be
selected as the deterministic nominal controls optimal control for
(7.124). The initial condition $\hat{X}(k + 1/k,u(k))$ depending on a
control $u(k)$ is the predicted value of $X(k + 1)$ given Z^k for the
optimal control $u^*(k)$ yet to be found. It is obvious by the
definition of the approximation that it starts at $(k + 1)$ after
$u(k)$ was applied. The cost to go is redefined to include
$L(X(k + 1),k)$ and is expanded about the nominal trajectory,
retaining only up to second-order terms.

$$V_{N-k-1}(X(k+1)) \triangleq L(X(k+1),k) + V_{N-k-1}(X(k + 1)/_Z k+1)$$

$$= V_{N-k-1}^0(X(k + 1)) + \Delta V_{N-k-1}^0 \qquad\qquad (7.125)$$

where, for $N(u_0(N),N) \neq 0$, $N_u(u_0(N),N) \neq 0$, $N_{uu}(u_0(N),N) \neq 0$

$$V^o_{N-k-1}(X(k+1)) \triangleq \sum_{j=k+1}^{N} [L(X_0(j),j-1) + N(u_0(j),j)] \qquad (7.126)$$

and

$$\Delta V^o_{N-k-1} \triangleq \underset{\delta U^{N-1}_{k+1}}{\text{Min}} \sum_{j=k+1}^{N} [L_X(X_0(j),j-1)\delta X(j) \qquad (7.127)$$

$$+ \frac{1}{2}\delta X^T(j)L_{XX}(X_0(j))\delta X(j)$$

$$+ N_u(u_0(j),j)\delta u(j) + \frac{1}{2}\delta u^T(j)N_{uu}(u_0(j),j)\delta u(j)]$$

In the above equations the quantities $\delta X(j)$ and $\delta u(j)$ are defined by

$$\delta X(j) \triangleq X(j) - X_0(j) \qquad \delta u(j) = u(j) - u_0(j)$$

and

$$L_X \triangleq \frac{\partial L}{\partial X} \quad L_{XX} \triangleq \frac{\partial^2 L}{\partial X^2}$$

$$N_u \triangleq \frac{\partial N}{\partial u} \quad N_{uu} \triangleq \frac{\partial^2 N}{\partial u^2}$$

the Jacobian and the Hessian matrices of $L(X,k)$ and $N(X,k)$ respectively. From the definition of the nominal trajectory (7.124) and the dynamics of the system (7.120), the perturbation δX and δu satisfy the following approximate dynamic equations, where the quadratic terms have been kept to balance the terms containing active information:

$$\delta X(j+1) = f_X(X_0(j),u_0(j),j)\delta X(j) + f_u(X_0(j),u_0(j),j)\delta u(j)$$

$$+ \frac{1}{2}[\delta X^T(j),\delta u^T(j)]F_{Xu}(X_0(j),u(j)) \begin{bmatrix} \delta X(j) \\ \text{----} \\ \delta u(j) \end{bmatrix} + w(j)$$

$$j = k+1,\ldots,N-1 \qquad (7.128)$$

$$\delta X(k+1) = X(k+1) - X_0(k+1) = X(k+1) - \hat{X}(k+1/k),u(k))$$

where $f_X \triangleq \frac{\partial f}{\partial X}$ $f_u = \frac{\partial f}{\partial u}$

and $F_{Xu} \triangleq \begin{bmatrix} f_{XX} & | & f_{uX} \\ ---&-&--- \\ f_{Xu} & | & f_{uu} \end{bmatrix} = \begin{bmatrix} \frac{\partial^2 f}{\partial x^2} & | & \frac{\partial^2 f}{\partial u \partial X} \\ ---&-&--- \\ \frac{\partial^2 f}{\partial X \partial u} & | & \frac{\partial^2 f}{\partial u^2} \end{bmatrix}$

The problem, now defined, consists of minimizations of (7.127)
subject to (7.128), which is solvable, in spite of the quadratic
structure of the system, by always ignoring the higher-order terms.
The active feedback solution yields

$$\Delta V^o_{N-k-1} = \gamma_0(k + 1) + E\{q_0^T(k + 1)\delta X(k + 1) + \tfrac{1}{2}\delta X^T(k + 1)K_0(k + 1)$$

$$\delta X(k + 1)/_z k + 1\} + \frac{1}{2} \sum_{j=k+1}^{N-1} tr[K_0(j + 1)Q(j)$$

$$+ A^o_{XX}(j)P_0(j/j)] \tag{7.129}$$

The above minimization is presented in detail in refs. [7.34] and [7.35]
where the following definitions are made:

$$H^o_X(j) \triangleq L(X_0(j),j) + N(u_0(j),j) + P_0^T(j + 1)f(X_0(j),u_0(j)j)$$

$$H^o_{XX}(j) \triangleq H^o_{XX}(j) + f_X^T(X_0(j),u_0(j),j)K_0(j + 1)f_X(X_0(j),u_0(j),j)$$

$$H^o_{uX}(j) \triangleq H^o_{uX}(j) + f_u^T(X_0(j),u_0(j),j)K_0(j + 1)f_u(X_0(j),u_0(j),j)$$

$$H^o_{uu} \triangleq H^o_{uu}(j) + f_u^T(X_0)(j),u_0(j),j)K_0(j + 1)f_u(X_0(j),u_0(j),j)$$

$$A^o_{XX}(j) \triangleq H^{oT}_{uX}(j)H^o_{uu}{}^{-1}(j)H^o_{uX}(j) \tag{7.130}$$

$$q_0(j) = H^o_X(j) - H^{oT}_{uX}(j)H^o_{uu}{}^{-1}(j)H^o_u(j) \qquad q_0(N) = L_X(X_0(N))$$

$$k_0(j) = H^o_{XX}(j) - A^o_{XX}(j) \qquad\qquad K_0(N) = L(X_0(N))$$

$$\gamma_0(j) = \gamma_0(j + 1) - H^{oT}_u(j)H^o_{uu}{}^{-1}(j)H^o_u(j) \qquad \gamma_0(N) = 0$$

$$j = N - 1,\dots,k + 1$$

The active feedback control is given by

$$\delta u^*(j) = -H_{uu}^{o^{-1}}(j)[H_{ux}^{o}(j)\delta\hat{X}(j/j) + H_{u}^{o}(j)] \tag{7.131}$$

where

$$\delta X(j/j) \triangleq E\{\delta X/_z j\}$$

and $S_{\alpha}(\alpha)$, $S_{\alpha\alpha}(\alpha)$ are the Jacobian and Hessians of the arbitrary quantities S with respect to α. The approximation of the active feedback is made obvious in (7.129) from the dependence of ΔV_{N-k-1}^{o} in $P_{o}(j/j)$, the covariance of the state along the nominal trajectory.

In order to obtain the optimal control $u^*(k)$, (7.126) and (7.129) are combined to yield

$$V_{N-k}(X(k)/_z k) = \underset{u(k)}{\text{Min}}\ E\{N(u(k),k) + V_{N-k-1}^{o}(X_{0}(k+1))$$

$$+\ \gamma_{0}(k+1) + q_{0}^{T}(k+1)\delta X(k+1)$$

$$+\ \frac{1}{2}\ \delta X^{T}(k+1)K_{0}(k+1)\delta X(k+1)$$

$$+\ \frac{1}{2}\ \sum_{j=k+1}^{N-1} \text{tr}[K_{0}(j+1)Q(j) + A_{xx}^{o}(j)P_{0}(j/j)]/_z k\} \tag{7.132}$$

Considering (7.124) and (7.127) from where

$$E\{\delta X(k+1)/_z k\} = 0$$

$$E\{\delta X^{T}(k+1)K_{0}(k+1)\delta X(k+1)/_z k\} = \text{tr}[K_{0}(k+1)P(k+1/k)] \tag{7.133}$$

the following *deterministic* expression must be minimized to yield

$$u^*(k): \underset{u(k)}{\text{Min}}\ [N(u(k),k) + V_{N-k-1}^{o}(X_{0}(k+1)) + \gamma_{0}(k+1)$$

$$+\ \frac{1}{2}\ \text{tr}[K_0(k+1)P(k+1/k)] + \frac{1}{2}\ \sum_{j=k+1}^{N-1} \text{tr}[K_0(j+1)Q(j)$$

$$+\ A_{xx}^{o}(j)P_{0}(j/j)]] \tag{7.134}$$

After $u^*(k)$ is computed from (7.134) and applied, and the new
measurement $z(k + 1)$ is obtained, the procedure (7.128)-(7.134) is
repeated. The prior estimate $P_0(j/j)$ of the revised state
covariance $P(j/j)$ entering in the control evaluation introduces
the active use of information by the control, in the sense of
Fel'dbaum, and justifies the name of the algorithm.

It is apparent that this algorithm is extremely complex to
implement, which is the trade off for the account of an approximate
active feedback control.

Simulation results have been obtained from the following
third-order example and are depicted in Fig. 7.17 and 7.18.

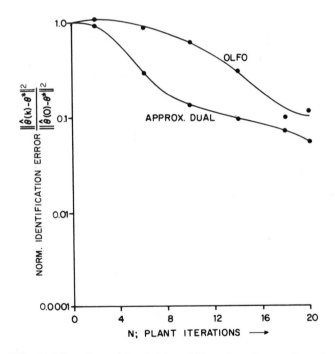

FIG. 7.17.   Normalized Identification Error for the
            Approximately Dual Optimal Control.

$$x(k + 1) = \begin{bmatrix} 0 & \vdots & 1 \\ --- & - & --- \\ \theta_1 & \theta_2 & \theta_3 \end{bmatrix} x(k) + \begin{bmatrix} \theta_4 \\ \theta_5 \\ \theta_6 \end{bmatrix} u(k) + w(k)$$

$$z(k) = [0,0,1]x(k) + v(k)$$

where

$$w(k) \sim N[0,1] \quad v(k) \sim N[0,1] \quad k = 0,\ldots,N - 1$$

$$= [1.8, -1.01, 0.58, 0.3, 0.5, 1]^T$$

$$\hat{\theta}(0/0) = [1, -0.6, 0.3, 0.1, 0.7, 1.5]$$

$$P_\theta(0/0) = \text{diag} [0.1, 0.1, 0.01, 0.01, 0.01, 0.1]$$

$$\hat{x}(0/0) = x(0) = 0$$

and

$$J(u) = \frac{1}{2} E\{ (x(N)-20)^2 + 10^{-3} \sum_{i=0}^{N-1} u^2(i)\} \quad N = 20$$

The normalized identification error is compared with the Athans and Tse OLFO algorithm extended by Ku and Athans [7.15], in Fig. 7.17. The performance of the controller is compared in Fig. 7.18 where it is obvious that the approximate dual control of Tse and Bar-Shalom consumes more energy. However, as it is drawn from Table 7.2, the total cost of reliability of results (standard deviation of cost) and average miss from the target point $x_3 = 20$ are superior in the dual control case and compare favorably with the LQG optimal.

7.8   DISCUSSION

Ten different algorithms qualifying as parameter-adaptive self-organizing have been presented in this chapter. They represent the various points of view of their creators on the problem of optimization of systems with unknown parameter vectors. They represent approximations by linearization, open-loop feedback optimal, upper- and lower-bound minimization, parallel identification and control, and actively adaptive, of the optimal dual solution.

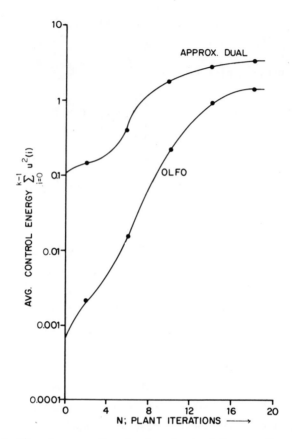

FIG. 7.18.   Average Cumulative Control Energy for the
Interception Example.

Most of these algorithms will be compared in Chap. 9 with each
other and with the performance-adaptive S.O.C. algorithms of
Chap. 8.   Therefore, little will be said here about their a
advantages and disadvantages.   It is clear, however, from the
above investigation that the most complicated ones are not always
the ones that yield the lowest cost.   The Saridis and Dao algorithm
and the Saridis and Lobbia algorithms represent the best combination
of simplicity and good approximation to the optimal.   They also are
the only ones in which a reasonable stability investigation which

TABLE 7.2.   Comparison of Final Results for Tse-Bar-Shalom Algorithm.

Control	LQG Optimal With Known θ	OLFO, Tse-Athans With Unknown θ	Approximate Dual Tse-Bar-Shalom
Average Cost $V_N(X(0))$	6.0	114.0	14.0
Standard Dev. of cost	6.0	140.0	16.0
Average square miss distance	12	225	22.00
Cumm. control energy, prior to final stage	0.1×10^{-3}	1.4×10^{-3}	3.2×10^{-3}

is so necessary for any practical design, can be performed off-line.
Practical application of several of these algorithms have been
reported [7.14],[7.19],[7.23],[7.24],[7.36], however due to space
limitations they have not been reported here. Variations of some of
the algorithms, like for instance the use of random search for
identification in the Saridis and Lobbia algorithm has not been
discussed in spite of its relative speed of convergence noted in
Sec. 6.7.   Such minor improvements are left as an exercise to the
reader.

The last remark is made in support of the author's selection
of the algorithms presented. Even though the account is not complete,
it is felt that the most representative ones have been discussed.

The next chapter will cover the performance-adaptive S.O.C.
algorithm and will represent a different philosophy of the problem
of designing effective controls for systems with uncertainties.

7.9 REFERENCES

[7.1]   Aoki, M., *Optimization of Stochastic Systems*, Academic Press, New York, 1967.

[7.2]   Åström, K. K., "On Self Tuning Regulators," *Automatica* - IFAC J. <u>9</u>, (2) 185-199 (1973).

[7.3]   Bellman, R., *Adaptive Control Processes: A Guided Tour*. Princeton University Press, Princeton, N.J., 1961.

[7.4]   Bar-Shalom, Y., and Sivan, R., "On the Optimal Control of Discrete-Time Linear Systems with Random Parameters," *IEEE Trans. Automatic Control,* <u>AC-14</u>, (1) 3-8 (1969).

[7.5]   Cooper, G. R., Gibson, J. E., *et al.*, "A Survey of the Philosophy and State of the Art of Adaptive System," Tech. Rept. PRF 2358, Purdue University, Lafayette, Ind., July 1960.

[7.6]   Cox, H., "On The Estimation of State Variables and Parameters for Noisy Dynamic Systems," *IEEE Trans. Automatic Control,* <u>AC-4</u> (1) 5-12 (1964).

[7.7]   Curry, R. E., "A New Algorithm for Suboptimal Stochastic Control," *IEEE Trans. Automatic Control,* <u>AC-14</u>, (5) 533-536 (1969).

[7.8]   Deshpande, J. G., Upadhyay, T. N., and Lainiotis, D. G., "Adaptive Control of Linear Stochastic Systems," *Automatica* - IFAC J. <u>9</u>, 107-115 (1973).

[7.9]   Eveleigh, V. W., *Adaptive Control and Optimization Techniques*, McGraw-Hill, New York, 1967.

[7.10] Farison, J. B., Graham, R. E., and Shelton, R. C., "Identification and Control of Linear Discrete-Time Systems," *IEEE Trans. Automatic Control,* <u>AC-12</u>, (4) 438-442 (1967).

[7.11] Fel'dbaum, A. A., *Optimal Control Systems*, Academic Press, New York, 1965.

[7.12] Florentin, J. J., "Optimal Probing, Adaptive Control of a Simple Bayesian System," *J. Electronics Control,* <u>13</u>, 165 (1962).

[7.13] Gorman, J., and Zaborsky, J., "Stochastic Optimal Control of Continuous Time Systems with Unknown Gain," *IEEE Trans. Automatic Control,* <u>AC-13</u> (6), 630-638 (1968).

[7.14] Jenkins, K., and Roy, R., "A Design Procedure for Discrete Adaptive Control Systems," Preprints JACC, 1966, p. 624.

[7.15] Ku, R., and Athans. M., "On the Adaptive Control of Linear Systems Using OLFO Approach," Proc. Conference on Decision and Control, New Orleans, La., Dec. 1972.

[7.16] Kopp, R. E., and Orford, R. J., "Linear Regression to System Identification for Adaptive Systems," *J. American Institute of Aeronautics and Astronautics*, (10), 2300-2306 (1962).

[7.17] Lainiotis, D., Deshpande, J. G., and Upadhyay, T. N., "Optimal Adaptive Control: A Nonlinear Separation Theorem," *Intern. J. Control*, 15, 877-888 (1972).

[7.18] Lee, R. C. K., *Optimal Estimation Identification and Control*, MIT Press, Cambridge, Mass., 1964.

[7.19] Lobbia, R. N., and Saridis, G. N., "Identification and Control of Multivariable Stochastic Feedback Systems," *J. Cybern.*, (1), 40-59 (1973).

[7.20] Mendel, J., and Fu, K. S., *Adaptive Learning and Pattern Recognition Systems: Theory and Applications*, Academic Press, New York, 1970.

[7.21] Mishkin, E., and Braun, Jr. L., *Adaptive Control Systems*, McGraw-Hill, New York, 1961.

[7.22] Murphy, W. J., "Optimal Stochastic Control of Discrete Linear Systems with Unknown Gain," *IEEE Trans. Automatic Control*, AC-13, (4), 338-343 (1968).

[7.23] Saridis, G. N., and Dao, T. K., "A Learning Approach to the Parameter Adaptive Self-Organizing Control Problem," *Automatica - IFAC J.* 9, (5) 589-598 (1972).

[7.24] Saridis, G. N., and Lobbia, R. N., "Parameter Identification and Control of Linear Discrete-Time Systems," *IEEE Trans. Automatic Control*, AC-17, 52-60 (1972).

[7.25] Saridis, G. N., and Lobbia, R. N., "A Note on Parameter Identification and Control of Linear Discrete Time System," *IEEE Trans. Automatic Control*, AC-20 (3) 442 (1975) (correspondence).

[7.26] Saridis, G. N., and Kitahara, R. T., "Comparison of Per-Interval and Overall Parameter-Adaptive SOC," Proc. 8th Symposium on Adaptive Processes, Penn. State, Dec. 1969.

[7.27] Schweppe, F., *Uncertain Dynamic Systems*, Prentice Hall, Englewood Cliffs, N.I., 1973.

[7.28] Simon, H., "Dynamic Programming under Uncertainty with a Quadratic Function, " *Econometrica*, 24, 74-81 (1956).

[7.29] Spang, H. A., "Optimum Control of an Unknown Linear Plant Using Bayesian Estimation of the Error," *IEEE Trans. Automatic Control*, AC-10, (1) 80-83 (1965).

[7.30]   Stein, G., and Saridis, G. N., "An Approach to the Parameter
         Adaptive Control Problem," Tech. Rept. TR-EE-68-35, School of
         Electrical Engineering, Purdue University, Lafayette, Ind.,
         Jan. 1969.

[7.31]   Stein, G., and Saridis, G. N., "A Parameter Adaptive Control
         Technique," Proc. 4th IFAC, Warsaw, 1969; *Automatica - IFAC J.*
         5, (6) 731-740, (1969).

[7.32]   Sworder, D. D., *Optimal Adaptive Control Systems*, Academic
         Press, New York, 1966.

[7.33]   Tse, E., and Athans, M., "Adaptive Stochastic Control for a
         Class of Linear Systems," *IEEE Trans. Automatic Control*,
         AC-15, (1) 38-51 (1972).

[7.34]   Tse, E., Bar-Shalom, Y., and Meier, L., "Wide-Sense Adaptive
         Dual Control of Stochastic Nonlinear Systems," *IEEE Trans.
         Automatic Control*, AC-18, (2), 98-108 (1973).

[7.35]   Tse, E., and Bar-Shalom, Y., "An Actively Adaptive Control
         for Discrete-Time Systems with Random Parameters," *IEEE
         Trans. Automatic Control*, AC-18, (2), 109-117 (1973).

[7.36]   Tse, E., and Bar-Shalom, Y., "Adaptive Dual Control for
         Stochastic Nonlinear Systems with Free End-Time," *Proc. 1974
         IEEE Conference on Decision and Control*, Phoenix, Arizona,
         Nov. 1974: also *IEEE Trans. Automatic Control*, AC-20, (5)
         670-674 (1975).

[7.37]   Tse, E., "On the Optimal Control for Linear Systems with
         Incomplete Information," Rept. ESL-R-412, Electronic Systems
         Laboratory at MIT, Cambridge, Mass., January 1970.

[7.38]   Ya, Tsypkin, *Adaptation and Learning in Automatic Systems*,
         translated by Z. J. Nikolic, Academic Press, New York, 1971.

Chapter 8

PERFORMANCE-ADAPTIVE SELF-ORGANIZING CONTROL

8.1 INTRODUCTION

In the preceding chapter the S.O.C. problem has been considered
from an "analytic" point of view, meaning that the controller was
designed to function on-line, the way a design engineer would have
approached the problem. The missing information in terms of system
parameters was accumulated and the controller would be updated or
redesigned every time that a better estimate of these parameters was
obtained. A behavioral and structural point of view will be taken
here to approach the S.O.C. problem. The function of the system is
described through a performance evaluator and the controller is
modified structurally or parametrically to improve the performance of
the system as the process evolves. This approach resembles, somehow, the function of a learning living organism, as defined by the
physiologists [8.6], [8.7]. As discussed in Chap. 1, "learning control
systems" have been defined from a similar point of view. The difference is that learning control systems represent a considerably
wider class including decision-making controllers of a higher degree
of sophistication demonstrating intelligent capabilities higher than
reflex imitations [8.12], [8.13], [8.53], [8.81], [8.83].

Therefore, trainable learning control systems have not been
included in S.O.C. and simple structural changes in the controller
are not covered by learning control. The concept of optimality is
also considerably widened in this chapter to accommodate more control problems with a certain degree of simplicity of implementation.
This approach can treat problems where dynamics are completely or
partially unknown, operating in an unknown stochastic environment
such as bioengineering or physiological systems, without limitations

309

on the dimensionality of the problem. The simplicity of an algorithm
that may be interpreted as decoupling subsystems and states, initial
value-search techniques, and successive improvement schemes, which
are obtained by relaxing the strict optimality conditions, are the
factors that permit handling of large-scale systems. Instead, the
ideas of asymptotic optimality or instantaneous improvement of a
suitable performance criterion are applied and yield satisfactory
results for problems of the complexity of S.O.C. systems. A word
of caution is appropriate at this point. The performance-adaptive
methods are developed to solve problems of higher degrees of complexity and should *not* be used for simpler problems that can be
handled by other control methods. It is only when these methods
fail to give reasonable results that S.O.C. is necessary. A typical
performance-adaptive S.O.C. system is demonstrated in Fig. 8.1.

In conclusion, performance-adaptive S.O.C.s represent the first
step toward an intelligent decision maker and can handle the lower-
level control function. On the other hand, they are on the top of
the control hierarchy since they can treat advanced situations with
considerable uncertainties without much sacrifice of the mathematical
rigor. Intuitively appealing procedures may replace a strictly optimal formulation whenever it is necessary [8.9], [8.68] for simplicity

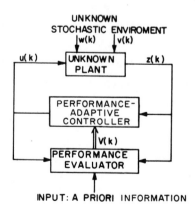

FIG. 8.1.   A Typical Performance-Adaptive S.O.C. System.

of implementation, but further mathematical investigation guarantees
satisfaction of the specifications needed for effective and safe
operation of the system.

The area is relatively new and therefore few investigations can
be accounted for. A historical review gives the proper chronological
perspective; it also helps relate the various methods to be discussed
and to account for certain algorithms not presented in the text. A
long list of references at the end of this chapter complements this
effort.

8.2 HISTORICAL REVIEW

Performance-adaptive S.O.C. systems evolved from various
independent attempts to produce advanced decision-making devices for
control purposes [8.12], [8.44], [8.68], [8.75]. One trend of thought
was the natural extension of the idea of parameter-adaptive S.O.C.s
which was popular in the 1950s. The work of Narendra and his colleagues [8.54], [8.55] is the most representative of this classical
approach to generate a controller based on the performance evaluation
of the system. A representative algorithm is presented in this
chapter. A second trend of thought in performance-adaptive S.O.C.
originated from the principles of learning systems and the applications of behavioral approaches such as reinforcement learning
algorithms to the control problem. The work of Fu, and his colleagues [8.25], [8.33], [8.101] is most representative in this area,
and a brief discussion of their most prominent algorithms is presented here. On the other hand, a variable structure stochastic
automaton [8.21], [8.24], [8.80], [8.93], [8.94], [8.96], [8.98] can be
used to implement a linear reinforcement algorithm and therefore generate
a performance-adaptive self-organizing controller. The idea in both
cases is that the selection of a control action is based upon the
improvement of a subgoal (e.g., per-interval performance criterion)
which is compatible with the overall performance of the system. The
work of Fu and McLaren [8.23], [8.48] is presented in this chapter as

the most representative in the area. It was preceded by the important investigations of Fu and McMurtry [8.24], while the work of
Riordon [8.65] and Narendra and his colleagues [8.56],[8.57],[8.98],
[8.99] followed and extended it. Mendel [8.44] and Mendel and
McLaren [8.46] summarized the state of the art of the reinforcement
techniques and learning automata algorithms while Mendel has applied
the concept to precise attitude control of a satellite [8.43].

A third school of thought has evolved through the application
of stochastic approximation as a performance-adaptive self-organizing
technique. Tsypkin in the U.S.S.R. [8.81],[8.82] and Nikolic and Fu
in the United States [8.58],[8.59] have proposed various formulations
of the solution to the performance-adaptive problem, using stochastic
approximation models. In spite of certain objections [8.76], these
methods represent new points of view to a successful solution and
will be described in the sequel. Another application of the stochastic approximation method treats the control problem as a class
of distributed systems with unknown dynamics. It was developed by
Badavas and Saridis [8.4] and represents one more aspect of the
flexibility of stochastic approximation methods to treat S.O.C.
[8.66] over the ones already described in Chap. 7.

Bayesian estimation techniques being used for S.O.C. [8.61],
[8.69] have been referred to the parameter-adaptive chapter 7 mainly
because of their parameter-searching nature.

Finally, the expanding subinterval algorithm, first developed
by Saridis and Gilbert for attitude control of an orbiting satellite
[8.70], and then generalized by Saridis [8.67],[8.68], represents the
last conceptually different approach to the performance-adaptive
S.O.C. problem to be discussed. It produces a global asymptotically
optimal solution to the control problem with completely unknown
plant dynamics by searching for a minimum of a per-interval cost
function defined over an expanding subinterval of time, thus relating
the subgoal to the overall goal, a concept of behavioral methods.
Structural considerations and stability investigation of the algorithm
have been discussed by Saridis [8.70], [8.72].

The algorithms to be presented in this chapter represent the major approaches to the performance-adaptive S.O.C. problem. Many other efforts to solve the problem have been made and are listed in the reference section of this chapter. They are not treated here in detail because they represent only small variations in philosophy from the major ones.

Many of the algorithms discussed were developed by their creators under the name *learning algorithms*. Since they still represent only the lowest level of an intelligent controller, they are presented in this chapter while the more advanced concepts of learning and intelligent controllers are discussed in Chap. 10. However, they still qualify as learning controls according to the definition in Chap. 1, since the class of learning systems is conceptually more general.

The plants considered in this chapter are assumed to have completely or partially unknown dynamics since the control effort is always confined to improve their performance directly. However, certain conditions must be imposed on the behavior of these plants in order that control is possible. In most cases it is assumed (1) that the plant does not contain any discontinuities in its single-valued function or, if it does, they are finite, and (2) that the plant is open-loop stable or that there is always a control loop with stabilizing effects on the plant. While the first condition is justified for most physical systems, the second imposes the most severe restrictions since stability investigation requires relative knowledge of the plant especially when it operates in a stochastic environment. The stability problem will be discussed further in Chap. 9.

In general the plant considered in all the methods mentioned in this section can be described mathematically by the following set of differential equations:

$$\frac{dx}{dt} = f(x(t), u(t), w(t)) \quad x(0) = x_0 \tag{8.1}$$

$$z(t) = h(x(t), v(t))$$

or the set of difference equations in the case of a sampled data
or discrete-time system

$$x(k + 1) = f(x(k),u(k),w(k)) \quad x(0) = x_0 \tag{8.2}$$

$$z(k) = h(x(k),v(k))$$

In both (8.1) and (8.2), $x(s)$, $s = t,k$, is the inaccessible
state vector of unknown dimension n, belonging to a space Ω_x,
$x(s) \in \Omega_x$, which may be composed of finite, countable, or continuously infinite elements; $u(s)$, $s = t,k$, is an m-dimensional control
vector belonging to the space of admissible controls Ω_u, also
possibly of finite, countable, or continuously infinite elements;
and $z(s)$, $s = t,k$, is the r-dimensional vector of all plant outputs
available for measurement and belongs to the measurement space
Ω_z which may contain a finite, countable, or infinite number of
elements in compliance with Ω_x to which it is one-to-one related
when $v(s) \equiv 0$, $s = t,k$. The initial state x_0 is also a random
variable of unknown probability density function. The process
noise $w(s)$, $s = t,k$, and the measurement noise $v(s)$, $s = t$, k representing the interaction of the plant with the environment, are samples of random processes of appropriate dimensions with unknown
density functions (Fig. 8.1).

The functions $f(\cdot,\cdot,\cdot)$ and $h(\cdot,\cdot)$ describing the plant and
measurement, respectively, are bounded functions of their arguments
partially or completely unknown. They are mapping Ω_x, Ω_u, and the
process noise space Ω_w and measurement noise Ω_v one-to-one onto the
output space Ω_z.

In other words, there is a concrete system operating in a
stochastic environment for which the control designer has little
information and he is interested in recovering only the part of
it which pertains to the improvement of its performance, characterized by a performance criterion of the form

$$J(u) = E\{\lim_{N\to\infty} J_N(z^N,v^N)\} \tag{8.3}$$

The solution of the problem is contingent upon the realizability of a performance evaluator depicted in Fig. 8.1.

In the case of a linear system where operational mathematics are applicable, Eq. 8.1 may be replaced by the unknown transfer function

$$\frac{Z(s)}{U(s)} = G(s)$$

and the noises are applied additively. More details are to be presented for each individual case.

8.3 THE CROSSCORRELATION METHOD OF NARENDRA

The first method to be presented is due to Narendra and Streeter [8.55]. It utilizes the correlation techniques discussed in Sec. 6.3 to construct the performance-adaptive self-organizing controller, shown in Fig. 8.2, for undefined linear processes.

It has been demonstrated by Narendra and McBride [8.54] that for a given unknown linear system $G(s)$, driven by a zero-mean random sequence $w(t)$, one may assume a feedback controller of the form

$$\frac{v(s)}{Z(s)} = \sum_{i=1}^{m} k_i \phi_i (s) \tag{8.4}$$

where $\phi_i(s)$ are known transfer functions and k_i are adjustable gains. Define a performance index as the mean-square error of the output from a desired output $D(s)$ prespecified for the system

$$J(k) = E\{(d(t) - z(t))^2\} = \lim_{T \to \infty} \frac{1}{2T} \int_{-T}^{T} (d(t) - z(t))^2 \, dt \tag{8.5}$$

where ergodicity and statistical independence have been assumed for all the stochastic processes involved. The optimum values k_i^*, $i = 1,\ldots,m$, may be sought to minimize $J(k)$. Since the form of the plant is unknown, one may search sequentially for the minimum by defining the rate of k_i to change in the direction of steepest descent of $J(k)$ [8.54]

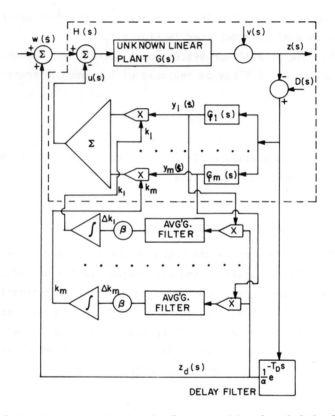

FIG. 8.2.   Crosscorrelation Performance-Adaptive S.O.C. System.

$$\frac{dk_j}{dt} = -\beta\frac{\partial J(k)}{\partial k_j} = -\frac{\beta}{\pi j}\int_{-\infty}^{\infty} [D(-s) - Z(-s)]$$

$$\cdot \left[\frac{G(s)\phi_j(s)}{1 + G(s)\sum_{i=1}^{m} k_i\phi_i(s)}\right] ds = -2\beta E\{(d(t) - z(t)) \cdot n_i(t)\}$$

$$i = 1,\ldots,m \qquad\qquad\qquad (8.6)$$

where

$$L[\ell(t)] = L(s) \text{ for some } \ell(t) \text{ and}$$

$$L[\eta_j(t)] = \frac{G(s)\phi_j(s)}{1 + G(s) \sum\limits_{i=1}^{m} k_i\phi_i(s)} \triangleq H(s)\phi_j(s)$$

Narendra and McBride showed that such a scheme is stable in the absence of inputs but depends on the system's unknown transfer function $G(s)$.

Narendra and Streeter showed that another crosscorrelation, not depending on $G(s)$ explicitly, may be used to obtain dk_j/dt. If the output $z(t)$ is attenuated by α and delayed by T_D, then

$$Z_D(s) = \frac{1}{\alpha} (D(s) - Z(s))e^{-T_D s} \tag{8.7}$$

The new error function $Z_D(t)$ can be produced by a delay filter, and for a sufficiently large T_D it can be made to be uncorrelated with $w(t)$ and $v(t)$.

Let

1. $|\phi_j(t)| < \epsilon \approx 0$, $t > a$, $j = 1,\ldots,m$, where $\phi_j(t)$ is the impulse response of $\phi_j(s)$.

2. $|h(t)| < \epsilon \approx 0$, $t > b$, where $h(t)$ is the impulse response of $H(s)$ defined in (8.6).

3. $R_{ww}(\tau) < \epsilon \approx 0$, $t > c$, where R_{ww} is the autocorrelation function of $w(t)$.

Then by choosing

$$T_0 \geq \max [3b, a + 2b + c] \quad \alpha^2 \gg 1$$

the partial derivatives $\partial J(k)/\partial k_j$ needed may be approximately obtained from the crosscorrelation of $y_j(t) = L^{-1}[\phi_j(s)z(s)]$ and the output of the delay filter $z_D(t)$.

$$E\{y_j(t)z_D(t)\} \approx \frac{1}{2\pi\alpha^2} \int_{-\infty}^{\infty} [D(s) - Z(s)][D(-s) - Z(-s)]$$

$$\cdot \left[\frac{G(s)\phi_j(s)}{1 + G(s) \sum\limits_{i=1}^{m} k_i\phi_i(s)}\right] ds = -\frac{1}{2\alpha^2}\left[\frac{\partial J(k)}{\partial k_j}\right] \tag{8.8}$$

Then the algorithm which produces the optimal gains k_i, $i = 1,\ldots,m$, are approximately produced by the algorithm that generates the correlation of $y_i(t)$ and $z_D(t)$

$$\frac{dk_i}{dt} = 2\alpha^2\beta \quad E\{y_i(t)z_D(t)\} \tag{8.9}$$

The algorithm is depicted in Fig. 8.2.

The following third-order system is to be controlled as a regulator system

$$G(s) = \frac{0.5s}{s^3 + 1.305s^2 + 1.565s - 0.98}$$

The simulation results of the third-order system are presented in Fig. 8.3.   The results were obtained by Narendra and Streeter [8.55] for T_D = 75 time units, α = 3, β = 10, while the constraint $\Delta k_1^2 + \Delta k_2^2 \leq 0.2$ was used for stability.

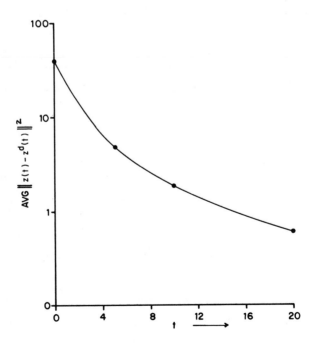

FIG. 8.3.   Simulation Results for Narendra's Crosscorrelation S.O.C.

The plot represents the average square error of the actual output from the desired output $d(t) \equiv 20$ units for $t > 0$.

The algorithm is simple enough and straightforward as a performance adaptive S.O.C. However, it assumes that the structure of the system is linear which may be restrictive for most applications.

8.4 LINEAR REINFORCEMENT AND STOCHASTIC AUTOMATA ALGORITHMS

This class of algorithms was developed by the "behavioral school of thought" which for some years has applied the scientific studies and experimentation on human and animal behavior to learning systems and control problems. Such applications have been made gradually and at various levels of sophistication from artificial intelligence to simple decision making and control and, therefore, cover a whole spectrum of their own which lies outside the scope of this book. However, the lowest-level systems developed from a "behavioral" point of view to be used as self-organizing controllers certainly belong here and will be discussed in the sequel. A brief discussion of higher-level learning control and intelligent hierarchical structures will also be given in Chap. 10 as a preview of the future of the field. The algorithms presented here summarize the pioneering work of Fu and his colleagues during the 1960s [8.10] - [8.29] and the applications of Mendel's discoveries to the aerospace industry [8.42] - [8.47].

8.4.1 Reinforcement Learning Models

In order to comprehend the function of the control algorithms a brief review of the mathematical learning models and theory developed for psychology is appropriate [8.6],[8.7],[8.53].

Learning has been considered by the psychologists in relation to a human or animal behavior as *any systematic change in a system's performance with a certain specified goal.*

One of the most important principles in learning theory is the *law of reinforcement*. It defines the effect of rewards or punishments that are generated as a result of the attainment or not of a goal, on the changes in behavior of a subject considered as a system. The law takes different forms in different learning theories and is implemented by the change of a set of *response probabilities* $P_i^{\ell}(n)$, in mathematical learning models. Such mathematical models have been experimentally verified. One of the most popular experiments used to establish the mathematical model is the problem of a hungry rat entering the ℓth T-maze, $\ell = 1,\ldots,s$, and deciding to turn right or left, e.g., $i = 1,2$, at the fork in the maze to find food which represents its reward (see Fig. 8.4). For each T-maze which represents one situation or event out of $E_1,\ldots,E_s$, the rat may learn to turn to the proper direction by n successive attempts. This experience with E is stored in its *short-term memory*. However, as he is moved from one T-maze to another, he has to repeat his learning process for each situation E_ℓ, $\ell = 1,\ldots,s$, and store the experience in its *long-term memory*. The rewards and punishments are not always of the same strength. However, it may be thought that the rat tries always to extremize the reward under similar circumstances.

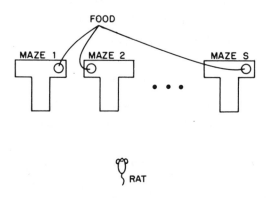

FIG. 8.4.   The s T-Maze Experiment.

A probabilistic model is based on the response probabilities $P_i^\ell(n)$ which represent the probability of the rat's response to the ith direction when in the ℓth T-maze after n attempts. The more general model, however, will be developed here in terms of control terminology to correlate it easier with the control algorithms that follow.

In a control problem the space of observable states is sub-divided to control situations $\ell = 1, \ldots, s$, similar to the s T-mazes of the behavioral experiment each of which has a different *short-term goal or subgoal*. At each control situation $\ell = 1, \ldots, s$, the control has to make a decision to select an *action* which forces the system to one of the mutually exclusive and exhaustive classes of response $\omega_1, \ldots, \omega_m$, and thus affect its performance. The performance change may be expressed by the change or reinforcement of the set of response probabilities P_i^ℓ where P_i^ℓ is the probability of occurrence of the ith class of response within the situation. The reinforcement of P_i^ℓ may be mathematically described by

$$P_i^\ell(n_\ell + 1) = \alpha_\ell P_i^\ell(n_\ell) + (1 - \alpha_\ell)\lambda_i^\ell(n_\ell) \quad i = 1, \ldots, m$$

$$\sum_{i=1}^{m_\ell} P_i^\ell(0) = 1 \tag{8.10}$$

where $P_i^\ell(n_\ell)$ the probability of ω_i at an instant n_ℓ of the control being in the ℓth control situation, $0 < \alpha_\ell < 1$ and $0 \le \lambda_i^\ell(n) \le 1$ for $\ell = 1, \ldots, s$. The $\lambda_i^\ell(n)$ are specified by the nature of the learning experiment and must satisfy

$$\sum_{i=1}^{m} \lambda_i^\ell(n) = 1$$

If α_ℓ and $\lambda_i^\ell(n) = \lambda_i$ are given, it is straightforward to show that

$$P_i^\ell(n_\ell) = \alpha_\ell^{n_\ell} P_i^\ell(0) + (1 - \alpha_\ell^{n_\ell}) \lambda_i^\ell \; , |\alpha_\ell| < 1 \tag{8.11}$$

and

$$\lim_{n_\ell \to \infty} P_i^\ell(n_\ell) = \lambda_i \qquad \begin{array}{l} i = 1,\ldots,m \\ \ell = 1,\ldots,s \end{array}$$

The results from (8.11) can be interpreted to mean that after repeated attempts the system always learns the quality λ_i^ℓ within a control situation, $\ell = 1,\ldots,s$. On the other hand, λ_i^ℓ is the limiting probability of $P_i^\ell(n)$ and should be related to the performance evaluated from the control situation ℓ at the instance n_ℓ to the controller. The control situation is usually related to the output of the plant or a function of it. Therefore, $\lambda_i^\ell(n)$ can be associated to the normalized index of performance associated with the ith class of control actions ω_i within the ℓth control situation.

The controller's task is to learn on-line the best control action, in the sense of minimizing the original index of performance, by applying control actions to the plant in the absence of complete information about it and the environmental disturbances. This can be accomplished by reinforcing the probabilities of occurrence of the class that yields an improvement of performance, while penalizing the rest.

A special case of association of the quantity $\lambda_i^\ell(n)$ to the reward or punishment outcome of the experiment within a control situation is obtained by assigning

$$\lambda_i^\ell(n) = \begin{cases} 1 & \omega_i = \text{reward} \\ 0 & \omega_j, \ j \neq i, \ \text{punishment} \end{cases} \qquad (8.12)$$
$$\ell = 1,\ldots,s$$
$$n_\ell = 1, 2,\ldots$$

It is obvious that the limiting probability vector becomes a $[0,\ldots,0, 1, 0,\ldots,0]$ with 1 at the ith column, demonstrating the probabilistic learning of the algorithm.

8.4.2 Reinforcement Algorithm for Control

The first reinforcement algorithm was produced as a result of Waltz and Fu's pioneering work [8.27], [8.100], [8.101] to learn the bang-bang solution of a minimum-time problem. Further research

on the subgoal problem was produced by Jones and Fu [8.33],[8.34]
and space applications were formulated by Mendel and his colleagues
[8.43], [8.44],[8.47].

   A generalized formulation will be presented here which highlights the learning properties of the algorithm. Consider the
nonlinear system with unknown dynamics described either by Eq. (8.1)
if it evolves continuously with time, or by Eq. (8.2) if it evolves
discretely. The state space Ω_x is assumed to be continuous in x and
is mapped into the measurement space Ω_z, e.g., $\Omega_x \rightarrow \Omega_z$, through
the measurement equations of (8.1) or (8.2). Let Ω_z be subdivided
in subregions $\Phi_\ell \subset \Omega_z$, $\ell = 1, 2,\ldots,s$, e.g., hyperspheres, where s
is finite if the region of operation in Ω_z is bounded, infinite if
it is not.

   A local cost function

$$J_\ell = J_\ell(Z,U) \quad Z \in \Phi_\ell \quad \ell = 1,\ldots,s \tag{8.13}$$

is assigned as a subgoal to each subregion Φ_ℓ which is related to the
overall performance cost to be minimized as the primary goal of the
control problem

$$J = J(Z,V) \quad Z \in \Omega_z \quad V \in \Omega_u \tag{8.14}$$

The control input to the system u belongs to a finite set of admissible controls corresponding to possible control actions and V is
some sequence in time of u's, all in Φ_ℓ,

$$u \in \Omega_u = \{u_1,\ldots,u_m\} \tag{8.15}$$

Now the control problem can be reformulated as follows: When the
system enters a subregion Φ_ℓ, a control situation is established
on Φ_ℓ with subgoal $J_\ell(Z,V)$ of (8.13). A control $u \in \Omega_u$ may be
generated that maximizes the probability of minimizing the subgoal
$J_\ell(Z,V)$

$$u^*(k) = u_i: \; P_i^\ell(k - 1) = \underset{j}{Max} P_j^\ell(k - 1) \quad j = 1,\ldots,m \tag{8.16}$$

The probabilities are generated after every control action u_i by the
reinforcement algorithm

$$P_i^\ell(k) = \alpha P_i^\ell(k - 1) + (1 - \alpha) \left[1 - \frac{J_\ell(k)}{\underset{k}{\text{Sup}} \, J_\ell(k)} \right]$$

$$P_j^\ell(k) = \alpha P_j^\ell(k - 1) + \frac{1 - \alpha}{1 - m} \left[\frac{J_\ell(k)}{\underset{k}{\text{Sup}} \, J_\ell(k)} \right] \qquad j \neq i \qquad (8.17)$$

for some $0 < \alpha < 1$.

In (8.17] λ_i have been replaced by "normalized" subgoals to satisfy the condition $\sum_{i=1}^{m} \lambda_i = 1$.

The probabilities are stored in the controller's short memory and are used as long as the system is in the control situation Φ_ℓ. When the system moves to another control situation, say Φ_n, the probabilities P^ℓ are stored in the long-term memory and the probabilities P^n, generated from past visits in Φ_n, are brought from the long-term memory and used in (8.17) and (8.17).

As an example, the problem treated by Waltz and Fu [8.100], [8.101] will be discussed. It is the application of the reinforcement algorithms (8.16) and (8.17) to the minimum-time problem.

The following differential equation represents a linear plant:

$$\dot{x}_1(t) = x_2(t)$$
$$\dot{x}_2(t) = -x_2(t) + 100 \, u(t) + w(t) \qquad |u| \leq 1 \qquad (8.18)$$

with discrete-time measurement

$$z(kT) = x(kT) + v(kT) = x(t) + v(t)$$

where $w(t)$ and $v(t)$ are the interactions with the environment. The following overall performance criterion is defined to be minimized:

$$J(u) = \sum_{j=1}^{n} j \, z^2(jT)$$

where T is the instant that the system arrives within a distance δ from the origin. Control situations are formed in the measurement space Ω_z by defining sets Φ_ℓ as circles with diameter D, as in Fig. 8.5. The control can take either of the two choices, +1 or -1 in each Φ_ℓ

$$u \in \Omega_u = \{+1,-1\} \tag{8.20}$$

All the state vectors and therefore the measurements in the same Φ_ℓ are assumed to have the same control choice. This particular control choice, e.g., +1 or -1, is learned by updating the probabilities $p_i^\ell(k)$ of (8.17) through the evaluation of subgoal J_ℓ in Φ_ℓ, defined as a quadratic form

$$J_\ell = \sum_{j=k}^{k+s} z^T(jT)G_\ell z(jT) \tag{8.21}$$

where s and $G_\ell = \text{diag}[g_1,g_2]$ were heuristically selected to minimize (8.19). The *a priori* probabilities for each control situation were selected to be equal.

It was shown [8.101] that, as the system was moving in each Φ_ℓ, the learning progressed and a control +1 or -1 was assigned to each Φ_ℓ. It was necessary to define circles with smaller diameter d as the system moves closer to the switching boundary in the state space (see Fig. 8.5). The learning process is demonstrated by plotting the overall performance index (8.21) versus time in Fig. 8.6 for fixed environment, e.g., $w(t) \equiv v(t) \equiv 0$. The method was applied

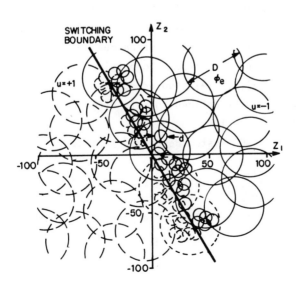

FIG. 8.5.   Distribution of Sets and Subsets in the Measurement Space.

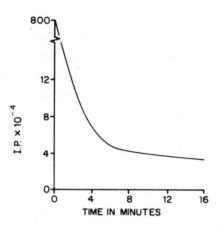

FIG. 8.6.   Performance of Index vs. Time for Plant in a
Fixed Environment w ≡ v ≡ 0.

on-line but it took too long to converge, e.g., 2-1/2 hr real time
for the initial conditions of Fig. 8.6, to be of any value as a
minimum-time solution.

Therefore, the selection of the problem to which the method
should be applied is crucial for the success of the method. The
other critical factor for the success of the reinforcement algorithm
is the selection of the subgoal. Waltz and Fu assigned it
heuristically and it may lead to erroneous results for some Φ_{ℓ}'s if,
for instance, the distance from the origin is selected as a subgoal.
Jones and Fu [8.33], [8.34] studied some special cases of LQG systems
where a subgoal may be defined that is directly related to the
overall performance criterion.

Consider the LQG system described in Sec. 4.5 with a performance
criterion

$$J(u) = E \left\{ \sum_{k=0}^{n-1} \left[x^T(k + 1)M(k)x(k + 1) + u^T(k)N(k)u(k) \right] \right\} \qquad (8.22)$$

To this performance criterion corresponds an optimal control, e.g.,
(4.45)

$$u^*(k) = -A_{N-k}^T \hat{x}(k/k) \qquad\qquad (8.23)$$

$$A_{N-k}^T = [B^T(k)(M(k) + S_{N-k-1})B(k) + N(k)]^{-1}B(k)(M(k) + S_{N-k-1})F(k)$$

where S_{N-k} is given also in (4.45). Define the subgoal as the
per-interval cost

$$J_\ell(u) = E\{x^T(k + 1)G_\ell(k)x(k + 1) + u^T(k)H_\ell(k)u(k)\} \qquad (8.24)$$

To this criterion corresponds the following optimal control,
$u^{**}(k)$, which was derived in (7.114):

$$u^{**}(k) = -K(k)\hat{x}(k/k)$$

$$K(k) = [B^T(k)G_\ell(k)B(k) + H_\ell(k)]^{-1}B^T(k)G_\ell(k)F(k) \qquad (8.25)$$

It is obvious that if

$$G_\ell(k) \equiv M(k) + S_{N-k-1} \quad H_\ell(k) \equiv N(k) \qquad (8.26)$$

$$k = 0, 1, \ldots, n-1$$

the two controls are identical and minimization of the subgoal yields
minimization of the overall goal. Unfortunately this result cannot
be generalized and therefore the selection of subjoals is still an
unanswered problem. Saridis [8.67], by defining the performance
criterion over the expanding subinterval, gave a satisfactory answer
to this problem, discussed in Sec. 8.6. There are several other
shortcomings to be discussed in the sequel that have seriously
limited the applications of the reinforcement algorithms. However,
they represent a very useful design for learning systems and
advanced intelligent controls.

8.4.3 The Reinforcement Controller as a Stochastic Automaton

The reinforcement learning control algorithms discussed in the
preceding section are typical examples of applications of "behavioral"
methods to control systems with unknown dynamics in a stochastic
environment. Since the reinforcement algorithms may be used to
describe the function of a variable structure stochastic automaton,
it is only natural to consider using such an automaton as a
controller [8.80], [8.93],- [8.97]. Fu and McMurtry's pioneering

investigation in the area [8.24] established the feasibility of
such a controller. McLaren's and Fu's subsequent work produced the
algorithms which are discussed here. First, however, a brief
discussion of stochastic automata is presented to familiarize the
reader.

A stochastic automaton is defined to be a quintuple $\{Z,Q,U,F,H\}$
where $Z = \{z\}$ is a finite set of inputs, $Q = \{q\}$ is a finite set of
states, $U = \{u\}$ is a finite set of outputs, F is the state-transition
function, and H is the output function.

$$q(n + 1) = F[q(n),z(n)] \qquad (8.27)$$
$$u(n) = H[q(n)]$$

In general the function F is stochastic and the function H may
be deterministic or stochastic, assumed to satisfy the Markov
property. For each input $z^k(n)$, applied at the nth instant, the
function F moves the automaton from state $q^j(n)$ to state $q^i(n + 1)$,
and an appropriate output $u(n + 1)$ is obtained. In a stochastic
automaton, F may be represented by a state-transition probability
matrix $T^k(n)$ with elements defined by

$$t^k_{ij}(n) = \text{Prob}\{q(n + 1) = q^j/q(n) = q^i; \; z(n) = z^k\} \geq 0 \quad i,j = 1,r \qquad (8.28)$$

$$\sum_{i=1}^{r} t^k_{ij}(n) = 1$$

If the random states are represented by their respective
probabilities

$$p^T(n) = [p_1(n),\ldots,p_r(n)], \; p_i(n) = \text{Prob}[q(n) = q^i], \; \sum_{i=1}^{r} p_i(n) = 1 \qquad (8.29)$$

then the probability transition equation may be written as

$$P(n + 1) = T^k(n)P(n) \qquad (8.30)$$

The transition matrix T^k depends on the $y(n)$ and $q(n)$, and is a
stochastic matrix, because it satisfies conditions (8.28). It is
known that such matrices satisfy also $|\lambda_i| \leq 1$, $i = 1,\ldots,r$, where
λ_i is the ith eigenvalue of T^k. The probabilities $p^k(n)$ are the
information states [8.5] for the automaton and describe its behavior

If the state transition probability matrices $T^1, T^2, \ldots$, corresponding to the inputs $z^1, z^2, \ldots$, respectively, the state transition probabilities for a sequence of inputs can be found by matrix manipulation. If the input z^k is applied n times, the n-step transition probability matrix is $[T^k]^n$. If the inputs are constant, the automaton is autonomous and can be interpreted as a Markov chain with the same state set. It can be shown that a deterministic finite automaton is a special case of a stochastic automaton, with the elements of the matrix T^k consisting of zeroes and ones, and each row of the matrix containing only one element equal to one. For this discussion, H will be considered to be deterministic even though this is not necessary.

The algorithm (8.27) may be interpreted as a nonlinear reinforcement algorithm attempting to attain a prespecified goal. Therefore, a stochastic automaton may serve as a suitable model for learning systems [8.21], [8.24], [8.96], [8.97]. Such a model suggests modification of the state probabilities p_i to improve the system's on-line performance. Such an automaton has a variable structure, that results in its learning behavior. The modification of these probabilities may be accomplished by reinforcement algorithms, the amount of reinforcement being the function of the automaton. The new sets of probabilities resulting from such a procedure reflect the information which the automaton has received from the input and consequently its ability to improve its performance.

The reinforcement algorithm applies a "reward" in the situation where improvement of the performance is observed, or a "penalty" when a performance deterioration is detected. The corresponding probabilities, generated from the random environment, are R_j for a penalty and $(1 - R_j)$ for a reward at the jth state. The overall measure of performance of the automaton is the mathematical expectation of the penalty

$$I = \sum_{j=1}^{r} R_j p_j \qquad (8.31)$$

where $p_j = \lim_{n \to \infty} p_j(n)$, the final probability of the state q_j. If

$$\underset{i}{\text{Min}} \ (R_1,\ldots,R_r) = I_{min} \leq I < \frac{1}{m} \sum_{j=1}^{r} R_j \qquad (8.32)$$

the performance of the automaton is called *expedient* where expediency is measured as the distance from I to I_{min}. In the case $I = I_{min}$, the automaton has *optimal* performance.

Expedient performance of an automaton, serving as a controller for an unknown system in a random environment, will be analyzed in the next section and will demonstrate the salient features of such an approach. A linear reinforcement algorithm will serve for the function of the automaton. Results with nonlinear reinforcement are scarce at the present time. The linear reinforcement algorithms to be discussed in the sequel are expedient but not optimal [8.25], [8.98].

8.4.4 McLaren's Variable Structure Automaton as a Controller

McLaren proposed [8.49] a finite state algorithm, i.e., a variable structure stochastic automaton used as a controller which uses a performance evaluation function $J(n) \ \epsilon \ \Omega = \{J_1,\ldots,J_s; \ 0 \leq J_i < 1; \ i = 1,\ldots s < \infty\}$ as an input, and produces $u(n) \ \epsilon \ U = \{u_1,\ldots,u_s; \ s < \infty\}$ as the output of the automaton. Then the McLaren automaton has the configuration $\{\Omega,U,U,F,I\}$, where F is the decision function and I is the identity operator. The complete control system is depicted in Fig. 8.7. The performance is evaluated in a separate unit, while the automaton comprises the

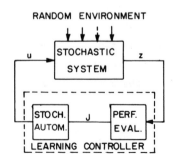

FIG. 8.7. The Stochastic Automaton as a Controller.

actual controller. Of course, under the present formulation the
learning controller may operate only on a finite set of controllers
and performance values as, for instance, $\Omega = \{1,0\}$.   In this case,
1 corresponds to a reward when improvement is obtained, while 0
corresponds to a penalty when no improvement is obtained. The
value of J in other performance evaluation algorithms should be a
normalized quantity, e.g., $0 \leq J_k \leq 1$, $k = 1,\ldots,s$.

Such a function J may be easily generated from an actual
performance criterion $J(u^\ell)$, corresponding to a subgoal as in (8.13)
by simple normalization

$$J_\ell(n) = \frac{J(u^\ell)}{\max_j J(u^j)} \qquad \ell = 1,\ldots,r \tag{8.33}$$

A linear reinforcement algorithm is now used to define the
structure of the r-state stochastic automaton where the outputs are
identical to the states, i.e., $H = I$.   For $0 < \alpha < 1$

$$p_i(n + 1) = \alpha p_i(n) + (1 - \alpha)(1 - J_i(n)) \quad u(n) = u^i$$

$$p_j(n + 1) = \alpha p_j(n) + (1 - \alpha)\frac{J_i(n)}{r-1} \quad j \neq i \tag{8.34}$$

The algorithm may be autonomized by taking the expectation of $J(n)$
and considering it as the performance criterion to be minimized

$$I(n) = E\{J(n)\} = \sum_{i=1}^{r} E\{J_i(n)|u^i\} \cdot E\{p_i(n)\} \triangleq \sum_{i=1}^{r} m_i \bar{p}_i(n)$$

where

$$E\{J_i(n)/u^i\} \triangleq m_i \tag{8.35}$$

$$E\{p_i(n)\} = E\{Prob\{u(n) = u^i\}\} \triangleq \bar{p}_i(n)$$

Then Eq. (8.34), which represents the transition equation of the
automaton, can be written as

$$\bar{P}(n + 1) = T \bar{P}(n) \tag{8.36}$$

where

$$
T = \begin{bmatrix} 1 - (1 - \alpha)m_1, \ldots, \dfrac{1 - \alpha}{r - 1} m_r \\ \vdots \qquad\qquad \vdots \\ \vdots \qquad\qquad \vdots \\ \dfrac{1 - \alpha}{r - 1} m_1, \ldots, 1 - (1 - \alpha)m_r \end{bmatrix} \qquad \bar{P}(n) = \begin{bmatrix} \bar{p}_1(n) \\ \vdots \\ \bar{p}_r(n) \end{bmatrix} \qquad (8.37)
$$

and T is a stochastic matrix. By defining the limiting probability

$$
\bar{P} = \lim_{n \to \infty} E\{\bar{P}(n)\} = \lim_{n \to \infty} T^n E\{P(o)\} = T \bar{P}
$$

it can be easily shown that

$$
\bar{p}_i = \sum_{k=1}^{r} \frac{1}{m_i/m_k} \qquad\qquad\qquad (8.38)
$$

and

$$
I = \lim_{n \to \infty} I(n) = \sum_{k=1}^{r} m_k \bar{p}_k = I_{min} \qquad\qquad (8.39)
$$

which is independent of the initial choice $\bar{P}(o)$. The convergence of this algorithm is given in (8.49).

This particular algorithm, in spite of its generality and its independence from the particular structure of the system to which it corresponds, is limited by the assumption that the control outputs belong to a limited finite set. The assumption that the inputs and the performance values belong to a finite set is not so limiting to the procedure previously discussed, since they may correspond to some reward-penalty transformations of a more conventional criterion.

Relation (8.39) indicates also that the algorithm is only asymptotically optimal in the sense of convergence discussed in Sec. 8.4.3.

8.4.5 McLaren's Growing Automaton as a Controller

To overcome the limitation of the previous algorithms created by the requirement of a finite number of control choices, McLaren proposed a modified algorithm [8.51] of a growing automaton where

the control outputs may take zero value from a countable set,
$s = 1, 2,\ldots,$.

In this case, the stochastic automaton used as a controller is
again composed of $\{\Omega, U, U, F, I\}$, but U is a countable set. The
performance evaluator is designed as

$$J(k) = J(z(k), u(k - 1)) \quad 0 \leq J(k) < 1 \tag{8.40}$$

The overall performance criterion to be minimized by $\{u\}$ is defined
as

$$I(k) = \frac{1}{k + 1} \sum_{j=1}^{k} E\{J(j)\} \tag{8.41}$$

Assume that at time k only r_k control actions have been selected,
e.g., $\Omega_k = \{u^1, \ldots, u^{r_k}\}$, and a level of uniform probability $L(k)$
over the set Ω_k has been retained. If, as previously,

$$p_j(k) \triangleq \text{Prob}\{u(k) = u^j\} \tag{8.42}$$

and a control action $u(k)$ has produced a performance value $J_j(k + 1)$
which is observed by the automaton, the following two cases are
possible:

1. A new controller that minimizes $I(k)$ is selected
    $u(k) = u^{r_k+1} \in \Omega_k$. Then the transition probabilities are given by
    $$p_{r_k+1}(k + 1) = [(1 - \alpha)(1 - J(k + 1))]L(k)$$
    $$p_j(k + 1) = p_j(k) \tag{8.43}$$
    $$L(k + 1) = [\alpha + (1 - \alpha)J(k + 1)]L(k)$$

2. If a controller was selected among the previously evaluated,
    $u(k) = u^j$, $j \in \{1,\ldots,r_k\}$, then the transition probabilities
    are given by
    $$p_j(k + 1) = \alpha p_j(k) + [(1 - \alpha)(1 - J(k + 1))](1 - L(k + 1))$$
    $$p_\ell(k + 1) = \alpha p_\ell(k) + (1 - \alpha) \frac{J(k+1)}{r_k - 1}(1 - L(k + 1)) \tag{8.44}$$
    $$L(k + 1) = L(k)$$

These probabilities must satisfy for every k

$$\sum_{j=1}^{r_k} p_j(k) + L(k) = 1 \tag{8.45}$$

This algorithm is identical to the preceding one when the control
is selected from the previously used output values u^j, $j \in [1,\ldots,r_k]$,
while when a new controller is selected u^{r_k+1}, its probability of
occurrence is drawn from the latent uniform distribution $L(k)$. In
accordance with the preceding algorithm the performance criterion
(8.41) can be rewritten as

$$I(k) = \sum_{j=1}^{r_k} E\{J(j)\} = \sum_{j=1}^{r_k} m_j \, E\{p_j(k)\} + \bar{m} \, E\{L(k)\} \tag{8.46}$$

where

$$m_j = E\{J(k)/u^j\} \quad \bar{m} = E\{m_j\}$$

As $k \to \infty$, the number of new control values $r_k \to \infty$ but slower than k.
Therefore, for a finite k, $L(k) > 0$ and there exists a finite probability of selecting a controller arbitrarily close to $u^* \in U$.

The convergence of the growing automaton algorithm can be
established by referring to the preceding algorithm (8.51) but its
importance from the control point of view is much more general. It
usually applies to stationary processes or steady-state nonstationary
ones. However, the assumption is made implicitly that the system
remains stable under all possible control actions. This is a strong
requirement in view of the difficulty of investigating stability of
unknown nonlinear systems. The growing automaton algorithm as well
as the preceding one is only asymptotically optimal and expedient.

Computational results have been obtained on various examples by
Saridis and Kitahara [8.73] properly adapted for this algorithm. The
method will be illustrated here by the applications to the following
control problem:

$$\dot{x}(t) = \begin{bmatrix} 0 & 1 \\ -10 & -1.264 \end{bmatrix} x(t) + \begin{bmatrix} 0 \\ 1 \end{bmatrix} u(t) + \begin{bmatrix} 0 \\ 1 \end{bmatrix} w(t) \quad x(0) = x_0$$

$$z(t) = (1,0)x(t) + v(t)$$

where $w(t) \sim N[0,0.01]$, $v(t) \sim N[0,0.01]$, $x_0 \sim N\left[\begin{pmatrix} 5 \\ 5 \end{pmatrix}, \begin{pmatrix} 9 & 0 \\ 0 & 4 \end{pmatrix}\right]$

The overall performance criterion is given by

$$J(u) = \lim_{t_f \to \infty} \frac{1}{t_f - t_0} E\left\{\int_{t_0}^{t_f} [(z_d(t) - z(t))^2 + 0.01 \, u^2(t)] \, dt\right\}$$

$$z_d(t) = 1 \quad t \geq t_0$$

The optimal solution requires a dynamic state estimator and control gains described by a set of n-differential equations with u unknowns, and the appropriate output equation with adjustable coefficients

$$\dot{\hat{x}}(t) = A \, \hat{x}(t) + gz(t)$$

$$u(t) = -c^T\hat{x}(t) + \alpha z_d(t)$$

where

$$A = \begin{bmatrix} 0 & 1 \\ a_1 & a_2 \end{bmatrix} \quad g = \begin{bmatrix} g_1 \\ g_2 \end{bmatrix} \quad c = \begin{bmatrix} 1 \\ 0 \end{bmatrix} \quad \alpha = 20$$

In this case a_1, a_2, g_1, and g_2 are the adjustable coefficients confined in the following regions for stability reasons:

$$\left\{\begin{array}{l} -10 \leq a_1 \leq -1 \\ -20 \leq a_2 \leq -2 \\ 0 \leq d_1 < 1 \\ 0 \leq d_2 < 1 \end{array}\right\} = \Omega_p$$

The adjustable parameter space Ω_p is equivalent to the automaton output space Ω_u and is searched, after it is discretized, to minimize the subgoals defined by

$$\text{SUBGOAL} = \frac{1}{(t_{i+1} - t_i)C_N} \int_{t_i}^{t_{i+1}} [(z_d(t) - z(t))^2 + 0.01 \, u^2(t)] \, dt$$

C_N is a normalizing constant used, in absence of information about SUBGOAL_{max}, to keep $0 \leq \text{SUBGOAL} \leq 1$. Plots of the average accumulated performance cost and its standard deviations are given in Figs. 8.8 and 8.9 for various C_N. The properties of the algorithms are further discussed in Chap. 9.

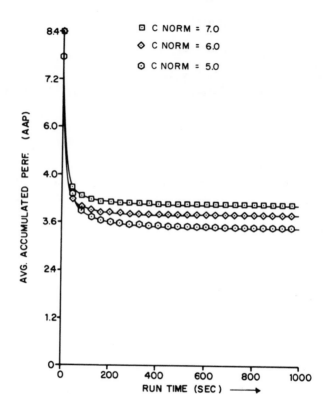

FIG. 8.8.   Average Performance Cost for McLaren's Growing Automaton
             Algorithm.

8.5 THE STOCHASTIC APPROXIMATION ALGORITHMS

Stochastic approximation has been used for parameter identification and control algorithms in Chap. 6 and 7 because of the simplicity of its implementation and the consistency of the estimation
involved.

Tsypkin, in the U.S.S.R. [8.81] - [8.90], and Nikolic and Fu
[8.58], [8.59] independently produced two algorithms which qualify
as performance-adaptive S.O.C. for systems with unknown dynamics
operating in a stochastic environment. Both algorithms demonstrate
a lower-level learning capability for updating the controller

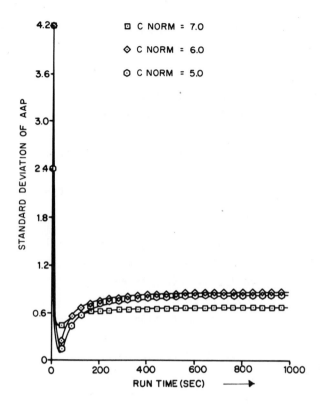

FIG. 8.9.   Standard Deviation of Cost for McLaren's Growing Automaton
            Algorithm.

toward an asymptotically optimal solution.   Both have advantages and
disadvantages to be discussed.   In spite of its name, a modification
of Riordan's adaptive automaton algorithm using stochastic approximation is an extension of the algorithm of Nikolic and Fu and is
also discussed here.   These three algorithms cover a wide range of
systems, all described by ordinary differential or difference
equations.   Another algorithm that uses stochastic approximation for
S.O.C. is presented separately in Sec. 8.7 because it treats systems
described by partial differential equations.

8.5.1 Tsypkin's Stochastic Approximation Algorithm

The method introduced by Tsypkin [8.82] utilizes a procedure
where the unknown system dynamics as well as the control action are
modeled by linearly independent functions weighted by adjustable
coefficients. The learning scheme consists of updating the weighting
coefficients to match the approximate combination of stochastic
plant and optimal control. The problem may be formulated as
follows:

Let the unknown stochastic plant be described by

$$x(k + 1) = f(x(k),u(k)) \tag{8.47}$$

where $x(k)$ is the n-dimensional state vector, $u(k)$ is the m-
dimensional input vector to be chosen to minimize an appropriate
performance criterion J_1, and $f(..)$ is an unknown stochastic function satisfying appropriate smoothness conditions. Let $f(x,u)$ be
approximated by the linear combination

$$\hat{f}(x,u) = \sum_{i=1}^{N} C_i \, \psi_i(x,u) = \psi(x,u)C \tag{8.48}$$

where $\psi_i(x,u)$, $i = 1,...,N$, are *a priori* assumed vector functions of
x, u, and C_i, $i = 1,...,N$, are the adjustable plant coefficients to
be determined by minimizing a convex function of the error $J_2(C)$

$$J_2(C) = E\{F[x(k + 1) - \psi(x(k),u(k))C]\} \tag{8.49}$$

where F is a convex differentiable function of its arguments.

The optimal matrix C may be obtained successively from a
stochastic approximation procedure of the Robbins-Monroe type which
seeks the zero of the gradient of (8.49)

$$\hat{C}(k + 1) = \hat{C}(k) + \gamma(k + 1) \, \nabla \, F[x(k + 1)$$
$$- \psi(x(k),u(k))\hat{C}(k)] \tag{8.50}$$

The control action $u(k)$ should be chosen to drive system (8.47) to
minimize a performance criterion

$$J_1 = E\{L[x^d(k) - x(k)]\} \tag{8.51}$$

where $x^d(k)$ represents the desired state vector of the system and $L(.)$ is a convex differentiable function of its argument. Such a control action may be approximated by a linear combination of known functions of the state $x(k)$, e.g.,

$$u(k) = \sum_{j=1}^{M} g_j(x(k)) \, \alpha_j = G(x(k))\alpha \tag{8.52}$$

In (8.52), $g_j(x)$ are known m-dimensional vector functions of the state and α_j are the adjustable coefficients which are chosen to minimize $J_1(\alpha,C)$ in (8.53)

$$J_1(\alpha,C) = E\{L[x^d(k + 1) - \psi(x(k),G(x(k)\alpha)C]\} \tag{8.53}$$

A stochastic approximation algorithm is defined to yield the optimal values α^*

$$\begin{aligned}
\hat{\alpha}(k + 1) = \hat{\alpha}(k) &+ \gamma_2(k + 1)\nabla_\alpha L[x^d(k + 1) \\
&- \psi(x(k),G(x(k))\alpha)C] \tag{8.54}
\end{aligned}$$

Algorithms (8.52) and (8.54) will converge to c^* and α^*, the respective optimal values, if Dvoretzky's conditions are satisfied

$$\sum_{k=1}^{\infty} \gamma_i(k) = \infty \quad \sum_{k=1}^{\infty} \gamma_i^2(k) < \infty \quad j = 1, 2,\ldots \quad ||\alpha(0)|| < \infty$$

$$||c(0)|| < \infty \tag{8.55}$$

The algorithm is depicted in a block diagram in Fig. 8.10.

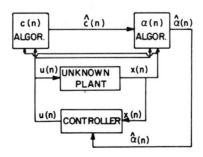

FIG. 8.10.   Block Diagram Representation of Tsypkin's S.O.C.
             Algorithm.

The implementation of algorithms (8.50) and (8.54) is complicated by the fact that the computation of the gradients of F(.) and
L(.) requires the computation of the partial derivatives $\partial x/\partial \alpha$ and
$\partial x/\partial c$, e.g., the "sensitivity functions" of the plant which are in
general unknown if the plant dynamics are unknown. In Tsypkin's
original work [8.81] these functions are either assumed known,
which contradicts the concept of controlling an unknown plant, or
estimated <u>off-line</u>, which violates the idea of feedback control.

There are more difficulties of theoretical as well as practical
nature related to this particular algorithm. The first one is the
generation of a biased estimate of the parameters because of the
idiosyncracies of the stochastic approximation algorithms. Unless
a good deal is known about the structure of the plant and more
restrictions are imposed upon the statistical properties of the
noise processes, the stochastic approximation estimates will, in
general, be biased. Stochastic approximation algorithms generating
unbiased estimates are produced for a limited class of systems for
which the knowledge of the plant structure is required. The second
difficulty arises with the choice of the appropriate matrix functions
$\psi(x,u)$ in (8.48) and $G(x)$ in (8.52), which should span the corresponding space and control spaces in order to approximate successfully
the plant and its optimal controller. Such a choice is extremely
difficult if the plant dynamics are unknown.

A discrete-time version of the second-order system used in
Sec. 8.4.5 without measurement noise will illustrate Tsypkin's
algorithm. After time discretization with $\Delta T = 0.1$ sec, the plant
may be written as a second-order difference equation in terms of
its output z

$$z(k + 2) = z^T_{k+1} \theta + \xi(k + 2)$$

where

$$z^{k+1} \triangleq \begin{bmatrix} z(k) \\ z(k+1) \\ u(k) \\ u(k+1) \end{bmatrix}, \quad \theta \triangleq \begin{bmatrix} f_1 \\ f_2 \\ b_1^0 \\ b_2^0 \end{bmatrix} \triangleq \begin{bmatrix} f \\ b^0 \end{bmatrix}$$

$$\xi(k+2) = g_2^0 w(k)$$

Hence, for purposes of identification, the model

$$\hat{z}(k+2) = z^{T^{k+1}} \hat{c}$$

is employed, where

$$\hat{c} \triangleq \begin{bmatrix} \hat{f} \\ \hat{b}^0 \end{bmatrix}$$

The criterion for determining the optimal identification parameters $\hat{c}^*$ was assumed to be

$$J_2(\hat{c}) = E\{[z(k+2) - z^{T^{k+1}} c]^2\}$$

Hence, the stochastic approximation algorithm of Sec. 6.4.1 for identification assumes the following form:

$$\hat{c}(k+1) = \hat{c}(k+1) + \gamma_1\left(\frac{k-1}{n+1}\right) \{z^{k+1}[z(k+2) - z^{T^{k+1}} \hat{c}(k-1)]\}$$

$$k = 1, n+2, 2n+3,\ldots$$

In this example

$$\hat{c}_0 = \begin{bmatrix} -0.8 \\ 0.9 \\ 0.0 \\ 0.05 \end{bmatrix}$$

and

$$\gamma\left(\frac{k-1}{n+1}\right) = \text{GAIN}/\left(\frac{k-1}{n+1} + 1\right)$$

No bias term appears in the identification algorithm since $g_1 = 0$ and $g_2 = g_n$ and the measurement noise $v(k)$ is assumed identically equal to zero. Otherwise the algorithm (8.50) should anticipate this situation and use the algorithms of Sec. 6.4.1 to eliminate the bias.

The criterion to be minimized at the end of each identification cycle, $k = 1, n + 2, 2n + 3, \ldots$ was chosen to be

$$J_1(\alpha, \hat{c}) = E\{(1 - z^{T_{k+1}} \hat{c})^2 + 0.01u^2(k + 1)\}$$

The stochastic approximation scheme for the control action is

$$\alpha(k + 2) = \alpha(k - 1) + \gamma_2 \left(\frac{k - 1}{n + 1}\right)\left\{-(1 - z^{T_{k+1}} \hat{c})\frac{\partial \hat{z}(k + 2)}{\partial \alpha}\right.$$

$$+ 0.01u(k + 1)\frac{\partial u(k + 1)}{\partial \alpha}\right\}$$

$$= \alpha(k - 1) + \gamma_2\left(\frac{k - 1}{n + 2}\right)\left[[z(k + 2) - 1]s(k + 2)\right.$$

$$+ 0.01u(k + 1)\frac{\partial u(k + 1)}{\partial \alpha}\right\}$$

where

$$\alpha^T = [a_1, a_2, g_1, g_2] \quad \frac{\partial \hat{z}(k)}{\partial \alpha} \triangleq s(k)$$

The parameters α appear in the dynamic feedback controller described by

$$\hat{x}(k + 1) = A\hat{x}(k) + gz(k)$$

$$u(k) = h^T x(k) + \alpha z^d(k)$$

$$A = \begin{bmatrix} 0 & 1 \\ a_1 & a_2 \end{bmatrix} \quad g = \begin{bmatrix} d_1 \\ d_2 \end{bmatrix} \quad h = \begin{bmatrix} 1 \\ 0 \end{bmatrix} \quad \alpha = 20 \quad z^d \equiv 1$$

while $s(k)$ satisfies the following equation and is used in the stochastic approximation scheme recursively to obtain $s(j)$, $j = 0$, $1, \ldots$

$$s(k + 2) = \left[s(k), s(k + 1), \frac{\partial u(k)}{\partial \alpha}, \frac{\partial u(k+1)}{\partial \alpha}\right]\hat{c}(k - 1)$$

$$s(0) = s(1) = 1$$

In order to complete the stochastic approximation scheme, it is required to compute

$$\frac{\partial u(k)}{\partial \alpha} = -T_k c$$

where

$$T_k = \frac{\partial \hat{x}(k)}{\partial \alpha} = \begin{bmatrix} \dfrac{\partial \hat{x}_1(k)}{\partial \alpha_1} & \dfrac{\partial \hat{x}_2(k)}{\partial \alpha_1} \\[3mm] \dfrac{\partial \hat{x}_1(k)}{\partial \alpha_2} & \dfrac{\partial \hat{x}_2(k)}{\partial \alpha_2} \end{bmatrix}$$

The rows of T_k can, in this case, be computed by

$$t_i(k) = \frac{\partial \hat{x}(k)}{\partial \alpha_i} = At_i(k - 1) + gs_i(k) + \psi_i(k) \quad t_i(0) = 0$$

where $\psi_i(k)$ is the ith row of $\Psi(k)$ defined by

$$\Psi(k) = \begin{bmatrix} 0 & 0 & z(k - 1) & 0 \\ \hat{x}_1(k - 1) & \hat{x}_2(k - 1) & 0 & z(k - 1) \end{bmatrix}$$

It is fortunate in this case that the $\partial z/\partial \alpha$ and $\partial \hat{x}/\partial \alpha$ are available for computation. This is true because it was assumed that the plant and the controller were linear. If measurement noise $v(k)$ were present, it can be shown that

$$E\{z^{k+1} \xi(k+2)\} \neq 0$$

and the stochastic approximation scheme should subtract the bias term as in Sec. 6.4.1.

These two facts, the assumption of linearity of the plant in order to compute $\partial z/\partial \alpha$ and $\partial u/\partial \alpha$ and the bias in the stochastic approximation algorithms, are the major drawbacks of the algorithm which otherwise converges very fast. This is demonstrated in Figs. 8.11 and 8.12 for the previous example. The plots are drawn for various constant-gain matrices used in $\gamma_i\left(\dfrac{k - 1}{n + 1}\right)$.

$$\gamma_i\left(\frac{k - 1}{n + 1}\right) = \text{GAIN} \ \frac{1}{\dfrac{k - 1}{n + 1} + 1} \qquad \begin{array}{l} k = 1, \ n + 1, \ 2n + 3 \\ i = 1, 2,\ldots \end{array}$$

to demonstrate the dependence of convergence on such gains. Further

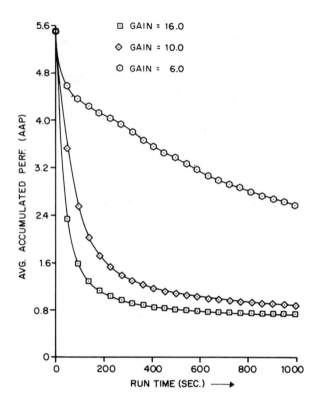

FIG. 8.11.   Average Performance Cost for Tsypkin's Stochastic
             Approximation Method.

discussion and comparisons will be given in Chap. 9.

8.5.2   The Algorithm of Learning Without External Supervision of Nikolic and Fu

The stochastic automaton formulation given in Sec. 8.4 has its
own limitations. To overcome these limitations, a stochastic
approximation algorithm that utilizes "unsupervised learning" was
developed by Nikolic and Fu [8.59]. The term "unsupervised" may
be interpreted in a wide sense as <u>on-line</u>. It requires a minimum
amount of *a priori* information about the unknown stochastic plant-
environment relation. The only assumptions necessary are a first-

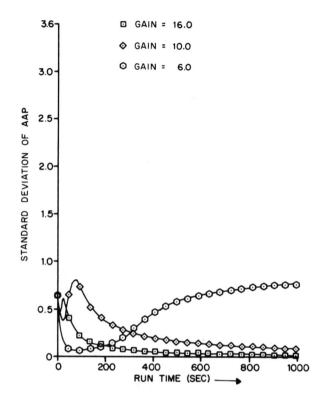

FIG. 8.12.   Standard Deviation of Cost for Tsypkin's Stochastic
             Approximation Method.

order Markov property about the plant and the stability of the
overall system under all possible control actions in order to
guarantee convergence of the algorithm to the optimal solution.

    The performance evaluation of an unknown continuous plant and
the optimal control prediction are impossible with the currently
available methods and therefore a discretization of the state and
control spaces is absolutely necessary. The proposed method is
implemented by two stochastic approximation algorithms, the first
to estimate the current performance of the system and the second
to make the selection for the next control action. In the first
one, an instantaneous evaluation of the performance is required.
Therefore, a *subgoal* minimization policy is pursued which eventually

should lead to the optimization of the overall performance of the
system.   In the second algorithm, the controller selects a control
action from a finite set of allowable actions via a pure random
strategy.   Subjective probabilities of applying a given control
action from the admissible set are altered in accordance with the
specified performance evaluation of the observed response of the
system.   Specifically, consider a stochastic plant-environment
relation described by

$$x(k + 1) = f(x(k),u(k)) \tag{8.56}$$

where $x(k)$ is the n-dimensional observed response of the system at
instant k, k = 0, 1, 2,..., and $f(..)$ is an unknown stochastic
function due to the influence of the environment.   It is assumed
for this algorithm that $x(k) \in \Omega_x = \{x^j; j = 1, 2,...,s < \infty\}$, a
finite set.   This may result from the discretization of the output
space and retaining a finite region.   The control input to the
plant $u(k) \in \Omega_u = \{u^i; i = 1, 2,...,A < \infty\}$ belongs to a finite set
of control actions.   At every instant k after an event $[u^i, x^j]$ has
occurred, e.g., after a control action u^i has driven the system to
the state x^j, a new control action $u(k + 1) \in \Omega_u$ is sought to
minimize the performance criterion

$$V(u(k + 1)) = E\{J(k + 1)/u(k) = u^i, x(k) = x^j; u(k + 1)\} \tag{8.57}$$

the instantaneous value of which is given by

$$J(k + 1) = L[x(k + 1),u(k),x(k)] \tag{8.58}$$

where $L(.,.,.)$ is a prespecified positive definite function of its
arguments.   Furthermore, assuming that the overall system is stable
for all control action

$$0 < J(k) < \infty \text{ for every } k \tag{8.59}$$

The probability distribution function of $J(k + 1)$ is unknown, but
it is assumed stationary,

$$p(J(k + 1)/u(k) = u^i, x(k) = x^j, u(k + 1) = u^\ell)$$
$$= p(J/u^i, x^j, u^\ell) \text{ for all } k \tag{8.60}$$

Because of the uncertainties in the probability distribution and the plant the optimal control u^* cannot be directly evaluated from

$$V(u^*) = E\{J/u^i, x^j, u^*\} = \min_{\ell=1,\ldots,A} E\{J/u^i, x^j, u^\ell\} \qquad (8.61)$$

One stochastic approximation algorithm is set up to yield recursive estimates of the unavailable performance criterion

$$\hat{V}_{n(ij)+1}(u^\ell) = \hat{V}_{n(ij)}(u^\ell) + \gamma_{n(ij)+1}[J(n(ij)+1) - \hat{V}_{n(ij)}(u^\ell)]$$

$$\hat{V}_0(u^\ell) = 0 \qquad (8.62)$$

where $n(ij) < k$ is the number of times in k that the event $[u^i, x^j]$ has occurred.

Algorithm (8.62) converges with probability one to the true value of the performance criterion corresponding to the ℓ control action

$$\text{Prob}[\lim_{n(ij) \to \infty} (\hat{V}_{n(ij)}(u^\ell) - V(u^\ell))] = 1 \quad \ell = 1,\ldots,A \qquad (8.63)$$

if the equivalents of Dvoretzky's conditions are satisfied

$$||V_0(u^\ell)|| < \infty \quad \lim_{k \to \infty} \gamma_n = 0, \quad \sum_{n=1}^{\infty} \gamma_n = \infty, \quad \sum_{n=1}^{\infty} \gamma_n^2 < \infty$$

$$\ell = 1,\ldots,A \qquad (8.64)$$

The subjective probabilities necessary for the selection of the next control action are defined by

$$p_{k+1}(u^*/u^i, x^j) = \max_{\ell} p_{k+1}(u^\ell/u^i, x^j) \qquad (8.65)$$

It is obvious that if a pure random strategy is used as a control, Eq. (8.65) represents the optimal strategy.

$$u(k+1) = u^* \qquad (8.66)$$

should be selected as the optimal control action, if $p(u^*/u^i, x^j) = 1$. However, because of (8.65), (8.66) will be selected as an approximation of the optimal solution at time k. In that case a stochastic

approximation algorithm is chosen to update sequentially the
necessary subjective probabilities.

$$P_{n(ij)+1}(u^\ell) = P_{n(ij)} + \beta_{n(ij)+1} \left[\xi_{n(ij)+1}(u^\ell) \right.$$
$$\left. - P_{n)ij)}(u^\ell) \right] \quad \ell = 1,\ldots,A \tag{8.67}$$

where

$$\xi_{n(ij)}(u^\ell) = \begin{cases} 1, & \text{if } \hat{V}_{n(ij)}(u^\ell) = \underset{r}{\text{Min}} \; \hat{V}_{n(ij)}(u^r) \\ 0, & \text{otherwise} \end{cases} \tag{8.68}$$

Then $0 \le p_n(u^\ell) \le 1$ and if

$$\lim_{n \to \infty} \beta_n = 0 \quad \sum_{n=1}^{\infty} \beta_n = \infty, \quad \sum_{n=1}^{\infty} \beta_n^2 < \infty \tag{8.69}$$

the algorithm (8.67) converges with probability one to the optimal
pure strategy, e.g.,

$$\text{Prob}\left[\lim_{n(ij) \to \infty} P_{n(ij)}(u^*) = 1 \right] = 1 \tag{8.70}$$

The combined stochastic approximation algorithms yield a global
minimum in the limit, provided that all outputs of the system are
visited by applying all the inputs of the system when

$$n(ij) = \underset{r}{\text{Min}} \; [n_r(ij); \; r = 1,\ldots,A] \tag{8.71}$$

The algorithm is depicted in Fig. 8.13. Extension of this algorithm
to a countable output set Ω_x has not been successful [8.59]. However,
a shifting discretization grid of the output set Ω_x has been successfully applied by Riordon [8.65] in a modification of the method
which handles nonstationary plants as well and will be discussed in
the sequel. However, the algorithm presented in this section has
been developed in more vigorous mathematical terms, and measures of
the reduction of the uncertainties of the plant-environment
relation have been shown to converge monotonically to one.

The problem used in the previous section was simulated here to
illustrate the method.

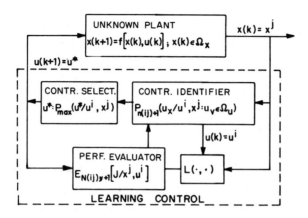

FIG. 8.13. Block Diagram of the Algorithm of Learning Without Supervision of Nikolic and Fu.

$$x(k + 1) = \begin{bmatrix} 0 & 1 \\ f_1 & f_2 \end{bmatrix} x(k) + \begin{bmatrix} 0 \\ b \end{bmatrix} [u(k) + w(k)]$$

$$z(k) = [1,0]x(k) + v(k)$$

where $f_1 = -0.974$, $f_2 = 1.874$, and $b = 0.01$

$$x(0) = x_0 \sim N\left[\begin{pmatrix} 0 \\ 0 \end{pmatrix}, \begin{pmatrix} 9 & 0 \\ 0 & 4 \end{pmatrix} \right]$$

$$w(k) \sim N(0,0.1)$$

$$v(k) \sim N(0,0.1) \qquad \text{for all } k$$

The performance criterion for this system is defined as

$$J(u) = \lim_{N \to \infty} \frac{0.1}{N} E \left\{ \sum_{k=1}^{N} [(z^d(k) - z(k))^2 + 0.01 \, u^2(k)] \right\}$$

and $z^d(k)$ the desired response is chosen to be a unit-step function.

The output space was discretized into 10 regions z_i, $i = 1,\ldots,$ 10, for which the lower-boundary value is assigned if the output lies in the region. The boundary values are given by the set

{2.5, 2.0, 1.6, 1.3, 1.1, 0.9, 0.6, 0.2, -0.3}. The control values
for the linear feedback controller can be chosen from one of the
two combinations

$$
\begin{bmatrix} f_1 \\ f_2 \\ f_3 \\ f_4 \end{bmatrix} = \begin{bmatrix} -2.0 \\ -3.0 \\ 0.1 \\ 0.1 \end{bmatrix} \text{ or } \begin{bmatrix} -1.0 \\ -9.1 \\ 1.0 \\ 0.9 \end{bmatrix}
$$

appropriately chosen to match the growing automaton and expanding
subinterval results. A similar quantity to the *subgoals* selected
for the other algorithms is defined here. This subgoal is appropriately adjusted to meet the requirements of the algorithm of
learning without supervision

$$
J_s(k) = [z^d(k) - z(k)]^2 + [z^d(k + 1) - z(k + 1)]^2 + 0.01 \ u^2(k)
$$

The results are plotted in Figs. 8.14 and 8.15. The algorithm
was rather slow and consumed a large computer memory space.

8.5.3 Riordon's Adaptive Automaton Algorithm

Riordon presented in [8.65] an automaton with performance-
adaptive capabilities which can be considered as a direct extention
of the learning algorithm of Nikolic and Fu by minimizing a multistage performance criterion.

The scheme recursively combines process model identification,
optimal feedback control policy estimation, and decision making to
synthesize the S.O.C. automaton. It also provides for automaton
structure adjustment through appropriate quantizations of the
control and state spaces. As will be noted in the forthcoming
development of the algorithm, the states of the system are assumed
to be directly measurable with no measurement error.

Consider a stochastic process of long duration described by
the following stationary state equation:

$$
x(k + 1) = f[x(k),u(k)] + g[x(k),u(k)] \tag{8.72}
$$

where $x(k)$ is the state variable at time k and is assumed to be

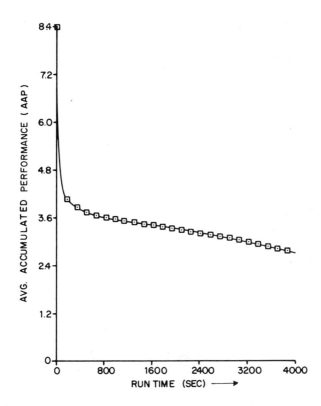

FIG. 8.14. Average Performance Cost for the Algorithm of Learning Without Supervision of Nikolic and Fu.

completely observable without error; $\{u(k) \in \Omega_u = u_{iq}$, $i = 1,...,N$, $q = 1,...,r\}$ is the control variable at time k; $h(.)$ is a continuous and twice differentiable function of x and u; and g is a random variable whose density function is smooth in x and u. It is also assumed that for fixed x and u, $g(k)$ for $k = 1, 2,...$ is an independent sequence. It is assumed that the functions $f(.)$ and $g(.)$ are initially completely unknown and that the process dynamics are, in general, nonlinear and that state- and control-dependent noises may be present. The object of control is to determine the optimal feedback control policy, $u^*(x)$, belonging to the admissible set Ω_u, which minimizes the following criterion for the long duration process:

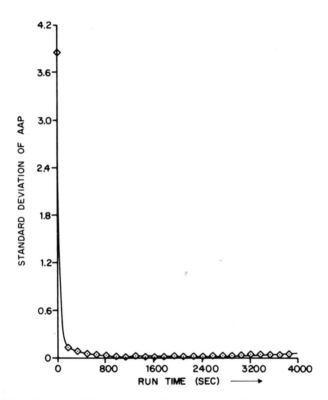

FIG. 8.15.   Standard Deviation of the Cost of the Algorithm of
Learning Without Supervision of Nikolic and Fu.

$$J = \lim_{T \to \infty} \frac{1}{T} E \left\{ \sum_{k=0}^{T-1} L[x(k),x(k + 1),u(k)] \right\} \qquad (8.73)$$

where L is a given convex function of x and u.   If L is separable in
x and u, then discrete-state process costs are defined by cost
matrices B and C derived from L.   If L is not separable, B and C are
replaced by a single 3 x 3-matrix of costs; B is an N x r control-
cost matrix, each of whose elements b_{iq} is the cost of using control
u_{iq}; and C is an N x N transition-cost matrix, each of whose elements
c_{ij} is the cost of a probabilistic transition from *interval state*
ϕ_i to *interval state* ϕ_j.

The interval states q_i are formed by first quantizing the admissible range of the state x into a set of N-discrete intervals

$$\Phi = \{\phi_1, \phi_2, \ldots, \phi_N\} \qquad (8.74)$$

each one being designated as a process state. For each process state ϕ_i there are assigned a set of r discrete alternative control inputs $\{\psi_{iq}, q = 1, 2, \ldots, r\}$. Define as a decision state $\xi_{iq}(k) = [\phi_i(k), u_{iq}(k)]$ where this implies the event (process state at stage k is ϕ); a decision has been made to apply control alternative q during the interval $[k, k + 1]$. Note that ξ_{iq} belongs to the product space $\Xi = \phi x \psi$, where ψ is the set of all inputs u_{iq}; $\xi_{iq} = [\phi_i, u_{jq}]$ is admissible only if $i = j$.

Using the above quantizations, the discrete-state dynamics are then modeled by an $N \times N \times r$, 3×3-matrix P, whose elements are un-known but stationary.

$$P = \{p_{ijq} = Pr[\phi(k + 1) = \phi_j | \xi(k) = \xi_{iq}]\} \qquad (8.75)$$

The feedback control policy is defined by an $N \times r$ decision matrix D, whose element d_{iq} is the probability that control alternative ψ_{iq} will be applied when the process state is ϕ_i. The object of control is to determine a stationary optimal policy $D = D^*$ which minimizes (8.73) over a long duration of operation. Clearly, this process model is a finite Markov chain whose transition probabilities are to be altered during self-organization by appropriate control decisions. It is assumed that the process model is an ergodic chain.

The identification part of the algorithm is comprised of the estimation of the transition probabilities. For that a stochastic approximation algorithm may be used at the $n = (n_{ijq} + 1)$-transition from state q_i to state q_j using the control action u_{iq}

$$\hat{p}_{ijq}(n + 1) = \hat{p}_{ijq}(n) + \gamma_{n+1}[1 - \hat{p}_{ijq}(n)] \qquad (8.76)$$

$$\hat{p}_{i\ell q}(n + 1) = \hat{p}_{i\ell q}(n) - \gamma_{n+1} \hat{p}_{i\ell q}(n) \qquad \ell \neq j$$

$$\hat{P} = \{\hat{p}_{ijq}\}$$

Using conventional methods, Riordon develops the algorithm for computing the estimated optimal policy

$$\hat{D}^*(k) = [D^*|P=\hat{P}(k)]$$ (8.77)

where the conditioning denotes the fact that $\hat{D}^*(k)$ is the optimal policy computed using $\hat{P}(k)$ for P. Generally speaking, $\hat{D}^*(k)$ is determined by

$$\hat{D}^*(k) = \underset{D}{\text{Min}} \left\{ \sum_{q=1}^{r} d_{iq}\hat{n}_{iq} \right\} \text{ for } i = 1, 2,\ldots,N$$ (8.78)

where

$$\hat{n}_{iq} = \hat{\mu}_{iq} + \sum_{j=1}^{N} \hat{P}_{ijq}\hat{v}_j$$ (8.79)

$$\hat{v}_i = \hat{\ell}_i + \sum_{i=1}^{N} \hat{r}_{ij}\hat{v}_j - \hat{g}$$ (8.80)

$$\hat{\ell}_i = \sum_{q=1}^{r} d_{iq}\hat{\mu}_{iq}$$ (8.81)

$$\hat{r}_{ij} = \sum_{q=1}^{r} d_{iq}\hat{P}_{ijq}$$ (8.82)

and

$$\hat{\mu}_{iq} = b_{iq} + \sum_{j=1}^{N} \hat{P}_{ijq}c_{ij}$$ (8.83)

Using further known results, Riordon states that since for a given $\hat{P}$ the optimal policy $\hat{D}^*$ contains elements which are either zero or unity, i.e., the optimal policy is deterministic. Hence, instead of (8.78), the ith element $\hat{d}^*_{is}$ of the estimated optimal policy where $\hat{s} = \hat{s}(i)$ is determined by

$$\hat{n}_{i\hat{s}} = \underset{q}{\text{Min}} \{\hat{n}_{iq}\}$$ (8.84)

When a new observation changes $\hat{P}(k)$ to $\hat{P}(k + 1)$, $\hat{D}^*(k)$ is updated to obtain $\hat{D}^*(k + 1)$. Riordon does present a modified method that accomplishes the computation of (8.79) through the use of various

transformations. This eliminates the necessity for the on-line
inversion of certain matrices. Since it is not necessary for the
basic understanding of the S.O.C. algorithm, it is omitted here
(see Ref. [8.65]).

The next function of the performance-adaptive S.O.C. is to
decide which control action u_{iq} to apply after state ϕ_i has been
observed. Riordon employs known results for single-state processes
and extends them to the multistage by considering the latter as a
series of single-stage decisions which are modified by, and interact
through, the variables $\hat{v}_j$. He uses a mixed strategy to seek at each
stage

$$n_{is} = \underset{q}{\text{Min}} \ \{n_{iq}\} \tag{8.85}$$

The strategy applies the control action, which incurs the minimum
expected cost for a given reduction of the single-stage error
probability Ω_i, where

$$\Omega_i = \Pr[\hat{s}(i) \neq s(i) | M, B, C] \tag{8.86}$$

Alternative $s(i)$ is the minimum cost control input for the observed
state ϕ_i, calculated when P is known. Riordon defines the following
strategy for control decision:

1.   For a given ϕ_i, the probability θ_i of choice of the estimated
     optimal control action $\psi_{i\hat{s}}$ is given by

$$\theta_i = 1 - \gamma \ \Omega_i \quad \text{for } \gamma \ \Omega_i < 0.5 \tag{8.87}$$

$$\theta_i = 0.5 \quad \text{for } \gamma \ \Omega_i \geq 0.5$$

     where γ is a constant fixed by the designer.

2.   If $\psi_{i\hat{s}}$ is not chosen then ψ_{iq}, $q \neq \hat{s}(i)$, is chosen so that

$$\frac{\exp(-\hat{\rho}_{iq}^2/2)}{n_{iq}\hat{\sigma}_{iq}} = \underset{\substack{h=1,2,\ldots,r \\ h\neq\hat{s}(i)}}{\text{Max}} \frac{\exp(-\hat{\rho}_{ih}^2/2)}{n_{iq}\hat{\sigma}_{ih}} \tag{8.88}$$

where

$$\hat{p}_{ih} = \frac{\hat{\eta}_{ih} - \hat{\eta}_{is}}{\hat{\sigma}_{ih}} \qquad (8.89)$$

$$\hat{\sigma}^2_{iq} = \frac{1}{n_{iq}} \sum_{j=1}^{N} \hat{p}_{ijq}(1 - \hat{p}_{ijq})c^2_{ij} \qquad (8.90)$$

Riordon has shown that

$$\Omega_i = \sum_{\substack{q=1 \\ q \neq \hat{s}}}^{r} F_{iq}(\hat{\eta}_{i\hat{s}}) \qquad (8.91)$$

where F_{iq} is a Gaussian cumulative distribution function

$$F_{iq}(x) = \frac{1}{(2\pi)^{1/2}\hat{\sigma}_{iq}} \int_{-\infty}^{x} \exp\left[-\frac{1}{2}\left(\frac{y - \hat{\eta}_{iq}}{\hat{\sigma}_{iq}}\right)^2\right] dy \qquad (8.92)$$

The value of Ω_i is determined on-line by interpolation from a table of normalized values of $F_{iq}(x)$. For γ constant, all values Ω_i, $i \neq s$, tend to zero with this strategy, so that the asymptotic probability of a correct control choice for every state ϕ_i tends to unity.

The above represents a general outline of the self-organizing technique proposed by Riordon. He also provides for adjustment of the basic structure of the automaton model itself through a proposed requantization of the control space based upon prespecified threshold levels for Ω_i. An analogous procedure is also suggested for requantization of the state space based upon the steady-state probabilities of occupancy. These probabilities are estimated on-line from previously determined variables. However, the cost (performance index) is usually less sensitive to state quantization than to control quantization.

The adaptive automaton method is very interesting for attacking multistage minimization problems for relatively low-order Markov processes. However, it is only applicable to Markov processes with directly measureable states, although the process dynamics themselves may be completely unknown.

This method is rather difficult to implement and for higher-order systems requires a large computer memory. It has additional theoretical shortcomings in that the convergence of the scheme to the optimal control policy is ultimately determined by the convergence of $\hat{P}$ to P. This convergence, in turn, is inherently connected with the overall stability of the self-organizing system, which, in general, is not guaranteed for the scheme proposed. However, like all the self-organizing methods discussed thus far, the stability of the overall system must be assumed. When such an assumption is valid, Riordon has shown that the estimate $\hat{P}$ does converge to P and therefore, for the long-duration process, the estimated optimal control law does converge to the optimal.

The method was tested on the following heat-treatment process presented by Riordon [8.65]. The physical process is nonlinear and first order, involving an endothermic reaction for temperatures below 800°K and an exothermic reaction for higher temperatures. The process dynamics are

$$z(k + 1) = z(k)(1.005 + 0.015 \tanh[0.1 (z(k) - 803.446)])$$

$$+ 0.00333 u(k) + w_1(k) + w_2(k) + w_3(k) + w_4(k)$$

where

$z(k)$ = temperature (°K) at stage k

$u(k)$ = heat input (kcal) at stage k

$w(k)$, i = 1,...,4, are independent samples drawn from normal
      zero-mean distributions with the following respective
      standard deviations: σ_i, i = 1,...,4

$\sigma_1 = 0.0002 z(k) y|z(k) - 800|$

$\sigma_2 = 0.005 z(k)[1 + z(k) - 800|^{1/2}]^{-1}$

$\sigma_3 = 0.0005 |u(k)|$

$\sigma_4 = 1$

$$y = \begin{cases} 1; & z(k) > 800 \\ \\ 0; & z(k) \le 800 \end{cases}$$

The uncertainties relative to the process for this particular example are so that the exact form of the process dynamics is unknown, thereby casting it into the framework of self-organization. The object of control in this situation is to maintain the operating temperature near 800°K, which is actually a point of unstable equilibrium. The state $z(k)$ is assumed to be exactly measureable with negligible error.

Deviations from the desired temperature of 800°K are assigned a cost (in cents) given by

$$L_1(z(k),z(k+1)) = \begin{cases} 0.015 \ [(z(k) - 800)^2 + (z(k+1) - 800)^2] \\ \qquad\qquad \text{for } z(k) < 850 \\ 2880 \qquad \text{for } z(k) \geq 850 \end{cases}$$

If $z(k)$ exceeds or equals 850°K, then a shutdown of the process occurs at a cost of \$28.80, and the process is restarted at $z(k) = 775°$K. In addition, penalties on the amount of control energy expended are added to the total cost. Heating and cooling effort results in a cost of 2 cts. per 1000 kcal. If extreme energy is needed, a reduction in the expected lifetime of the control equipment follows, hence such extremes are heavily penalized. In any event, $u(k)$ cannot exceed 10,000 kcal. The total cost (in cents) of control can then be formulated as

$$L_2(u(k)) = 0.002 \ |u(k)| + 60 \ (u(k)/10^4)^6$$

where

$$|u(k)| < 10^4$$

The object of control in this situation is to find a feedback control policy that minimizes the expected cost of operation per unit of time over a long period as described by the following index of performance:

$$J = \lim_{T \to \infty} \frac{1}{T} E \left\{ \sum_{k=0}^{T-1} [L_1[z(k),z(k+1),u(k)] + L_2[u(k)]] \right\}$$

The results are plotted in Figs. 8.16 and 8.17.

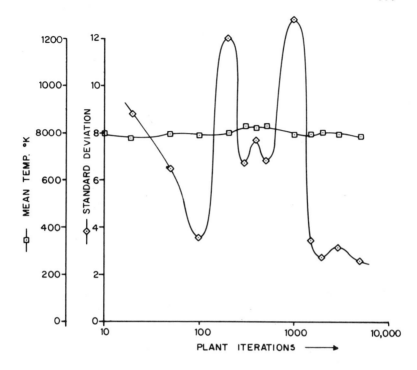

FIG. 8.16.   Single-Run Results for Riordon's Adaptive Automaton.

8.6   SARIDIS' EXPANDING SUBINTERVAL ALGORITHM

This particular algorithm was conceived when Gilbert and Saridis were developing a control algorithm with asymptotically optimal properties for an orbiting satellite [8.71].   These basic ideas developed into the *expanding subinterval control algorithm* which is believed to be the most general and complete of the on-line learning control algorithms that treat systems with unknown dynamics operating in a stochastic environment.   The advantages of this algorithm are pointed out by summarizing its outstanding features:

1.   The controller does not depend structurally or parametrically on the explicit form of the plant dynamics or the noise statistics.   Therefore, minimal information about the plant is required and continuous or discrete linear or nonlinear systems can be treated.

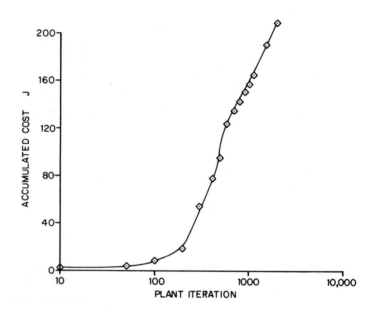

FIG. 8.17.   Accumulated Cost for Riordon's Plant Iteration Adaptive
             Automaton.

2.   There is a definite relation established between the *subgoal*
     used and the *overall performance criterion* of the system.   There
     the algorithm obtains the asymptotic optimality for the system.
3.   The search technique possesses global properties while it is
     very easy to implement.
     The assumptions required for the algorithm are that the system
stays finite for all control actions, that the controller is properly
structured to effectively drive the plant, and that the length of the
process is infinite as for all the processes treated by a learning
algorithm.

8.6.1   The Random Search Algorithm

     The algorithm will be developed here for both a continuous- and
a discrete-time by either

$$\dot{x}(t) = f(x(t),u(t),w(t),t) \quad u(t) \ \varepsilon \ \Omega_u \qquad x(0) = x_0 \tag{8.93}$$

$$z(t) = h(x(t),v(t),t)$$

or

$$x(k + 1) = f(x(k),u(k),w(k),k) \quad x(0) = x_0 \quad u(k) \ \varepsilon \ \Omega_u \tag{8.94}$$

$$z(k) = h(x(k),v(k),k)$$

where $x(t)$ or $x(k)$ is an n-dimensional state vector in the state space Ω_x; $u(t)$ or $u(k)$ is an m-dimensional control vector in the closed and bounded set of *admissible* controls Ω_u; x_0, $w(t)$ or $w(k)$, and $v(t)$ or $v(k)$ are vector valued random variables belonging to independent random processes with unknown statistics; $z(t)$ or $z(k)$ is an r-dimensional output vector $r \leq n$; $f(.)$ and $h(.)$ are unknown nonlinear continuous and bounded functions of their arguments of appropriate dimensions. The vectors $w(t)$, $w(k)$, $v(t)$, and $v(k)$ represent the plant-environment relationship.

Define the performance criterion for the overall process for the continuous-time case

$$I(u) = E\{J(u)\} = E\left\{\lim_{T \to \infty} \frac{1}{T - t_0} \int_{t_0}^{T} L(z(t),u(t),t) \ dt \right\} \tag{8.95}$$

or the discrete-time case

$$I(u) = E\{J(u)\} = E\left\{\lim_{N \to \infty} \frac{1}{N} \sum_{k=1}^{N} L(z(k),u(k),k)\right\} \tag{8.96}$$

where L is a known nonlinear continuous bounded operator. A *specific* control sequence $\{u_c(t)\}$ or $\{u_c(k)\}$ is sought to minimize $I(u)$ of (8.95) or (8.96)

$$u_c(t) \triangleq F[c^T\phi(z(t),t)t] \ \varepsilon \ \Omega_u \tag{8.97}$$

or

$$u_c(k) = F[c^T\phi(z(k),k)] \ \varepsilon \ \Omega_u \tag{8.98}$$

so that $I(u_c)$ is finite for all controls $u_c(k) \ \varepsilon \ \Omega_u$, i.e., the admissible set and,

$$0 \leq I_{min} \leq I(u_c) < \infty, \text{ for all } u_c \in \Omega_u \tag{8.99}$$

In the above specific controller $\phi(.)$ is a p-dimensional vector of linearly independent continuous and bounded functions of the measureable outputs properly chosen to span Ω_u, c is a p-dimensional vector of adjustable coefficients, and F is an m-dimensional nonlinear bounded operator in Ω_u. Since there is not enough information for the controller to minimize $I(u)$ directly, the following "subgoal" $J(c_i)$ may be defined to be sequentially improved by the controller so that $I(u)$ will be asymptotically minimized

$$J(c_i) = \frac{1}{t_i - t_{i-1}} \int_{t_{i-1}}^{t_i} L(z(t), F[c_i^T \phi(z(t),t)]) dt \tag{8.100}$$

or

$$J(c_i) = \frac{1}{t_i - t_{i-1}} \sum_{k=t_{i-1}}^{t_i} L(z(k), F[c_i^T \phi(z(k),k)]) \quad i = 1,2,\ldots,\infty \tag{8.101}$$

The length of the time interval $T_i = t_i - t_{i-1}$, $i = 1,2,\ldots,$ satisfies the conditions

$$T_{i-1} \leq T_i \quad \lim_{i \to \infty} T_i = \infty \text{ for all } i \tag{8.102}$$

The c_i^* is sought through an appropriate search technique applied once at every interval to improve sequentially the value of $J(c_i)$. A global random search technique is used for this purpose at every subinterval T_i

$$c_{i+1} = \begin{cases} \rho_i, J(c_i) - J(\rho_i) > 2\mu \\ \\ c_i, J(c_i) - J(\rho_i) \leq 2\mu \end{cases} \quad \text{at } T_i \tag{8.103}$$

where ρ_i is the ith sample value of a vector random process defined at every subinterval T_i on $\Omega_c = \{c/u_c(i) \in \Omega_u\}$ with a probability density function $p(\rho) \neq 0$ over Ω_c, and μ is some arbitrary positive number. Define the cost "accrued" during the search

$$V(n) = \frac{\sum\limits_{i=1}^{n} T_i \, E_{wv}\{J(c_i)\}}{\sum\limits_{i=1}^{n} T_i} \qquad n = 1, 2, \ldots \qquad (8.104)$$

where T_i is defined by (8.102).

Define by c^* the value of c that minimizes $I(u_c)$

$$c^* = \{c/\text{Min}_c \, I(u_c) = I_{min}\} \qquad (8.105)$$

It can be easily shown that after successive applications of c^* for c_i in $V(n)$, the "cost" is the accumulative performance criterion over the interval $[0, t_n]$

$$V(n) = \frac{1}{t_n} E \left\{ \sum\limits_{k=1}^{n} L(z(k), F[c^{*T}\phi(z(k), k)], k) \right\} \qquad (8.106)$$

and in the limit

$$\lim_{n \to \infty} V(n) = I(u_c^*) = \text{Min}_c \, I(u_c) = I_{min} \qquad (8.107)$$

Then the following theorem establishes the asymptotic optimality of the algorithm:

Theorem 8.1.

Given an arbitrary number $\delta > 0$ and the random search (8.103) with the following properties:

A1.   The function $J(c)$ satisfies (8.101) and is such that $\Omega_c^* \triangleq \{c/J(c) - J_{min} < \delta, \forall \delta, c \in \Omega_c\}$ has a positive measure.

A2.   Conditions (8.102) are satisfied.

A3.   For any n_1, n_2, and n_3,

$$\lim_{\frac{n_3}{n_2} \to 0} \frac{\sum\limits_{i}^{n_3} T_i}{\sum\limits_{i=n_1+1}^{n_1+n_2} T_i} = 0$$

where $\sum\limits_{i}^{n_3}$ denotes a sum containing n_3 terms of the denominator.

Then there exists a number $\mu > 0$ such that

$$\lim_{n \to \infty} \text{Prob}[V(n) - l_{min} < 3\,\delta] = 1 \tag{8.108}$$

Theorem 8.1 and the convergence in probability implied in it have been proved [8.67]. Furthermore, it was found necessary for computational improvement to evaluate $J(c_i)$ at every iteration of the algorithm in order to use updated values of the performance criterion for comparison, during the transient period of the system. Since this procedure is performed on-line, the additional cost may be included in the accumulative cost function $V(n)$ which is redefined as

$$\bar{V}(n) = \frac{\sum\limits_{i=1}^{n} E_{wv}[T_i\,J(c_i) + \bar{T}_i\,J(\rho_i)]}{\sum\limits_{i=1}^{n} [T_i + \bar{T}_i]} \qquad \bar{T}_i = \begin{cases} T_i, c_{i+1} = \rho_i \\ 0, c_{i+1} = c_i \end{cases} \tag{8.109}$$

For this accumulative cost function Theorem 8.2 has been proved in [8.67].

Theorem 8.2.

Given an arbitrary $\delta > 0$, the random search (8.103) satisfying properties A1 to A3 of Theorem 8.1 and the additional property

A4. $\lim\limits_{n \to \infty} = \dfrac{\sum\limits_{i=1}^{n} T_i}{\sum\limits_{i=1}^{n} [T_i + \bar{T}_i]} = 0$

Then there exists a number $\mu > 0$ so that

$$\lim_{n \to \infty} \text{Prob}[\bar{V}(n) - J_{min} \leq 4\,\delta] = 1 \tag{8.110}$$

The usage of the above algorithm is limited by condition (8.107) which requires that the trajectories of the system remain finite

for the duration of the process. A general class of nonlinear
systems for which such conditions can be established without the
exact knowledge of the nonlinearity has been investigated [8.70].
These conditions can be translated into conditions of the parameters
c_i of the controller so that the appropriate admissible sets Ω_c are
defined before the process starts; they will be discussed in the
next chapter.

A block diagram of the expanding subinterval S.O.C. algorithm
is depicted in Fig. 8.18.

8.6.2 The Selection of the Structural Model

The two major problems facing the designer of an expanding
subinterval self-organizing controller as well as any other self-
organizing controller are stability of the system and the selection
of the $\phi(.)$ function to approximate as close as possible an
asymptotically optimal solution. The stability problem will be
discussed in the next chapter, but the selection of the structure of
the feedback control is discussed here. The problem is far from
being resolved in its generality. Therefore, only certain ideas

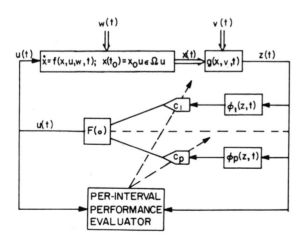

FIG. 8.18.   Block Diagram of the Expanding Subinterval S.O.C.
Algorithm.

applicable to a limited class of problems will be presented.

It is obvious that the structure of the feedback control, e.g., the selection of the $\phi(.)$ functions, depend on the plant's non-linearities. If it was known that the plant was linear, the feedback control should be a linear function of the states, and the optimal solution, also linear, would be asymptotically reached.

In general the structure of the plant is not known and without such a knowledge the structure of the control can only be speculative. It is conjectured that such a control should be described by a non-linear dynamic system of at least the same dimension as the plant. Such a system may be represented by an integral expression, e.g., a Volterra series expansion to represent the nonlinearities and unknown coefficients to be determined by the expanding subinterval algorithm. Such representations have been discussed in Chap. 2; they were found impractical because of the size of the number of terms in the expansion required to give a reasonable approximation to the structural form of the optimal control. Therefore, even though one may use this general approach for certain special cases, some knowledge of the nonlinearities and their location in the plant would reduce the difficulties of obtaining an effective feedback control.

This may be obtained through *pattern recognition method for the classification of nonlinearities* of Saridis and Hofstadter, using estimates of input-output crosscorrelation. A diagram of the method is depicted in Fig. 8.19 [8.72]. The method may be applied either off-line or whenever an intermediate stabilizing control is available to keep the system within bounds during the on-line identification period. It classifies nonlinearities to general classes of similar nonlinearities, e.g., the saturation class depicted in Fig. 8.20, located at different parts of the plant, e.g., feedforward or feedback loop etc. It is conjectured that the representative nonlinearity in the feedback control for most practical purposes. This is the equivalent of the engineering approximation of replacing a saturation curve by its piecewise linear approximation.

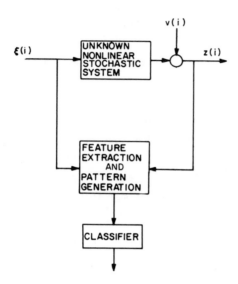

FIG. 8.19.   The Saridis-Hofstadter Nonlinear System Classification
Method.

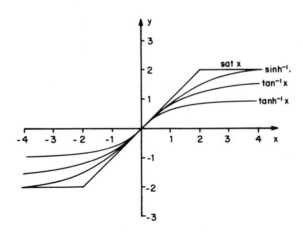

FIG. 8.20.   The Class of Saturation Nonlinearities.

Once the nonlinearities of the plant have been identified within their classes, the selection of the $\phi(.)$ functions must be accomplished. There is no known general method to relate the $\phi(.)$ functions to the plant's nonlinearities. Of course, heuristically one may assign $\phi(.)$ functions containing the same nonlinearities as the plant, thus having a controller that can always compensate for such structures. This, however, does not represent an optimal solution of the problem and may lead to unstable cancellations of functions. Another approach may be generated from the approximation of some algebraic expression of the optimal formulation of the system. For instance, the nonlinear system may be considered

$$\frac{dx}{dt} = Fx + f(x) + b(x)u \tag{8.111}$$

with a performance criterion

$$J(u) = \frac{1}{2} \int_0^T u^2(t)\ dt \tag{8.112}$$

where the state vector $x(t)$, control input $u(t)$, and the corresponding nonlinearities and coefficients are of appropriate dimension as previously defined.

For a time-independent system the optimal control is generated by

$$u^*(t) = -b^T(x)p \tag{8.113}$$

$$\frac{dp}{dt} = -F^T p - \frac{\partial f}{\partial x}^T p - \frac{\partial b}{\partial x}^T pu^*(t)$$

The constant Hamiltonian of the system becomes

$$H = \text{const} = p^T Fx + p^T f(x) + \frac{1}{z} p^T b(x)b^T(x)p \tag{8.114}$$

Since one is not interested in the coefficients, it may be easy, for some cases, to find $p(t)$ as a function of the states $y(x)$ so that it satisfies (8.114) which is quadratic in p. For instance, if the plant is scalar

$$p \triangleq y(x) = \frac{Fx + f(x) \pm \sqrt{(Fx + f(x))^2 - 2 \, Hb^2(x)}}{b^2(x)} \qquad (8.115)$$

and the structure of the controller is given by

$$u^*(x) = -b^T(x)y(x) \qquad (8.116)$$

which is structurally known. This may not be always possible to obtain, but it represents an interesting idea to follow up in future research. A successive approximation $y_i(x)$ to determine the function $y(x) = \sum_{i=1}^{w} a_i y_i(x)$ may also be considered.

However, from the above demonstration it has been shown that the optimal feedback controller is not necessarily a linear function of the plants nonlinearities.

The expanding subinterval algorithm is illustrqted by applying it to the following linear system:

$$x(k + 1) = \begin{bmatrix} 0 & 1 \\ f_1 & f_2 \end{bmatrix} x(k) + \begin{bmatrix} 0 \\ \beta \end{bmatrix} [u(k) + w(k)]$$

$$z(k) = [1,0]x(k) + V(k)$$

where $f_1 = -0.974$, $f_2 = 1.874$, and $\beta = 0.01$

$$x(0) = x_0 \sim N\left[\begin{pmatrix} 0 \\ 0 \end{pmatrix}, \begin{pmatrix} 9 & 0 \\ 0 & 4 \end{pmatrix} \right]$$

$$w(k) \sim N(0,0.1)$$

$$v(k) \sim N(0,0.1) \qquad \forall k$$

The performance criterion for this system is defined as

$$J(u) = \lim_{N \to \infty} \frac{0.1}{N} E\left\{ \sum_{k=1}^{N} [(z^d(k) - z(k))^2 + 0.01 \, u^2(k)] \right\}$$

and $z^d(k)$ the desired response is chosen to be a step function.

A linear controller was structured to drive the plant

$$\hat{x}(k + 1) = A\hat{x}(k) + dz(k)$$

$$u(k) = h^T\hat{x}(k)$$

with

$$A = \begin{bmatrix} 0 & 1 \\ a_1 & a_2 \end{bmatrix}, \quad d = \begin{bmatrix} d_1 \\ d_2 \end{bmatrix}, \quad h = \begin{bmatrix} 1 \\ 0 \end{bmatrix}$$

with self-organizing coefficients

$$c^T = [a_1, a_2, d_1, d_2]$$

"Subgoal" is defined as

$$J_s(k) = \frac{1}{T_i} \sum_{k=t_{i-1}}^{t_i} [(z^d(k) - z(k))^2 + 0.01 \, u^2(k)]$$

where $T_i = t_i - t_{i-1} = \Delta T = 1$ sec for $i = 1, \ldots, 1000$, and $T_i = (i-1000)$ ΔT for $i > 1000$. This satisfies the conditions of the expanding subinterval algorithm (8.101).

The simulation results are given in Figs. 8.21 and 8.22. The algorithm was stable for all runs and converged relatively fast probably due to the knowledge of the linearity of the system. Another application to a nonlinear plant is given in Chap. 9.

8.7 STOCHASTIC APPROXIMATION FOR CONTROL OF A CLASS OF DISTRIBUTED SYSTEMS

Stochastic approximation has been applied by Badavas and Saridis [8.4] to control distributed systems of the elliptic type with unknown coefficients in the presence of measurement noise. The method is treated separately to emphasize the difficulties involved in treating distributed systems operating in a stochastic environment. The problem can be stated as follows: Let the model of the system be defined on a bounded domain Ω_x with its boundary Ω_b,

$$\sum_{ij}^{M} a_{ij}(x) \frac{\partial^2 S(x)}{\partial x_i \, x_j} + \sum_{i=1}^{N} b_i(x) \frac{\partial S(x)}{\partial x_i} + h(x) \cdot S(x) = u(x) \quad x \in \Omega_x$$

$$(8.117)$$

with boundary condition, the Dirichelet condition

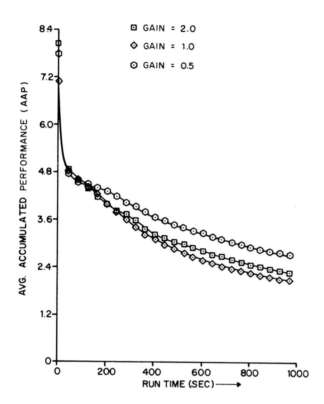

FIG. 8.21.   Average Performance for the Expanding Subinterval
             Algorithm.

$$S(x^b) = g(x^b), \quad x^b \ \epsilon \ \Omega_b, \text{ and noisy measurement} \qquad (8.118)$$

$$z(x) = S(x) + v$$

In (8.117) and (8.118), x is the N-dimensional spatial coordinate
vector, x^b is its boundary equivalent, S(x) is the scalar valued
response of the system, u(x) is a scalar distributed control function,
g(x) is a scalar valued boundary control function, v is a zero-mean
finite variance measurement noise, and coefficients $a_{ij} = a_{ji}$, b_i,
and $h \leq 0$ are assumed to be unknown to the designer. The problem
is to select a distributed and/or a boundary feedback controller to
track a given desired output $S^d(x)$ by minimizing a performance index
of the form

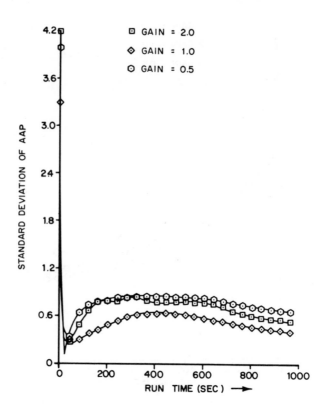

FIG. 8.22.   Standard Deviation of the Performance for the Expanding
Subinterval Algorithm.

$$J(u,g) = E \int_{\Omega_x} [S^d(x) - z(x)]^T M[S^d(x) - z(x)] \, dx \qquad (8.119)$$

It has been shown [8.4] that under certain conditions there
exists a specific controller of the form

$$u(x) = \sum_{m=1}^{M_1} \alpha_m \phi_m(x) = \alpha^T \phi(x) \qquad (8.120)$$

and/or

$$g(x^b) = \sum_{m=1}^{M_2} \beta_m \phi_m^b(x) = \beta^T \phi^b(x^b) \qquad (8.121)$$

and that $S(x)$ belongs to the same class of functions as $S^d(x)$ and has a similar complete Fourier series expansion. In this case, $\phi(x)$ and $\phi^b(x^b)$ are linearly independent functions belonging in the same class as $S^d(x)$, and α_m and β_m are adjustable parameters. Then it has been shown that

$$S(x) = \sum_{m=1}^{M_1} \alpha_m \theta_m(x) + \sum_{m=1}^{M_2} \beta_m \theta_m^b(x) \triangleq c^T \theta(x) \qquad (8.122)$$

$$\theta(x)^T = [\theta_1(x),\ldots,\theta_{M_1}(x),\theta_1^b(x),\ldots,\theta_{M_2}^b(x)]$$

are the unknown responses of the system to the distributed and boundary controls with adjustable coefficients $c^T = [\alpha^T, \beta^T]$.

8.7.1   The Selection of a Finite Number of Measurement Points

In order to account for the function of the distributed systems, its response must be measured at infinite points on each direction of the system coordinates. This is practically not feasible and a scheme is needed that requires only a finite number of measurement points. However, the response function $S(x)$ needs to be measured only at finite points at every coordinate direction, if a multidimensional version of the sampling theorem is satisfied. Such an n-dimensional sampling theorem has been developed and proved [8.4], and is repeated here.

Theorem 8.3.

Consider the function $q(x_1,\ldots,x_N)$ which has an n-dimensional Fourier transform $F(\omega_1,\ldots,\omega_N)$ defined over the frequency domain $G_\omega \triangleq [-\tilde{i}_i \le k_i \le \tilde{i}_i;\ i = 1,\ldots,N]$ such that $F(\omega_1,\ldots,\omega_N) = 0$ for all $\omega_1,\ldots,\omega_N \notin G_\omega$. Let

$$T_i = \frac{\pi}{\tilde{i}_i} = \frac{1}{2 f_i^0} \quad \text{or} \quad f_i^0 = \frac{\tilde{i}_i}{2\pi} \quad i = 1,\ldots,N \qquad (8.123)$$

$T_1, \ldots, T_N$ are the sampling intervals in the $x_1, \ldots, x_N$ coordinates, respectively. Then the function $q(x_1, \ldots, x_N)$ is completely determined by its values taken at a series of points separated at most by the sampling intervals $T_1, \ldots, T_N$ on the $x_1, \ldots, x_N$ spatial coordinates, respectively.

For this problem, let $S^d(\omega)$ be the Fourier transform of $S^d(x)$ and define by T_i the sampling intervals in each direction as

$$T_i = \frac{1}{2 f_1^0} \qquad i = 1, \ldots, N \tag{8.124}$$

where f_i^0 is the maximum admissible frequency in $S^d(\omega)$. Then the minimum number I_i of measuring points in the ith spatial coordinate sufficient to reproduce exactly the response is given by

$$I_i = \frac{L_i}{T_i} \tag{8.125}$$

where L_i is the length of Ω along x_i. Under those circumstances it is sufficient to know $S^d(x)$ and $\theta(x)$ at these discrete points only denoted by $S^d(i_1, \ldots, i_N)$ and $\theta(i_1, \ldots, i_N)$, respectively. The performance index (8.119) using the specific controller and the above discretization is rewritten as

$$J(c) = E\left\{ \sum_{i_1=1}^{I_1}, \ldots, \sum_{i_N=1}^{I_N} [S^d(i_1, \ldots, i_N) - c^T \theta(i_1, \ldots, i_N) \right. \tag{8.126}$$
$$\left. - v(i_1, \ldots, i_N)]^2 \right\}$$

If

$$\det \left| \sum_{i_1=1}^{I_1}, \ldots, \sum_{i_N=1}^{I_N} \theta(i_1, \ldots, i_N) \, \theta^T(i_1, \ldots, i_N) \right| \neq 0 \tag{8.127}$$

then $J(c)$ is unimodal in c.

8.7.2   The Self-Organizing Control Algorithm

Once the systems response can be reproduced exactly by a finite number of measurement points, a S.O.C. algorithm that improves the performance criterion directly can be established.

In order to do that, it is assumed that the measurement noise $v(i_1, \ldots, i_N)$ is a function of the spatial coordinates, and that it satisfies the following conditions:

$$E\{v(i_1, \ldots, i_N)\} = 0, \quad E\{v^2(i_1, \ldots, i_N)\} = \sigma^2(i_1, \ldots, i_N) < \infty \quad (8.128)$$

$$i_1 = 1, \ldots, I_1 \quad i_N = i_N = 1, \ldots, I_N$$

and

$$E\left\{ \sum_{i_1=1}^{I_1}, \ldots, \sum_{i_N=1}^{I_N} v^2 \ (i_1, \ldots, i_N) \right\} = \sigma^2 < \infty$$

Since $\theta(i_1, \ldots, i_N)$ is unknown, a stochastic approximation algorithm of the Kiefer-Wolfowitz type may be defined to find recursively the value c^* that minimizes $J(c)$, and therefore yields an optimal controller.   Define the observable function

$$Q(c_n) \triangleq \sum_{i_1=1}^{I_1}, \ldots, \sum_{i_N=1}^{I_N} [s^d(i_1, \ldots, i_N) - z(i_1, \ldots, i_N, c_n)]^2 \quad (8.129)$$

and generate the approximation of the jth component of its gradient

$$G_j(c_n) = \frac{Q_j(c_n + \beta_n) - Q_j(c_n - \beta_n)}{2\beta_n} \quad j = 1, 2, \ldots, M_1 + M_2 \quad (8.130)$$

With β_n and γ_n being real sequences satisfying

$$\lim_{n \to \infty} \beta_n = 0 \quad \lim_{n \to \infty} \gamma_n = 0 \quad \sum_{n=1}^{\infty} \gamma_n = \infty \quad \sum_{n=1}^{\infty} \left(\frac{\beta_n}{\gamma_n}\right)^2 < \infty \quad (8.131)$$

and M_1 and M_2 being the number of terms in $u(x)$ and $g(x^b)$, respectively.   The algorithm

$$c_{n+1} = c_n - \gamma_n G(c_n) \quad (8.132)$$

yields estimates of c that converge to c^* with probability one and in the mean-square sense

$$Pr[\lim_{n \to \infty} c_n = c^*] = 1$$

$$\lim_{n \to \infty} E\{||c_n - c^*||^2\} = 0 \tag{8.133}$$

The method was implemented on the following distributed system:

$$\frac{\partial^2 s}{\partial x_1^2} + \frac{\partial^2 s}{\partial x_2^2} = u(x_1, x_2)$$

with boundary conditions

$$S(x_1, 0) = S(x_1, \pi) = 0$$

$$S(0, x_2) = S(\pi, x_2) = 0$$

with a desired response

$$s^d(x_1, x_2) = \sum_{m=1}^{2} \frac{4}{m} \sin(mx_1) \sin(mx_2)$$

The controller was chosen to be in the same class as $s^d(x)$

$$u(x_1, x_2) = \sum_{m=1}^{2} \sum_{n=1}^{2} \alpha_{mn} \sin(mx_1) \sin(nx_2)$$

The maximum sampling intervals were $T_1 = T_2 = \pi/2$ and therefore only four measurements were required.

The convergence properties of this algorithm are demonstrated in Figs. 8.23 to 8.25.

8.8 DISCUSSION

Eight basic algorithms, which qualify for stochastic performance adaptive S.O.C., have been presented in this chapter. Even though they represent various philosophies of self-organization, they all operate by improving the performance of the system directly by appropriately updating a feedback controller. Very little

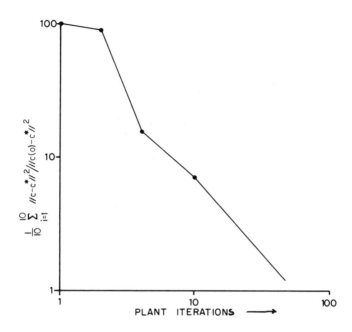

FIG. 8.23.   Normalized Average Parameter Error for Distributed System.

information is usually required about the plant and the stochastic
environment in which it functions. Therefore, systems such as
these demonstrate a more advanced learning capability than previously
described methods. However, the level of learning is still quite
primitive compared with any living system.

The two major problems involved with this formulation and
treated rather lightheartedly in this chapter, are the problems of
overall systems stability and the selection of the structure of
the feedback controller. The first problem is discussed in more
detail in Chap. 9, while the second one was briefly discussed in
context with the expanding subinterval algorithm for which it is
mostly pertinent. A quantitative comparison and evaluation of
these algorithms is given in Chap. 9.

In summary, the performance-adaptive S.O.C. algorithm have
demonstrated superior control capabilities in cases where the
knowledge of the plant dynamics is limited and the interactions with

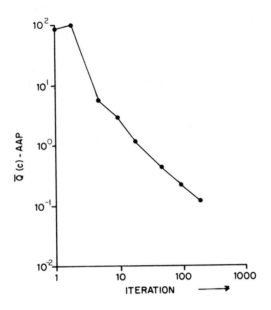

FIG. 8.24.   Accumulated Performance for Distributed System.

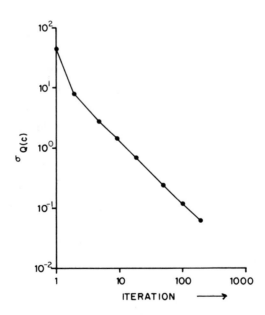

FIG. 8.25.   Standard Deviation of Average Accumulated
Performance (AAP) for Distributed System.

stochastic environment is unknown. Specifically, the crosscorrelation algorithm, which represents the classical approach, demonstrated good convergence properties, which, however, were very sensitive to the initial state of the plant. The method is limited, since it assumes a linear plant and requires some knowledge of its order to select the feedback transfer functions.

The reinforcement methods, which are based on the properties of behaviorial learning are usually more suitable for higher-level decision making from a finite discrete number of control alternatives. When interpreted as stochastic automata, the reinforcement algorithms have been extended to cover more general control problems. Nonlinear reinforcement could make up for the nonoptimality of the convergence of the algorithms. One of them, the growing automaton method, extends to continuous control spaces by the "growing" state and control spaces. It is relatively simple to implement and has been shown to converge to the optimal solution without bias, given a sufficiently long (infinite) time. It can handle a rather general class of problems with very little *a priori* knowledge of the plant structure. As with any of the schemes, enough information about the plant must be known *a priori*, of course, to be able to specify an appropriate performance criterion and to select the appropriate control action space in which to search for the optimal performance. One of its major implementation problems is the normalization of the performance index $J(k)$.

The stochastic approximation method, though formulated for systems with completely unknown dynamics is, practically speaking, considerably less general. In order to specify reasonable approximation functions to model the plant and structure the controller, a considerable amount of *a priori* knowledge of the plant struct must be known. It is applicable to problems with continuous valued state and control spaces, but is extremely sensitive to the stochastic approximation gain constants. The method is hampered by the problem of instability of the coupled identification and control parameter algorithms and sensitivity functions. However, the instantaneous

evaluations of the gradients of mean estimation error criterion F
and the mean performance criterion L in 8.50 and 8.54, respectively,
require the evaluation of the sensitivity matrices $\partial x/\partial b$ and $\partial x/\partial c$,
as well as $\partial u/\partial b$ and $\partial u/\partial c$. Such sensitivity functions can be obtained only when the plant is known. In the case of a linear plant
the sensitivity matrices are constant and depend on the unknown
system parameters. Implementation of this algorithm requires either
the a priori knowledge or the on-line evaluation of these sensitivity
functions. Unpredictable results may occur, however, with the biases
resulting from the application of the stochastic approximation
algorithm to a dynamic system without correction. Since the guaranteed
stability of the system is practically unknown, there is no way to
eliminate this bias or to guarantee stability of the system. It can
be said, however, that both will greatly depend on the strength of
the noises introduced by the environment. No proof of convergence
of the overall combined self-organizing scheme exists to date.
Hence, any application of the technique should be made with caution.
However, in the performed simulations, it was observed that the
method performed well during the finite operation of the learning
control as long as it could be kept stable. But the method may be
unappealing because there is no a priori indication if it will converge for a given self-organizing situation.

The approach of learning without external supervision of Nikolic
and Fu appears to be best suited for problems involving finite state
and control spaces. It requires very little a priori knowledge of
the plant except for the first-order Markov property. Additionally
enough information must be known a priori to specify an appropriate
control-action space and subgoal criterion to direct the search for
the optimal. In the example simulated here the convergence rate was
observed to be slow compared to the rest of the methods, but converged with relatively good confidence. The algorithm has been
shown to converge theoretically to the optimal solution with bias.
It is most effective for situations involving relatively good confidence. The algorithm has been shown to converge theoretically to the
optimal solution without bias. It is most effective for situations

involving relatively few control parameters choices, since for large-scale systems the amount of computer memory and associated learning time increases correspondingly.

Riordon's algorithms may be considered as an extension of the stochastic automaton of McLaren's and the learning without external supervision of Nikolic and Fu. It utilizes dynamic programming to handle the dynamics of the system and is highly cumbersome to implement. But by shifting the state and control grids it accommodates the relative minimum of the performance criterion where it exists outside the original grid of values. An attempt by Nikolic, to consider countable output states yielded biased results.

The random search technique seems to be very general in its formulation. It is limited by the need for a *priori* information to be able to specify an appropriate control-action space in which to search. Such a restriction is severe in cases where a stability analysis of the system is not possible. It can accommodate continuous state and control spaces. Like random search schemes in general, it is most efficient for problems with a relatively large number of control parameters. Hence, it is expected to be less effective than the other methods tested for low-order problems, as treated here. Nevertheless, the method has been shown to yield workable learning controllers. The random search has been shown to converge to the optimal solution without bias in a global sense, hence confidence in using the method is relatively high. It is easy to implement with reasonable computer requirements.

Finally, the stochastic approximation method of Badavas and Saridis for distributed systems has been presented to demonstrate the feasibility of S.O.C. to this class of systems.

In conclusion, the performance-adaptive philosophy for S.O.C. represents a new direction for research in areas of control systems when the plant and the environment are not known. The next chapter is devoted to the evaluation and comparison of the algorithms in order to strengthen and enhance the claim of usefulness of the methods. However, considerable research is still needed in the field.

8.9 REFERENCES

[8.1]   Aizerman, M. A., Braverman, E. M., and Rosonoer, L. I.,
         "Potential Functions Technique and Extrapolation in
         Learning Systems Theory," Preprints 3rd IFAC, London, 1966.

[8.2]   Aizerman, M. A., Braverman, E. M., and Rosonoer, L. I.,
         "Theoretical Foundations of the Potential Function Method
         in Pattern Recognition Learning," *Automation Remote Control*
         (Russian), 25, (6), June, 917-935 (1964).

[8.3]   Aoki, M., *Optimization of Stochastic Systems*, Academic
         Press, New York, 1967.

[8.4]   Badavas, P. C., and Saridis, G. N., "A Performance Adaptive
         Self-Organizing Control for a Class of Distributed Systems,"
         IEEE Trans. Systems Man Cybern., SMC-1, (2), 105-110 (1971).

[8.5]   Bellman, R., *Adaptive Control Processes*, A Guided Tour,
         Princeton University Press, Princeton, N. J., 1961.

[8.6]   Bush, R. R., and Estes, W. K (eds.), *Studies in*
         Mathematical Learning Theory, Stanford University Press,
         Stanford, Calif., 1959.

[8.7]   Bush, R. R., and Mosteller, F., *Stochastic Models for*
         Learning, Wiley, New York, 1955.

[8.8]   Fel'dbaum, A. A., "Dual Control Theory," *Automation Remote*
         Control, Part I: 21 (9), 874-880 (1960); Part II: 21 (11),
         1033-1039; (1960); Part III: 22 (1), 1-12 (1961);
         Part IV: 22 (2), 109-121 (1961).

[8.9]   Fel'dbaum, A. A., *Optimal Control Systems*, Academic Press,
         New York, 1965.

[8.10]   Fu, K. S., "Learning Control Systems," in *Advances in*
         Information Systems Science (J. T. Tou, ed.), Plenum
         Press, New York, 1969.

[8.11]   Fu, K. S., "Learning Control Systems and Intelligent
         Control Systems: An Intersection of Artificial Intelli-
         gence and Automatic Control," *IEEE Trans. Automatic*
         Control, AC-16, (1) 70-72 (1971).

[8.12]   Fu, K. S., "Learning Control Systems," in *Computer and*
         Information Sciences, (J. T. Tou and R. H. Wilcox, eds.)
         New York, Spartan Books, 1964.

[8.13]   Fu, K. S., "Learning Control Systems-Review and Outlook,"
         IEEE Trans. Automatic Control, AC-15, (7) 210-221 (1970).

[8.14] Fu, K. S., "Learning Techniques in System Design-A Brief Review," Fifth World Congress of IFAC, Paris, June 12-16, 1972.

[8.15] Fu, K. S., "Learning System Theory," in *System Theory*, (L. A. Zadeh and E. Polak, eds.), Chap. 11, McGraw-Hill, New York, 1969.

[8.16] Fu, K. S. (ed.), *Pattern Recognition and Machine Learning*, Plenum Press, New York, 1971.

[8.17] Fu, K. S., *Sequential Methods in Pattern Recognition and Machine Teaching*, Academic Press, New York, 1968.

[8.18] Fu, K. S. (ed.), *Learning Systems*, ASME Publ., New York, June 1973.

[8.19] Fu, K. S. (ed.), *Pattern Recognition and Intelligent Robots*, Plenum Press, New York, 1974.

[8.20] Fu, K. S., "A Class of Learning Control Systems Using Statistical Decision Functions," *Proc. Second IFAC (Teddington) Symposium on Theory on Self-Adaptive Control Systems*, Sept., 1965.

[8.21] Fu, K. S., "Stochastic Automata as Models of Learning Systems," *Computer and Information Sciences*, Vol. II (J. T. Tou, ed.), Academic Press, 1967.

[8.22] Fu, K. S., and Li, T. J., "Formulation of Learning Automata and Automata Games," *Inform. Sci.*, 1, (3), 237-256 July (1969).

[8.23] Fu, K. S., and McLaren, R. W., "An application of Stochastic Automata to the Synthesis of Learning Systems," Tech. Rept. TR-EE 65-17, School of Electrical Engineering, Purdue University, Lafayette, Ind., Sept. 1965.

[8.24] Fu, K. S., and McMurtry, G. J., "A Study of Stochastic Automata as Models of Adaptive and Learning Controllers," Tech. Rept. TR-EE 65-8, Purdue University, Lafayette, Ind., June 1965.

[8.25] Fu, K. S., and Nikolic, Z. J., "On Some Reinforcement Techniques and Their Relations with Stochastic Approximation," *IEEE Trans. Automatic Control*, AC-11, (2) 756-757(1966).

[8.26] Fu, K. S., and Nikolic, Z. J., "On the Stochastic Approximation and Related Learning Techniques," Tech. Rept. TR-EE-66-6, School of Electrical Engineering, Purdue University, Lafayette, Ind., April 1966.

[8.27]   Fu, K. S., and Waltz, M. D., "A Heuristic Approach to
         Reinforcement Learning Control Systems," *IEEE Trans.*
         Automatic Control, AC-10, (4), 390-398 (1965).

[8.28]   Gibson, J. E., Fu, K. S., et al., "Philosophy and State
         of the Art of Learning Control Systems," Tech. Rept. TR-EE
         63-7, Purdue University, Lafayette, Ind., Nov., 1963.

[8.29]   Hill, J. D., and Fu, K. S., "Learning Control System Using
         Stochastic Approximation for Hill-Climbing," Preprints JACC,
         1965.

[8.30]   Ho, Y. C., and Lee, R. C. K., "A Bayesian Approach to
         Problems in Stochastic Estimation and Control", *IEEE Trans.*
         Automatic Control, AC-9 (5) 333-338 (1964).

[8.31]   Ivakhnenko, A. G., and Lapa, V. G., "Cybernetic Predicting
         Devices," Translation TR-EE-66-4, School of Electrical
         Engineering, Purdue University, Lafayette, Ind., April 1966.

[8.32]   Jarvis, R. A., "Adaptive Global Search in a Time-Varying
         Environment Using a Probabilistic Automaton with Pattern
         Recognition," *IEEE Trans. Systems Sci. Cybern.,* SSC-6, (3)
         209-217 (1970).

[8.33]   Jones, L. E., III, "On the Choice of Subgoals for Learning
         Control Systems," *IEEE Trans. Automatic Control,* AC-13, (6)
         613-620 (1968).

[8.34]   Jones, L. E., III, and Fu, K. S., "On the Selection of
         Subgoals and the Use of a Priori Information in Learning
         Control Systems," *Automatica - IFAC J.,* 5, (6) (1969).

[8.35]   Konakovsky, R., and Binder, Z., "A Multimodal Searching
         Technique Using a Learning Controller," Third IFAC
         Symposium on Sensitivity, Adaptivity and Optimality,
         Ischia, Italy, June 18-21, 1973.

[8.36]   Krylov, V. Y., "On An Automaton That is Asymptotically
         Optimal in a Random Media," *Avtomatika i Telemekhanika,*
         24, (9) (1963).

[8.37]   Kushner, H., *Stochastic Stability and Control,* Academic
         Press, New York, 1967.

[8.38]   Lakshmivarahan, S., "Learning Algorithms for Stochastic
         Automata," Ph.D. thesis, Indian Institute of Science,
         Bangalore, Jan. 1973.

[8.39]   Lakshmivarahan, S., and Thathacher, M. A. L., "Optimal
         Nonlinear Reinforcement Schemes for Stochastic Automata,"
         Inform. Sci., 4 (2) 121-128 (1971).

[8.40]   Lakshmivarahan, S., and Thathacher, M. A. L., "Bayesian
         Learning and Reinforcement Schemes for Stochastic Automata,"
         Proc. International Conference on Cybernetics and Society,
         Washington, D. C., October 9-12, 1972.

[8.41]   Lee, R. C. K., *Optimal Estimation, Identification, and
         Control*, MIT Press, Cambridge, Mass., 1964.

[8.42]   Leondes, C. T., and Mendel, J. M., "Artificial Intelligence
         Control," in *Survey of Cybernetics* (R. Rose, ed.), ILLIFE
         Press, London, 1969.

[8.43]   Mendel, J. M., "Applications of Artificial Intelligence
         Techniques to a Spacecraft Control Problem," NASA CR-755,
         U.S. Govt. Print. Off., Washington, D. C., 1967.

[8.44]   Mendel, J. M., "Survey of Learning Control Systems for
         Space Vehicle Applications," Preprints JACC, Aug. 1966.

[8.45]   Mendel, J. M., and Fu, K. S. (eds.), *Adaptive, Learning
         and Pattern Recognition Systems: Theory and Applications*,
         Academic Press, New York, 1970.

[8.46]   Mendel, J. M., and McLaren, R. W., "Reinforcement-Learning
         Control and Pattern Recognition Systems," *Adaptive, Learning,
         and Pattern Recognition Systems: Theory and Applications*
         (J. M. Mendel and K. S. Fu, eds.), Academic Press, New York,
         1970, pp. 287-318.

[8.47]   Mendel, J. M., and Zapalac, J. J., "The Application of
         Techniques of Artificial Intelligence to Control System
         Design," in *Advances in Control Systems: Theory and
         Applications* (C. T. Leondes, ed.), Academic Press, New
         York, 1968.

[8.48]   McLaren, R. W., "A Stochastic Automaton Model for the
         Synthesis of Learning Systems," *IEEE Trans. Systems Sci.*,
         Cybern., SSC-2 (2), 109-114 (1966).

[8.49]   McLaren, R. W., "A Stochastic Automaton Model for a Class
         of Learning Controllers," *Proc. 1967 Joint Automatic Con-
         trol Conference*, University of Pennsylvania, Philadelphia,
         pp. 267-273, 1967.

[8.50]   McLaren, R. W., "Application of a Continuous-Valued Control
         Algorithm to the On-line Global Optimization of Stochastic
         Control Systems," *Proc. Nat. Electron. Conf.*, 25, 2-7
         (1969).

[8.51]   McLaren, R. W., "A Continuous-Valued Learning Controller for the Global Optimization of Stochastic Control Systems," Seminar on Learning Processes in Control Systems, Nagoya, Japan, pp. 77-81, August 1970.

[8.52]   McMurtry, G. J., and Fu, K. S., "A Variable Structure Automaton Used as a Multimodal Searching Technique," *IEEE Trans. Automatic Control*, $\underline{AC-11}$ (3) 379-387 (1966).

[8.53]   Mosteller, H. W., "Learning Control Systems," Rept. No. R64, ELC37, GE Advanced Electronics Center, Ithaca, New York, March 1964.

[8.54]   Narendra, K. S., and McBride, L. E., "Multiparameter Self-Optimizing Systems Using Correlation Techniques," *IEEE Transaction Automatic Control*, $\underline{AC-9}$ (1) 31-39 (1964).

[8.55]   Narendra, K. S., and Streeter, D., "An Adaptive Procedure for Controlling Undefined Linear Processes," *IEEE Transactions Automatic Control*, $\underline{AC-9}$, October 1964.

[8.56]   Narendra, K. S., and Viswanathan, R., "A Two Level System of Stochastic Automata for Periodic Random Environment," *IEEE Trans. Systems Man Cybern.*, $\underline{SMC-2}$, (2) 292-294 (1972).

[8.57]   Narendra, K. S., and Viswanathan, R., "Learning Models Using Stochastic Automata," *Proc. 1972 International Conference on Cybernetics and Society*, Washington, D. C. October 9-12, 1972.

[8.58]   Nikolic, Z. J., and Fu, K. S., "A Mathematical Model of Learning in an Unknown Random Environment," *Proc. Nat. Electron. Conf.*, $\underline{22}$ 607-613 Oct. (1966).

[8.59]   Nikolic, Z. J., and Fu, K. S., "An Algorithm for Learning Without External Supervision and its Application to Learning Control Systems," *IEEE Trans. Automatic Control*, $\underline{AC-11}$, (3) 414-423 (1966).

[8.60]   Patrick, E. A., "On A Class of Unsupervised Estimation Problems," *IEEE Trans. Information Theory*, $\underline{IT-14}$, 407-415 (1968).

[8.61]   Pugachev, V. S., "A Bayes Approach to the Theory of Learning Systems," Third World Congress of IFAC, London, England, June 1966.

[8.62]   Pugachev, V. S., "Optimal Learning System," *Dokl. Akad. Nauk. SSSR*, $\underline{175}$, (5), 762-764 (1967).

[8.63]   Pugachev, V. S., "Optimal Training Algorithms for Automatic
          Systems in Case of Non-Ideal Teacher," *Dokl. Akad. Nauk.
          SSSR*, <u>172</u> (5) 1039-1042 (1967).

[8.64]   Raible, R., and Gibson, J., "A Computer Study on a
          Learning Control System," Preprints 3rd IFAC Congress,
          London 1966, Paper 14E.

[8.65]   Riordon, J. S., "An Adaptive Automaton Controller for
          Discrete-Time Markov Processes," *Automatica-IFAC J.*, <u>5</u>
          (6) 721-730 (1969).

[8.66]   Saridis, G. N., "Stochastic Approximation Methods for
          Identification and Control - A Survey," *IEEE Trans. Auto-
          matic Control*, <u>AC-19</u>, (6), 798-809 (1974).

[8.67]   Saridis, G. N., "On a Class of Performance-Adaptive Self-
          Organizing Control Systems," in *Pattern Recognition and
          Machine Learning* (K. S. Fu, ed.), Plenum Press, New York,
          204-220 (1971).

[8.68]   Saridis, G. N., "On-line Learning Control Algorithms," in
          Learning Systems, (K. S. Fu, ed.) ASME Publ., New York,
          June 1973.

[8.69]   Saridis, G. N., and Dao, T. K., "A Learning Approach to the
          Parameter Adaptive Self-Organizing Control Problem,"
          Automatica, IFAC J., <u>8</u> (5) 589-598, Sept. (1972).

[8.70]   Saridis, G. N., and Fensel, P., "Stability Considerations on
          the Expanding Subinterval Performance Adaptive S.O.C.
          Algorithms," *J. Cybern.*, <u>3</u> (2), 26-39 (1973).

[8.71]   Saridis, G. N., and Gilbert, H. D., "Self-Organizing
          Approach to the Stochastic Fuel Regulator Problem," *IEEE
          Trans. Systems Sci. Cybern.*, <u>SSC-6</u> (3), 187-191 (1970).

[8.72]   Saridis, G. N., and Hofstadter, R. N., "A Pattern Recogni-
          tion Approach to the Classification of Nonlinear System,"
          IEEE Trans. Systems Man Cybernetics, <u>SMC-4</u> (4), 362-371
          (1974).

[8.73]   Saridis, G. N., and Kitahara, R. T., "Computational Aspects
          of Performance-Adaptive S.O.Control Algorithms," Tech.
          Rept. TR-EE 71-41, Purdue University, West Lafayette, Ind.,
          Dec. 1971.

[8.74]   Shapiro, I. J., and Narendra, K. S., "Use of Stochastic
          Automata for Parameter Self-Optimization with Multimodal
          Performance Criteria," *IEEE Trans. System Sci. Cybern.*,
          SSC-5, (5) 352-360 (1969); Witten, I. H. et al, "Comments
          on the 'Use of Stochastic Automata for Parameter Self-Opti-
          mization with Multimodal Performance Criteria," *IEEE Trans.
          Systems Man Cybern.*, SMC-2 (2) 289-290 (1972).

[8.75]   Sklansky, J., "Learning Systems for Automatic Control,"
          IEEE PGAC, AC-11 (1), 6, January (1966).

[8.76]   Stratonovich, R. L., "Does there exist a Theory of Synthesis
          of Optimal Adaptive, Self-Learning, and Self-Adaptive
          Systems?," *Avtomatika Telemechanika*, (1) 83-92 (1968).

[8.77]   Sworder, D. D., *Optimal Adaptive Control Systems*, Academic
          Press, New York, 1966.

[8.78]   Tou, J. T., "Systems Optimization via Learning and Adapta-
          tion," *Intern. J. Control*, 2 (1) 21-32 (1965).

[8.79]   Tou, J. T., and Hill, J. D., "Steps Toward Learning Control,"
          Preprints JACC, August 1966.

[8.80]   Tsetlin, M. L., "On the Behavior of Finite Automata in
          Random Environments," *Avtomatika Telemekhanika*, 22 (10)
          1345-1354 (1961).

[8.81]   Tsypkin, Ya. Z., *Adaptation and Learning in Automatic
          Systems*, Nauka, Moscow, 1968; (Transl. Z. J. Nikolic)
          Academic Press, New York, 1970.

[8.82]   Tsypkin, Ya. Z., "Adaptation, Learning and Self-Learning in
          Automatic Systems," *Avtomatika Telemekhanika*, 27 (1) 23-61
          (1966).

[8.83]   Tsypkin, Ya. Z., "All the same, does a Theory of Synthesis
          of Optimal Adaptive Systems Exist?," *Avtomatika Tele-
          mekhanika*, 29, (1) 93-98 (1968).

[8.84]   Tsypkin, Ya. Z., *Foundations of the Theory of Learning
          Systems*, Nauka, Mowcow, 1970; (Transl. Z. J. Nikolic)
          Academic Press, New York, 1973.

[8.85]   Tsypkin, Ya. Z., "Generalized Algorithms of Learning,"
          Automatika Telemekhanika, 31 (1) 86-92 (1970).

[8.86]   Tsypkin, Ya. Z., "Principles of Dynamic Adaptation in
          Automatic Systems," Fifth World Congress of IFAC, Paris,
          June 12-17, 1972.

[8.87] Tsypkin, Ya. Z., "Self-Learning-What is it?," *IEEE Trans. Automatic Control*, <u>AC-13</u>, (6) 608-612 (1968).

[8.88] Tsypkin, Ya. Z., "Smoothed Randomized Functionals and Algorithms in Theory of Adaptation and Learning," *Automatika Telemekhanika*, <u>32</u> (8) 29-50 (1971).

[8.89] Tsypkin, Ya. Z., Kaplimsky, A. I., and Larionov, K. A., "Adaptation and Learning Algorithms in Nonstationary Conditions," *Eng. Cybern.*, USSR (5), 9-21 (1970).

[8.90] Tsypkin, Ya. Z., and Kelmans, G. K., "Recursive Algorithms of Self-Learning," *Eng. Cybern.*, USSR, (5), 70-80 (1967).

[8.91] Tsypkin, Ya. Z., Kelmans, G. K., and Epstein, L. E., "Learning Control Systems," Preprints, Fourth World Congress of IFAC, Warsaw, Poland, June 16-21, 1969.

[8.92] Ula, N., and Kim, M., "An Empirical Bayes Approach to Adaptive Control," *J. Franklin Inst.*, <u>280</u> (3), 189-204, Sept. (1965).

[8.93] Varshavsky, V. I., "Automata Games and Control Problems," Fifth World Congress of IFAC, Paris, June 12-17, 1972.

[8.94] Varshavsky, V. I., "Collective Behavior and Control Problems," *Machine Intelligence*, <u>3</u>, Edinburgh University Press, Scotland, 1968.

[8.95] Varshavsky, V. I., "The Organization of Interaction in Collectives of Automata," *Machine Intelligence*, <u>4</u>, Edinburgh University Press, Scotland, 1969.

[8.96] Varshavsky, V. I., and Vorontsova, I. P., "On the Behavior of Stochastic Automata with Variable Structure," *Avtomatika Telemekhanika*, <u>24</u> (3), 353-360 (1963).

[8.97] Viswanathan, R., and Narendra, K. S., "A Note on the Linear Reinforcement Scheme for Variable-Structure Stochastic Automata," *IEEE Trans. Systems Man Cybern.*, <u>SMC-2</u>, (2) 292-294 (1972).

[8.98] Viswanathan, R. and Narendra, K. S., "Comparison of Expedient and Optimal Reinforcement Schemes for Learning Systems," *J. Cybern.*, <u>2</u> (1), 21-23 (1972).

[8.99] Viswanathan, R., and Narendra, K. S., "Stochastic Automata Models with Applications to Learning Systems," *IEEE Trans. Systems Man Cybern.*, <u>SMC-3</u>, (1) (1973).

[8.100] Waltz, M. D., and Fu, K. S., "A Computer-Simulated Learning
         Control System," IEEE *International Convention Record*, Pt.
         1, pp. 190-201 (1964).

[8.101] Waltz, M. D., and Fu, K. S., "A Heuristic Approach to Re-
         inforcement Learning Control Systems," IEEE *Trans. Automatic
         Control*, <u>AC-10</u> (4), 390-398 (1965).

Chapter 9

EVALUATION AND STABILITY CONSIDERATIONS OF
SELF-ORGANIZING CONTROL ALGORITHMS

9.1   INTRODUCTION

The various S.O.C. algorithms discussed in the previous two
chapters were selected for presentation because they meet the requirements of the definition 1.18 of stochastic S.O.C., "to reduce
the *a priori* uncertainties pertaining to the effective control of
the process, through subsequent observations of inputs and outputs as
the process evolves." They cover a wide spectrum of methods and
ideas, having in common the self-organizing philosophy for control
systems which was discussed extensively in Chap. 1. The result is a
collection of control techniques, highly sophisticated, sometimes
very complex, and always requiring a fairly large digital computer
for implementation.

It is obvious that such algorithms are not meant to solve
trivial control problems, e.g., with first- or second-order linear
dynamics with one adjustable parameter in the single feedback loop,
for which easier and many times ingenious "engineering" solutions
are possible [9.7]. Self-organizing control may oversolve such
problems. These algorithms represent systematic general approaches
to the solution of modern multidimensional, multiloop, multinon-
linear, control problems with considerable amount of uncertainties
about the plant dynamics and the stochastic environment likely to be
found in bioengineering, economic, societal, environmental, and
other processes of current interest. This should be brought to mind
whenever an evaluation of S.O.C. is attempted.

An evaluation of the S.O.C. algorithms within the framework of
their design is necessary at this point, in order to assess their

391

relative qualities and to justify their applicability to large-
scale uncertain processes [9.4], [9.21]. Some such applications of
S.O.C. requiring highly sophisticated algorithms are discussed in
Chap. 10, qualitative and quantitative evaluation of the various
algorithms will be given in this chapter. These will include an
analysis of the experiences gained by several simulation problems with
suggestions for the applicability to various types of plants, as
well as comparisons of actual simulation studies on speed of convergence, computer capabilities, biases and sensitivities, and ease
of implementation. The studies are performed on simple first- to
fourth-order linear or nonlinear plants, in order to demonstrate to
the reader the feasibility and the step-by-step procedure of the
algorithms, which may be impossible if higher-dimensional plants
were used. The extension to higher dimensions which will undoubtedly
generate more interesting problems is left to the reader with a
specific problem to apply. Chap. 10 presents the author's favorite
large-scale systems applications.

Finally, the problem of stability of the resulting systems which
was only superficially touched in the previous chapters will be
discussed since it is of imperative importance to the control designer. To the author's best knowledge, this is the first comprehensive study of this important problem and it is far from being
complete. It is based on a generalization of the investigation
presented in Sec. 7.6.5 where deterministic stability investigation
may be applied to an "engineering" approximation of the stochastic
system and a more rigorous study of the problem from the stochastic
point of view [9.23]. The first presents excellent results at the
sacrifice of mathematical rigor; the second is bogged down by mathematical complexity and therefore limited to lower-order systems with
limited nonlinearity. Both studies need a considerable amount of
further investigation and the interested reader is encouraged to
venture into the problem along with his idea of a new approach.

9.2 QUALITATIVE EVALUATION OF SELF-ORGANIZING CONTROL

Parameter-adaptive and performance-adaptive S.O.C.s represent two different philosophies for the <u>on-line</u> control of plants with uncertain dynamics. In general they also apply to different types of plants, the first one when the structural information about the plant is known, the second one when the structure is unknown. When tested on similar systems, e.g., linear quadratic systems for which the stochastic optimal solution with known parameters was available, they demonstrated the expected behavior.

In other words, parameter-dadptive algorithms demonstrated on the average, speed of convergence and computer size requirements one order of magnitude lower than the performance adaptive. In addition they presented similar difficulties of implementation as it will be quantitatively demonstrated in the sequel. Typical parameter- and performance-adaptive S.O.C. systems are given in Figs. 9.1 and 9.2, respectively.

The basic differences between parameter- and performance-adaptive S.O.C. algorithms are therefore due to the actual limitations of the methods. Since the limitations within each class will be discussed separately, only the general differences will be presented here with the possible exception of the actively adaptive algorithm of Tse and Bar-Shalom where the optimal solution, which is

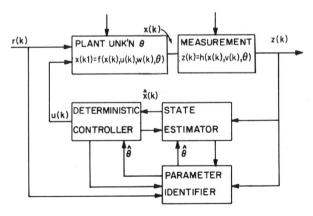

Fig. 9.1. A Typical Parameter-Adaptive S.O.C. System

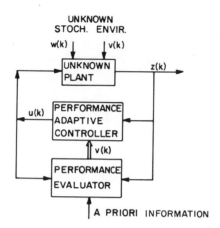

Fig. 9.2. A Typical Performance-Adaptive S.O.C. System

being approximated, is known only for the linear or linearized version of the plant. Therefore parameter-adaptive S.O.C. is limited
to such classes of plants for which this approximation is valid.
Performance-adaptive S.O.C. algorithms are not limited by the structure of the plant but by the structure of the controller which must
be assumed to approximate a globally optimal solution for the given
performance criterion. The optimal selection of such a controller,
either in the form of $\phi(.)$ functions in the crosscorrelation,
Tsypkin's stochastic approximation, or the expanding subinterval
algorithms, or the set of numbers in the control space Ω_u for the
reinforcement, automata, adaptive automata, and unsupervised learning algorithms has not yet been obtained. Some preliminary results
are given in Sec. 8.5.3, obtained by Riordon as a shifting grid of
points in the space, and in Sec. 8.6.2 for the expanding subinterval
algorithms. More research is needed in this area. Space discretization has been used both in parameter-adaptive (see Lainiotis in
Sec. 7.4.3, Stein and Saridis in Sec. 7.5.1, and Saridis and Dao in
Sec. 7.5.2), in performance-adaptive S.O.C. reinforcement, and
automata algorithms in Sec. 8.4; as well as unsupervised learning
in Sec. 8.5.2 and Riordon's work in Sec 8.5.3. The rest of the

algorithms in Chap. 8 treat continuous systems defined on state and
control spaces and/or continuous time.

The performance-adaptive S.O.C. algorithms being more sophisticated usually require on the average a larger-size computer memory
to store their various functions, while the parameter-adaptive S.O.C.
algorithm require more computational capabilities, since their
operating programs are longer. However, both classes produce rather
slowly learning algorithms and require fairly large-size digital
computers to treat a moderately high-dimensional process. These two
points, have been strongly criticized by potential users of the
algorithms and their improvement is imperative in order to make the
field more popular.

9.2.1 Qualitative Evaluation of Parameter-Adaptive S.O.C. Algorithms

Parameter-adaptive methods have been the more traditional S.O.C.s
and closer to the classical idea of "adaptation." The eight major
algorithms presented in Chap. 7 cover a wide spectrum of S.O.C.
algorithms with an explicit parameter identification. These algorithms may be classified into four major groups regarding their
approach to optimality: (1) the linearized stochastic optimal,
(2) the open-loop feedback optimal (OLFO), (3) the asymptotically
optimal, and (4) the approximate dual optimal. The linearized stochastic optimal of Sec. 7.3 with linearized or extended Kalman filter cover the first group. The algorithms of Farison, Tse and
Athans, and Lainiotis of Sec. 7.4 and Saridis and Dao of sec. 7.5.2
are the most representative of the second group, while the parallel
identification and control algorithms of Saridis and Lobbia of Sec.
7.6 belong to the third group. The algorithms of Stein and Saridis,
and especially the actively adaptive of Tse and Bar-Shalom are the
only ones known of group four.

The algorithms of the first group have been mostly popular in
the aerospace industry because of the theoretical speed of convergence of the identifier. However, they have demonstrated problems
with the convergence of the control scheme and the stability of the

overall system in the same manner as the identification algorithms
of Sec. 6.6. When they converge they do not behave asymptotically
optimal and they are difficult to implement.

The OLFO algorithms with fast convergence properties are simple
to implement, due to the speed of convergence of their information
states, e.g., the *a posteriori* probabilities which implement the
identifier. Their major shortcoming is the need of discretization
of the parameter space in most of them, in order to compute the
evolution of the information states which limits their application.
It was shown that for a finite-state process the Saridis and Dao
algorithm has a theoretical performance value lower than the
Lainiotis algorithm and that they are both converging to the optimal
control asymptotically and therefore are better than the Stein and
Saridis algorithm. The latter is an approximation to the dual optimal solution but does not converge to the asymptotically optimal and
is hard to implement.

The algorithms of parallel identification and control of Saridis
and Lobbia are the next best in convergence properties and better in
terms of implementation. Variations using per-interval performance
criteria may further improve performance and simplicity, sacrificing
the asymptotic optimality that the overall performance criterion
provides. It is the only algorithm for which a reasonable stability
analysis is available.

The actively adaptive algorithm of Tse and Bar-Shalom is the
closest approximation to the dual optimal. However, its complexity
has prevented implementation and therefore it has not been compara-
tively tested.

Table 9.1 gives a comparative qualitative evaluation of the
parameter-adaptive S.O.C. algorithm that describes best their
advantages and disadvantages.

9.2.2 Qualitative Evaluation of Performance-Adaptive S.O.C.
       Algorithms

Comparative evaluation of the performance-adaptive S.O.C. algorithms is harder to develop because they represent diverse

Table 9.1. Qualitative Comparison of Parameter-Adaptive S.O.C. Algorithms.

Parameter-adaptive S.O.C. algorithm	Type of System					Optimality			Parameter Ident.			General				Additional information
	Linear	Nonlinear	Time varying	Cont.-time	Discrete-time	Asymptotically optimal	Approximately dual optimal	OLFO	Structure assumed	Existence of bias	Discretization of Param. space	Stability assumed	Stability tested	A Priori information	Generality	
Linearized optimal of 7.3	X	X	X	X	X					X		X		X	X	Both linearized and ext. Kalman filter algs. stability probs. complex implement
Farison's OLFO of 7.4.1	X	X	X		X			X	X		X	X		X		All states exactly measurable
Tse and Athans OLFO of 7.4.2	X	X	X		X			X	X				X	X		Only control gains unknown
Lainiotis OLFO of 7.4.3	X		X	X	X	X		X			X	X			X	Difficult stability study
Stein and Saridis min. upper-bound of 7.5.1	X		X		X	X					X	X			X	Very complex implementation
Saridis and Dao min. lower-bound of 7.5.2	X		X		X			X			X	X	X		X	Stability study possible
Parallel ident. and control algorithms of 7.6	X		X	X	X	X							X	X	X	Overall and per interval P.I. per-interval not optimal
Actively adaptive of 7.7	X	X	X	X	X		X		X			X		X	X	Extremely complex implementation

philosophies and are designed for different types of systems.
However, comparative simulations have been obtained for some of them
on linear and nonlinear control systems [9.10] and they are used as
a basis for evaluation. The first eight algorithms of Chap. 8 have
been considered here, the stochastic approximation algorithm (Sec.
8.7) for distributed systems belonging to a separate class has not
been included in this comparison.

The classical control approach has been represented by Narendra's crosscorrelation algorithms limited to only linear time-
invariant systems. The "behavioral" algorithms, such as the reinforcement learning algorithms of Fu et al. (Sec. 8.4.2), Mc Laren's
automaton and growing automaton (Sec. 8.4.4 and 9.4.5, respectively),
the unsupervised learning of Fu and Nikolic (Sec. 8.5.2), and
Riordon's adaptive automaton (Sec. 8.5.3), all represent a classification of the performance-adaptive S.O.C. algorithms based on the
approach adopted for the reduction of the uncertainties pertaining
to the improvement of the performance of the process and the definition of the subgoal to be minimized. They are very elegant algorithms suitable for higher-level learning decision making. Their
main shortcoming is the need of discretization of the state and
control space which is not very appealing to the control engineer.
They also require complex implementation and large-size computers.

The stochastic approximation method of Tsypkin is the most
simpleminded algorithm and therefore easy to implement. It resembles conceptually the parallel identification and control algorithms
of Sec. 7.6 and would be the best "engineering" solution if it was
not hampered by a number of problems. The stochastic approximation
algorithms give biased estimates, as it was shown in Sec. 6.4, unless
the system is of a special form or the bias is subtracted. The
sensitivity functions $\partial z/\partial \theta$ are needed in order to compute the
stochastic gradient of the algorithms which are available only in
special cases, e.g., linear systems.

Finally, the expanding subinterval of Saridis is a compromise
to the above problems and at the same time it answers a number of

problems arising in performance-adaptive S.O.C. Asymptotically it
is a globally optimal algorithm if the feedback controll is appropriately structured. It minimizes a per-interval performance cost
which, in the limit, minimizes the overall criterion, thus relating
the subgoal to the goal, and it is easy to implement even for a completely unknown system. Its shortcomings are the size of computer
required for its implementation and the difficulties involved with
its stability investigation.

A comparative qualitative evaluation of the eight performance-
adaptive S.O.C. algorithms is given in Table 9.2.

9.3 QUANTITATIVE EVALUATION OF CERTAIN PARAMETER-ADAPTIVE S.O.C.
     ALGORITHMS

Comparative quantitative evaluation of the parameter-adaptive
S.O.C. algorithms may be obtained by testing the methods on a large
number of systems and comparing the results. However, this is an
extremely lengthy, tedious, and expensive process. On the other
hand, all the algorithms presented are not general enough for a
significant comparison to the rest. Therefore only the two linearized optimal, the Lainiotis OLFO, the minimum lower-bound of Saridis
and Dao, and the parallel identification and control of Saridis and
Lobbia were compared to the stochastic optimal solution with known
parameters. The actively adaptive algorithm of Tse and Bar-Shalom
and the minimum upper-bound algorithms of Stein and Saridis have not
been compared due to their complexity.of implementation. Specific
examples may be found, however, in Chap. 7. The comparative evaluation was performed on a representative linear second-order system
with quadratic performance criterion and Gaussian noises. The
linear system was chosen as a common denominator of the algorithms
since many of them are applicable to LQG systems for which the
stochastic optimal solution is available. A one-sample statistic is
not sufficient to permit sweeping statements. However, this representative example demonstrates the salient features of the algorithm
and lends itself to sufficient comparisons.

Table 9.2.   Qualitative Comparison of Performance-Adaptive S.O.C. Algorithms.

Performance-Adaptive S.O.C. Algorithm	Type of System							Optimality	General						Additional Information
	Linear	Nonlinear	Time-varying	Cont.-time	Discrete-time	Known structure	State vector needed	Performance Criterion to be improved	Finite Discr. output space	Countable output space	Finite discr. control space	Countable control space	Stability assumed	Apriori information	
Narendra's crosscorrelation of 8.3	X			X		X		Mean-square error					X		Sensitive to initial state
Reinforcement learning algorithm of 8.4.2	X	X	X	X	X		X	Subgoal	X	X	X	X	X	X	Measurable environmental interaction complex implementation most suitable off-line
McLaren's variable structure automaton of 8.4.3	X	X	X	X	X			Subgoal	X	X	X		X	X	Requires knowledge of PImax; steady state requires initialization
McLaren's growing automaton of 8.4	X	X	X	X	X			Subgoal	X		X	X	X		Requires knowledge of PImax requires initialization
Tsypkin's stochastic approximation of 8.5.1	X	X	X	X	X			Subgoal				X	X		No convergence proof. requires knowledge of sensitivity functions
Nikolic and Fu unsupervised learning of 8.5.2	X	X	X		X			Subgoal	X		X		X	X	Plant 1st-order Markov complex implementation
Riodorn's adaptive automation of 8.5.3	X	X	X		X			Overall P.I.		X		X	X		Complex implementation
Saridis' expanding subinterval algorithm of 8.6	X	X	X	X	X			Subgoal tending to overall					X	X	Stability study to follow

The system is described by the following difference equations:

$$x(k + 1) = \begin{bmatrix} 0 & 1 \\ f_1 & f_2 \end{bmatrix} x(k) + \begin{bmatrix} 0 \\ b \end{bmatrix} [u(k) + w(k) \qquad (9.1)$$

$$z(k) = [1,0]x(k) + v(k)$$

where $f_1 = -0.974$, $f_2 = 1.874$, and $b = 0.01$

$$x(0) = x_0 \sim N\left[\begin{pmatrix} 0 \\ 0 \end{pmatrix}, \begin{pmatrix} 9 & 0 \\ 0 & 4 \end{pmatrix} \right] \qquad (9.2)$$

$$w(k) \sim N(0,0.1)$$
$$v(k) \sim N(0,0.1) \qquad \forall k$$

The performance criterion for this system is defined as

$$J(u) = \lim_{N \to \infty} \frac{0.1}{N} E \left\{ \sum_{k=1}^{N} [(z^d(k) - z(k))^2 + 0.01u^2(k)] \right\} \qquad (9.3)$$

and $z^d(k)$, the desired response, is chosen to be a step function. This system may be considered as the discrete-time version of a continuous-time system with discretization interval $\Delta t = 0.1$ sec,[9.10]. The responses of the algorithms have been produced for different initial conditions of the unknown parameter set $\theta^T = [a_1, a_2, b]$; each averaged over 10 runs. The simulations were performed on Purdue University's CDC-6500 digital computer, and compared for computer time after 100 stages or plant iteration corresponding to 10 secs of real time and storage requirements. The accumulated average performance (AAP) and its standard deviation computed over 10 runs is plotted in Figs. 9.3 and 9.4. Numerical evaluations and components of the sensitivity to the initial conditions and implementation of the parameters are summarized in Table 9.3.

The expressions derived in Chap. 7 were used for the above simulations with appropriate modifications and adjustments.

For the linearized optimal algorithms both the linearized and extended Kalman filters were used; convergence problems due to plant instability were observed frequently.

Table 9.3.   Quantitative Comparison of Parameter-Adaptive S.O.C. Algorithms.

Parameter-adaptive S.O.C. algorithms	Computer time, sec, after 1000 stages	Storage location (octal)	Average performance cost		Initial state	Implementation
			After 100 stages	After 1000 stages		
Linearized optimal with linearized Kalman filter	20.1	17.3k	29.5	--	Close	Difficult
Linearized optimal with extended Kalman filter	16.2	17.6k	23.0	--	Close	Difficult
Lainiotis OLFO	6.3	25.0k	8.0	7.2	Anywhere	Less easy
Saridis and Dao	7.4	26.8k	7.0	6.6	Anywhere	Less easy
Parallel identification and control of Saridis and Lobbia	3.1	1k	20.5	18.0	Anywhere	Easy
Stochastic optimal with known parameter	0.95	1k	4.5	4.0	Anywhere	Easy

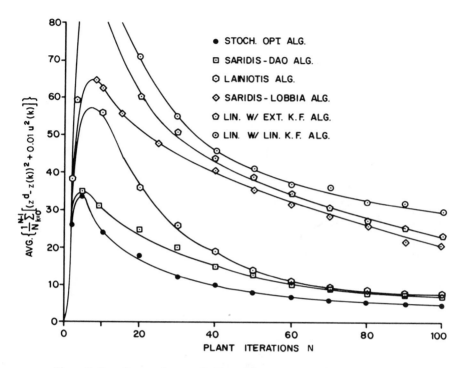

Fig. 9.3. Comparison of Plant Iterations for Parameter-Adaptive S.O.C. Algorithms

For the algorithm of Lainiotis and Saridis-Dao the parameter space Ω_θ was discretized as follows:

$$\Omega_\theta = \left\{ \begin{matrix} f_1 \\ f_2 \\ b \end{matrix} \right\} = \left\{ \begin{bmatrix} -0.1 \\ 0.1 \\ 0.05 \end{bmatrix}, \begin{bmatrix} -0.25 \\ -0.75 \\ 0.5 \end{bmatrix}, \begin{bmatrix} -0.5 \\ 1.0 \\ 0.1 \end{bmatrix}, \right.$$

$$\left. \begin{bmatrix} -0.6 \\ 1.5 \\ 0.15 \end{bmatrix}, \begin{bmatrix} -0.75 \\ -1.5 \\ 0.25 \end{bmatrix}, \begin{bmatrix} -0.974 \\ 1.874 \\ 0.01 \end{bmatrix} \right\} \qquad (9.4)$$

For each of these vectors the system was checked to be open-loop stable.

The overall performance criterion with the steady-state values for the state estimator and controller was used.

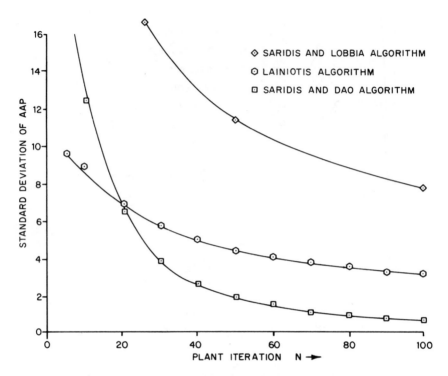

Fig. 9.4.   Comparison of Standard Deviation of AAP

Theoretical comparisons for the algorithms of Stein and Saridis, Lainiotis, and Saridis and Dao were given in Sec. 7.5.3 where the Saridis-Dao algorithm was proved to be superiorior to the other two at each stage of a finite-stage process.

$$V_{N-k}(k)_{S-D} \leq V_{N-k}(k)_{L}; V_{N-k}(k)_{S-D} \leq V_{N-k}(k)_{S-S} \qquad (9.5)$$

This first statement was experimentally verified in Fig. 9.3, while the second one was explained in Sec. 7.5.3.

If one would crudely classify the algorithms in the order of performance cost, one would put the two linearized optimal as the worst and the two OLFO as the best ones compared to the stochastic optimal, with the parallel identification and control in the middle.

Such an ordering has been expected from the arguments presented in Chap. 7.

9.4   QUANTITATIVE EVALUATION OF CERTAIN PERFORMANCE-ADAPTIVE S.O.C.
      ALGORITHMS

A careful selection of compatible performance-adaptive S.O.C.
algorithms from the eight presented in Chap. 8, was made [9.10].
Only Mc Laren's growing automaton, Tsypkin's stochastic approximation,
the unsupervised learning of Nikolic and Fu, Riordon's adaptive
automaton, and Saridis' expanding subinterval were chosen for comparative evaluation. The crosscorrelation, reinforcement, and stochastic automaton were not included because they apply to a limited
class of systems or they are special cases of tested algorithms.

The algorithms were tested on two plants, i.e., the linear plant
used in the preceding section so that a crosscomparison with parameter-
adaptive algorithms would be possible and a nonlinear system used by
Riordon in his paper [9.18]. These two examples give a better idea
about the behavior of these algorithms.

The linear discrete-time system used for the first example is
repeated here from Sec. 9.3 for completeness

$$x(k + 1) = \begin{bmatrix} 0 & 1 \\ f_1 & f_2 \end{bmatrix} x(k) + \begin{bmatrix} 0 \\ b \end{bmatrix} [u(k) + w(k)] \tag{9.6}$$

$$z(k) = [1,0]x(k) + V(k)$$

where $f_1 = -0.974$, $f_2 = 1.874$, and $b = 0.01$

$$x(0) = x_o \sim N\left[\begin{pmatrix} 0 \\ 0 \end{pmatrix}, \begin{pmatrix} 9 & 0 \\ 0 & 4 \end{pmatrix} \right]$$

$$w(k) \sim N(0,0.1)$$
$$v(k) \sim N(0,0.1) \qquad \forall k \tag{9.7}$$

The performance criterion for this system is defined as

$$J(u) = \lim_{N \to \infty} \frac{0.1}{N} E\left\{ \sum_{k=1}^{N} [(z^d(k) - z(k))^2 + 0.01u^2(k)] \right\} \tag{9.8}$$

and $z^d(k)$, the desired response, is chosen to be a step function.

The various algorithms were implemented according to the expressions derived in Chap. 8.

The growing stochastic automaton method (GAM) was implemented by defining the following heuristic "subgoal" to be minimized at time instant k:

$$J_s(k) = \frac{1}{NORM} \{[z^d(k) - z(k)]^2 + 0.01u^2(k)\} \tag{9.9}$$

when NORM was assigned originally an arbitrary value. Subsequently, if any evaluation of $J_s(k)$ exceeded NORM, the $J_s(k)$ was set equal to 1 and the algorithm allowed to continue. This merely indicated that the particular control action was "bad" and the probability of using it again was modified accordingly. An improvement of this method is to adapt NORM when $J_s(k)$ exceeds its value by setting

$$NORM(NEW) = J_s(k) \tag{9.10}$$

It was found that the algorithm was sensitive to the initial choice of NORM and the best values for this problem were 5 to 7. This procedure tended to bias the search toward the initial control choices. Therefore an initiation procedure based on randomly selecting a control among an initial set of control actions was used to overcome this difficulty.

The stochastic approximation method (SAMT) was implemented by assuming that for this example the plant is linear and second order, completely controllable, and completely observable.

The system equation may be written in a difference equation form

$$z(k + 2) = [z(k), z(k + 1), u(k + 1)] \begin{bmatrix} f_1 \\ f_2 \\ b \end{bmatrix} + \xi(k + 2) \tag{9.11}$$

where $\xi(k + 2)$ is the resulting correlated noise, a function of f_1, f_2, and b. Then the identification criterion may be set up along the lines of Sec. 8.5.1 with an error function

$$J_2(\hat{c}) = E\{(z(k + 2) - z^{k+1^T}\hat{c})^2\} \tag{9.12}$$

where $c^T = [f_1, f_2, b]$, ${z^{k+1}}^T = [z(k), z(k+1)]$

The instantaneous performance criterion or "subgoal" is defined as

$$J_1(a,c) = E\{(z^d(k+2) - {z^{k+1}}^T \hat{c})^2 + 0.01u^2(k+1)\} \qquad (9.13)$$

The rest of the algorithm evolves in exactly the same way as the example in Sec. 8.5.1 to which this example is almost identical. The only difference between this example and the example in Sec. 8.5.1 is that the noise $\xi(k+2)$ is correlated and is described by

$$\xi(k+2) = v(k+2) - f_1 v(k+1) - f_2 v(k) + bw(k) \qquad (9.14)$$

This means that the algorithm used by Tsypkin gives definitely biased estimates of c of the form discussed in Sec. 6.4.1. In this case, the sensitivity functions were readily compatible, but this is in general not true. Under the assumption of a linear plant and elimination of the bias this algorithm should function as fast as a parameter-adaptive algorithm and be very efficient, a fact which has been verified experimentally.

The learning without external supervision algorithm (LWES) is implemented along the lines of Sec. 8.5.2.

The output space was discretized into 10 regions z_i, i = 1,..., 10, for which the lower-boundary value is assigned, if the output lies in the region. The boundary values are given by the set [2.5, 2.0, 1.6, 1.3, 1.1, 0.9, 0.6, 0.2, -0.3]. The control values for the linear feedback controller can be appropriately chosen from one of the two combinations

$$\begin{bmatrix} f_1 \\ f_2 \\ g_1 \\ g_2 \end{bmatrix} = \begin{bmatrix} -2.0 \\ -3.0 \\ 0.1 \\ 0.1 \end{bmatrix} \text{ or } \begin{bmatrix} -1.0 \\ -9.1 \\ 1.0 \\ 0.9 \end{bmatrix}$$

to match the growing automaton and expanding subinterval results. A similar quantity to the "subgoals" selected for the other algorithms is defined here. This subgoal is appropriately adjusted to meet the requirements of the algorithm of learning without supervision.

$$J_s(k) = [z^d(k) - z(k)]^2 + [z^d(k + 1) - z(k + 1)]^2$$
$$+ 0.01u^2(k) \tag{9.15}$$

Finally, the expanding subinterval algorithm (random search technique, (RST) was implemented, as indicated in Sec. 8.6, to generate the appropriate learning controller. The "subgoal" was defined as

$$J_s(k) = \frac{1}{T_i} \sum_{k=t_{i-1}}^{t_i} \left[[z^d(k) - z(k)]^2 + 0.01u^2(k) \right] \tag{9.16}$$

where $T_i = t_i - t_{i-1} = \Delta T = 1$ sec for $i = 1,\ldots,1000$ and $T_i = (i - 1000) \Delta T$ for $i > 1000$. This satisfies the conditions for convergence of the algorithm.

The simulation of this example indicates that all the controllers produced stable responses for the system. The composite plots in Figs. 9.5 and 9.6 giving the average (AAP) over 10 runs of the performance index and its standard deviation indicate the convergence and confidence properties of the methods. The algorithms rank, SAMT, RST, LWES, and GAM.

Table 9.5 gives values of the performance (AAP), its standard deviation after 1000 stages, and the computer memory required for the simulation; RST is best and LWES is worst. The simulation was performed on a CDC-6500 digital computer.

A heat-treatment process discussed by Riordon [9.18] serves as the second example to the application of learning control algorithms. The physical process is nonlinear and first order, involving an endothermic reaction for temperatures below 800°K and an exothermic reaction for higher temperatures. This process seems to contain the generality of real-life systems with minimum computational requirements. The process dynamics are

$$z(k + 1) = z(k)(1.005 + 0.015 \tanh[0.1(z(k) - 803.446)])$$
$$+ 0.00333 u(k) + w_1(k) + w_2(k) + w_3(k) + w_4(k) \tag{9.17}$$

where

Table 9.4. Quantitative Comparison of Performance-Adaptive S.O.C. Algorithms on a Linear System.

Performance-adaptive S.O.C. algorithm	Computer time, after 1000 stages, Sec.	Storage location (octal)	Performance cost after 1000 stages		Initial state	Implementation
			Average	standard deviation		
McLaren's growing automation	~50	27.5K	34.5	0.85	Anywhere	Less easy
Tsypkin's stochastic approximation	~35	36.7K	8.3	~0	Almost anywhere	Easy, algorithm is biased
Nikolic and Fu, unsupervised learning	~65	42.0K	36.0	~0	Anywhere	Difficult
Riordon's adaptive automation	~100	--	--	--	Close	Very difficult
Saridis' expanding subinterval	~45	27.4k	19.5	0.4	Anywhere	Easy
Stochastic optimal with known parameters	9.00	1k	4.0	~0	--	--

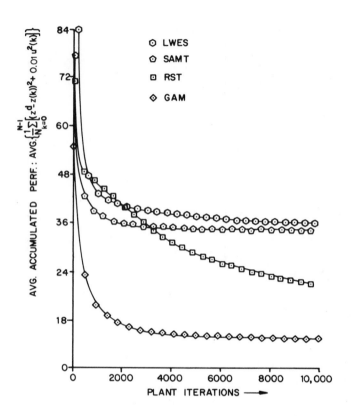

Fig. 9.5. Composite Results of Performance-Adaptive S.O.C.
           Average Accumulated Performance for Linear System

$z(k)$ = temperature (°K) at stage k

$u(k)$ = heat input (kcal) at stage k

$w_i(k)$, $i = 1,\ldots,4$, are independent samples drawn from normal
        zero-mean distributions with the following respective
        standard deviations, σ_i, $i = 1,\ldots,4$:

$\sigma_1 = 0.0002z(k)y|z(k)-800|$

$\sigma_2 = 0.005z(k)[1 + |z(k)-800|^{1/2}]^{-1}$

$\sigma_3 = 0.0005|u(k)$

$\sigma_4 = 1$

$$y = \begin{cases} 1; & z(k) > 800 \\ 0; & z(k) \leq 800 \end{cases}$$

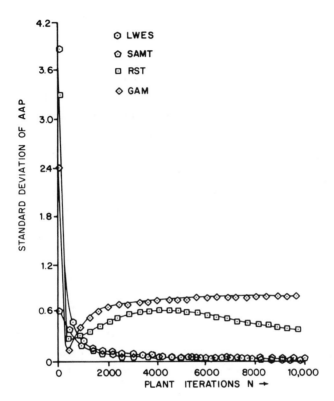

Fig. 9.6.   Composite Results of Performance-Adaptive S.O.C.
             Standard Deviation of AAP.

The uncertainties relative to the process for this particular example
are such that the exact form of the process dynamics, Eq. (9.17),
is unknown, thereby casting it into the framework of self-organization. The object of control in this situation is to maintain the
operating temperature near 800°K, which is actually a point of unstable equilibrium. The state $z(k)$ is assumed to be exactly measureable with negligible error.

Deviations from the desired temperature of 800°K are assigned
a cost given by

$$L_1(z(k),z(k + 1)) = \begin{cases} -0.015[(z(k) - 800)^2 + (z(k + 1) - 800)^2], \\ \qquad \text{for } z(k) < 850 \\ 2800 \quad \text{for } z(k) \geq 850 \end{cases} \qquad (9.18)$$

If $z(k)$ exceeds or equals $850°K$, then a shutdown of the process occurs at a cost of $28.80, and the process is restarted at $z(k) = 775°K$. In addition, penalties on the amount of control energy expended are added to the total cost. Heating and cooling effort results in a cost of 2 cts per 1000 kcal. If extreme energy is needed, a reduction in the expected lifetime of the control equipment is a result, hence such extremes are heavily penalized. In any event, $|u(k)|$ cannot exceed 10,000 kcal. The total cost of control can then be formulated as

$$L_2(u(k)) = 0.002|u(k)| + 60 \ (u(k)/10^4)^6 \quad \text{cts} \qquad (9.19)$$

where

$$|u(k)| < 10^4$$

The object of control in this situation is to find a feedback control policy which minimizes the expected cost of operation per unit of time over a long period as described by the following index of performance:

$$J = \lim_{T \to \infty} \frac{1}{T} E\left\{ \sum_{k=0}^{T-1} L[z(k)z(k + 1),u(k)] \right\} \qquad (9.20)$$

where

$$L[z(k),z(k + 1),u(k)] = L_1[z(k),z(k + 1)] + L_2[u(k)] \qquad (9.21)$$

The performance-adaptive S.O.C. algorithms were implemented, as in the linear example according to the rules derived in Chap. 8.

The growing automaton method (GAM) was applied to this problem using the same controller structure as in the RST and SAMT applications. The subgoal was modified in the following fashion:

$$J_s = \frac{1}{T_i \; NORM} \sum_{t_i}^{t_{i+1}} \{0.015[z(k) - 800]^2 + (z(k + 1) - 800)^2]$$

$$+ \; 0.002|u(k)| + 60\left[\frac{(u(k)]}{10^4}\right]^6 \}$$

where the normalizing factor NORM was added to insure the condition

$$0 \le J_s \le 1$$

as required by the algorithm. Due to the incomplete knowledge of the process, the choice of NORM is generally unclear. However, for this particular example a reasonable guess for NORM was found by assuming tolerable values for the state and control values. Using the heuristic procedure, a value of NORM = 10.0 was determined and found to yield good results.

To apply Tsypkin's stochastic approximation method (SAMT) to this problem the following simplified model was used to identify the plant, where at most quadratic terms were allowed in the structure:

$$\hat{z}(k + 1) = \phi^T(z(k),u(k))\hat{c} \tag{9.23}$$

where

$$\phi(z(k),u(k)) = \begin{bmatrix} z(k) \\ z^2(k) \\ u(k) \end{bmatrix} \quad c = \begin{bmatrix} c_1 \\ c_2 \\ c_3 \end{bmatrix}$$

Using the criterion

$$J_2(c) = E\{z(k + 1) - \phi^T (z(k),u(k))\hat{c})^2\} \tag{9.24}$$

for determining the optimal identification parameters, c^* the stochastic approximation algorithm assumes the following form:

$$\hat{c}(k + 1) = \hat{c}(k) + \gamma_k \phi(z(k),u(k))[z(k + 1) - \phi^T(z(k),u(k))\hat{c}(k)] \tag{9.25}$$

The controller was likewise given a relatively simple structure of the form

$$u(k) = -\hat{\psi}^T(z(k))\alpha \tag{9.26}$$

$$\psi(z(k)) = \begin{bmatrix} 200(z(k) - 800) \\ \\ 0.75(z(k) - 800)^3 \end{bmatrix}$$

Using the criterion

$$J_1(\underline{c}) = E\{L_1[z(k),\phi^T(z(k),\psi^T(z(k))\alpha,c)] + L_2[\psi^T(z(k))\alpha]\} \tag{9.27}$$

to guide the controller, the following stochastic approximation
algorithm was formed:

$$\hat{\alpha}(k + 1) = \hat{\alpha}(k) - \beta_k\Big\{-\psi(z(k)) \ [\hat{c}^T(k)\frac{\partial\phi(z,u)}{\partial u(k)}\cdot 0.03(\phi^T(z(k),u(k))\hat{c}(k)$$

$$- 800) + 0.036\left(\frac{u(k)}{10^4}\right)^5 + 0.002z(k)]$$

$$+ s(k) [\hat{c}(k)^T \frac{\partial\phi(z(k),u(k))}{\partial z(k)} \cdot 0.03(\phi^T(z(k),u(k))\hat{c}(k)$$

$$- 800) + 0.03(z(k) - 800) - \frac{\partial\psi(z(k))}{\partial z(k)} \hat{\alpha}(k) 0.036$$

$$\left(\frac{u(k)}{10^4}\right)^5 + 0.002z(k))]\Big\} \tag{9.28}$$

where

$$\hat{\alpha}_o = \begin{bmatrix} 0.5 \\ \\ 0.5 \end{bmatrix}$$

$$\frac{\partial\phi(z(k),u(k))}{\partial u(k)} = \begin{bmatrix} 0 \\ 0 \\ 1 \end{bmatrix}$$

$$\frac{\partial\psi(z(k))}{\partial z(k)} = \begin{bmatrix} 200 \\ \\ 2.25(z(k) - 800)^2 \end{bmatrix}$$

$$\frac{\partial\phi(z(k)u(k))}{\partial z(k)} = \begin{bmatrix} 1 \\ 2z(k) \\ 0 \end{bmatrix}$$

$$x(k) = \begin{cases} 1 & u(k) > 0 \\ 0 & u(k) = 0 \\ 1 & u(k) < 0 \end{cases}$$

The terms $s(k)$ represent the partials of $z(k)$ with respect to the parameters b, terms which are actually unavailable due to the lack of knowledge of the plant. Assuming that the identification is exact, (sensitivity) functions were generated by

$$s_i(k + 1) = \{\hat{c}_1(k) + 2\hat{c}_2(k)z(k) - \hat{c}_3(k)[200\hat{a}_1(k) + 2.25\hat{a}_2(k)$$

$$(z(k) - 800)^2]\} \cdot s_i(k) - \hat{c}_3(k)\psi_i(z(k)) \qquad (9.29)$$

where the double subscript denotes the component of the given vector and the time index, respectively, and $\psi_i(z(k))$ denotes the ith element of the vector $\psi(z(k))$.

In applying the learning techniques of Nikolic and Fu to this particular problem, basically the same approach was used as in the application of Riordon's Adaptive Automaton (AAM). Due to the computer memory limitations, a somewhat cruder control quantification of 1000 kcal deviations about the initial guess was implemented, again allowing five control choices for each of the process states. The subgoal necessary for the LWES simulation was heuristically chosen to be

$$J_s(k) = 0.015[z(k) - 800)^2 + (z(k + 1) - 800)^2] + 0.002|u(k)|$$

$$+ 60 \left[\frac{u(k)}{10^4}\right]^6 \qquad (9.30)$$

The random search technique (RST) was applied to this problem, using the same controller structure as in the SAMT application, i.e., Eq. (9.26). The subgoal or the expanding subinterval performance index was chosen to be

$$SUBGOAL = \frac{1}{T} \sum_{t_i}^{t_{i+1}} \{0.015[(z(k) - 800)^2 + (z(k + 1) - 800)^2]$$

$$+ 0.002|u(k)| + 60 \left[\frac{u(k)}{10^4}\right]^6\} \qquad (9.31)$$

where $T_i = t_{i+1} - t_i$ was held constant at 10 for this particular application.   The search for the control parameters b was implemented with
random variables generated from a zero-mean Gaussian distribution.
The standard deviation of the random step vector ρ was chosen to be

$$\sigma_\rho = \text{GAIN*} \begin{bmatrix} 1.0 \\ 1.0 \end{bmatrix}$$

where a GAIN of 0.1 was found to yield the best results.

The simulations performed for this 2nd example indicate that all
of the performance-adaptive S.O.C. methods tested produced controllers
which could stabilize the process temperature.   The composite plots
of Figs. 9.7 and 9.8 illustrate the performance of each method
relative to the others for this particular application.   As far as
the *rate of convergence* is concerned, Fig. 9.7 indicates that the
methods may be ranked in order of decreasing speed of convergence as
follows: SAMT, RST, GAM, and LWES.   Looking at the confidence in the
convergence as indicated by the *standard deviation of the* AAP over
the 10 test runs, the methods may be ranked in order of decreasing
strength as follows: SAMT, RST, LWES, and GAM.   Table 9.5 presents
a summary of the quantitative results obtained for this nonlinear
heat-treatment example.   As indicated by Table 9.2, the various S.O.C.
schemes, simulated on a CDC-6500 digital computer, may be ranked in
order of increasing storage requirements as follows: RST, SAMT,
GAM, and LWES.   Table 9.5 indicates that for a typical single run,
each S.O.C. system is operating near 800°K after 1500 stages of
operation in a fairly stable manner.   The sole exception is the
LWES S.O.C. which did, however, achieve a performance comparable to
the others after 3000 stages.

Using the two relatively simple but realistic examples, it was
demonstrated that practical controllers may be produced to drive
systems with various degrees of uncertainty in their dynamics.   Comparisons were performed among five major on-line performance-adaptive
S.O.C. algorithms with the stochastic optimal and were reported here.

Table 9.5. Quantitative Comparison of Performance-Adaptive S.O.C. Algorithms on a Nonlinear System.

Performance Adaptive S.O.C. Algorithm	Storage Location (octal)	Performance cost after 1500 stages		Temperature, K°, after 1500 stages		Implementation
		Average	Standard deviation	Average	Standard deviation	
McLaren's growing automation	33.6K	8.37	3.021	792.6	2.31	Easy
Tsypkin's stochastic approximation	25.6K	3.82	0.793	803.4	3.83	Easy
Nikolic and Fu unsupervised learning	63.2K	15.31	1.724	804.5	6.60	Difficult
Riordon's adaptive automation	~75K	4.27	0.75	794.3	3.5	Difficult
Saridis' expanding subinterval	25.4K	5.84	1.497	803.1	5.87	Less easy
Stochastic optimal with known parameter	---	---	---	---	---	Unknown

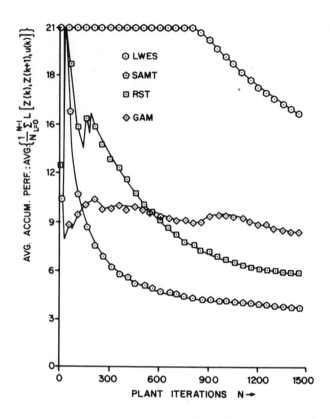

Fig. 9.7.   Composite Results of Performance-Adaptive S.O.C.
             Average Accumulated Performance for Nonlinear System

They demonstrate, among other features, the feasibility and the
shortcomings of these algorithms, and seem to rank them which was
the main goal of this section.

A record of the programs in FORTRAN language used for the
quantative evaluation of the performance-adaptive S.O.C. algorithm
can be found in Ref. [9.10].

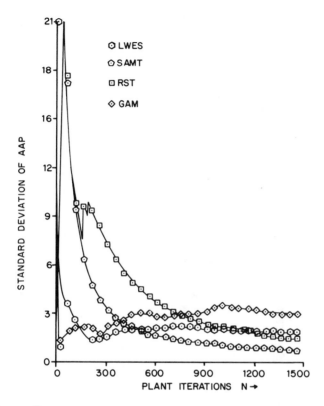

Fig. 9.8.   Composite Results of Performance-Adaptive S.O.C.
Standard Deviation of AAP

9.5   CROSSCOMPARISON OF PARAMETER-ADAPTIVE AND PERFORMANCE-ADAPTIVE
S.O.C. ALGORITHMS

After examining the parameter-adaptive and performance-adaptive
S.O.C. Algorithms both qualitatively and quantitatively, the next
question is how do they compare to each other.

Even though the two classes have been defined to be different,
comparison is possible in view of the linear example commonly used
for individual comparison of each of the classes. The results are
available in Fig. 9.3 to 9.6, as well as in Tables 9.3 and 9.4.
However, because a different number of runs were obtained for each

of the classes, the average accumulation performance and its standard
deviation for three parameter-adaptive and four performance-adaptive
algorithms were plotted for 10,000 plant iterations or stages in
Figs. 9.9 and 9.10, respectively.

The conclusion was drawn that the parameter-adaptive algorithms
are about one order of magnitude faster in convergence, they require
about one order of magnitude faster and bigger digital computers, and
they are more reliable (standard deviation) and rather easier to
implement since they are more amenable to analytic computations than
the performance-adaptive S.O.C. algorithms.

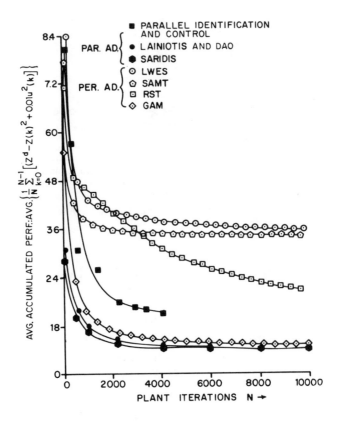

Fig. 9.9.   Composite Results for Crosscomparison of Parameter-
            Adaptive and Performance-Adaptive S.O.C. Average
            Accumulated Performance

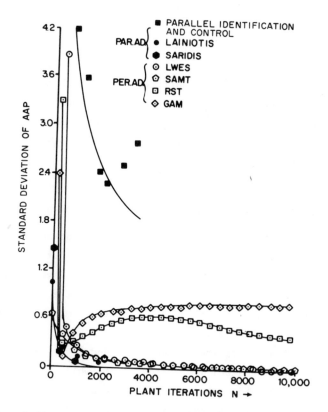

Fig. 9.10. Composite Results for Crosscomparison of Parameter-
Adaptive and Performance-Adaptive S.O.C.: Average
Deviation of AAP

This conclusion was almost expected, because, as it was previously stated, the performance-adaptive S.O.C. algorithm are designed to anticipate uncertainties in the structure of the plant as
well as the parameters. A verification of the above statement can
be provided by Tsypkin's stochastic approximation algorithm which
was designed with both the parameter-adaptive and performance-
adaptive points of view. When the structure of the system is known,
either because it is linear or simple nonlinear, the models of the
plant and the controller are structurally exact and the algorithm is

reduced to a parameter-adaptive algorithm with the expected properties. This was clearly demonstrated in Figs. 9.9 and 9.10.

In a final remark it is recommended here, as it was done
throughout this book, that an algorithm should be selected for its
own merits of simplicity and suitability to the particular problem
at hand without trying to overdesign the controler merely in order
to look elaborate. It is hoped that the plots and tables in this
chapter help the designer to select the proper algorithm that fits
best to his problem.

9.6 STABILITY CONSIDERATIONS FOR SELF-ORGANIZING CONTROLS

The stability problem in S.O.C. has been overlooked or by-
passed by most of the investigators in the area not out of negligence
but because of the difficulties associated with its investigation.
Nevertheless, stability is of vital importance for the implementation
of any S.O.C. algorithm and appropriate conditions must be established
before any design application.

In the algorithms described in Chaps. 7 and 8 very little mention was made about the stability of the plant, while most of the
effort was devoted to the stability, e.g., the convergence, of the
identification on performance improvement algorithms. In the case
of the parameter-adaptive S.O.C. techniques such a study was avoided
most of the time because of the analytic difficulties of such an
investigation. As a result, the linearized optimal methods,
especially the ones using a linearized Kalman filter, were plagued
with unstable runs. In the OLFO algorithms a stability study is
possible, though tedious. Especially in Lainiotis's algorithm as
well as in the algorithms of Stein and Saridis, and Saridis and Dao,
which were originally developed for open-loop stable plants, it is
easy to select the unknown parameter set to yield stable systems,
along with the knowledge that the asymptotically optimal solution
is always stable [9.17]. The parallel identification and control
algorithm of Saridis and Lobbia was originally designed for open-loop

stable systems with a saturation element in the feedback loop to
keep the control signal within the reasonable bounds that would
keep the system with the stable region, a common practice in aerospace industry [9.29]. The stability investigation associated with
the convergence of the identification algorithm [9.30] discussed in
Chap. 7 will be generalized and further explored in the next section.
No attempt has been made yet to investigate the stability of the
actively adaptive algorithm of Tse and Bar-Shalom.

The stability of the performance-adaptive S.O.C. algorithm is
a more difficult task because of the partial or complete lack of
knowledge of the plant. In the case of the algorithms studied in
Chap. 8, it was always assumed that the overall system was stable.
This was accomplished either by selecting the control space to contain only stable values for their admissible control vectors, as in
the case of the automata algorithms or the unsupervised learning
algorithms (Fu and Nikolic), or by selecting the control parameter
space to yield a stable system as in Tsypkin's stochastic approximation and Saridis' expanding subinterval algorithms. Investigation
of stability in such cases would require the use of methods of stochastic stability [9.14], [9.15] which requires a highly advanced
mathematical treatment and yields information sometimes questionable
from the designer's point of view.

In order to face the stability problem one has to understand
its nature in the S.O.C. case. By definition such a system has
partially or completely unknown dynamics operating in a stochastic
environment which may or may not be stable in some sense. The
problem would be to design the S.O. controller to improve its performance but at the same time keep the system stable in the same
sense as before. From the designers point of view, the stability
required for this problem is equivalent to the boundedness of the
output of the system for the duration of the process. This condition would permit any of the S.O.C. algorithms to function and
eventually converge to its steady state value which may assume the
asymptotic optimal value of the stochastic control problem postulated

to exist and be stable. Based on the above remark, one may narrow
the stability investigation down to bounded input-bounded output
(BIBO) stability or more general stability in large.

In the sequel two aspects of this problem will be pursued. The
first one considers the random disturbances from the environment as
inputs to the system. Since the rest of the plant and the controller
are assumed to be deterministic, the stability problem may be treated
by deterministic methods. Since this approach assumes somehow the
knowledge of the structure of the plant, it will be developed in
conjunction with the parameter-adaptive S.O.C. problem, even though
this restriction is only superficial.

The second approach treats the whole system as a stochastic
process with a linear or nonlinear structure being identified to
within a certain class of nonlinear structures of an assumed library
[9.26]. Then an investigation is possible to find the regions of
the control coefficients for which the system stays finite. This
approach is more suitable for performance-adaptive S.O.C. systems
and will be developed for such cases.

9.6.1 Deterministic Stability Investigation

This approach may be considered as a direct generalization of
the stability study of Sec. 7.6.5 performed in conjunction with the
convergence investigation of the identification algorithm. Its
purpose would be go obtain the regions of the adjustable parameter
of the system for which the system will stay bounded for any bounded
input.

In its most general form this procedure would treat the following continuous-time system:

$$\dot{x}(t) = f(x,u,v,w,t) \tag{9.32}$$

$$z(t) = h(x,\theta,v,t)$$

or discrete-time system

$$x(k + 1) = f(x,u,\theta,w,k) \tag{9.33}$$

$$z(k) = h(x,\theta,v,k)$$

where x is an n-dimensional state vector, u is an m-dimensional control, z is an r-dimensional output vector, w and v are random disturbances emanating from the environment of appropriate dimension,
and θ is a set of unknown parameters belonging to a closed and bounded
set Ω_θ. For this problem a feedback controller has been designed
using some S.O.C. algorithm for the continuous-time case

$$\hat{x}(t) = a(\hat{x},z,\alpha,t) \tag{9.34}$$

$$u(t) = c(\hat{x},\alpha,t)$$

or the discrete time case

$$\hat{x}(k + 1) = a(\hat{x},z,\alpha,k) \tag{9.35}$$

$$u(k) = c(\hat{x},\alpha,k)$$

where $\hat{x}(t)$ is an n-dimensional vector corresponding to some kind of
state estimate, and α is a set of adjustable parameters depending on
the unknown parameters θ of the system, e.g., $\alpha(\theta)$, and which are
obtained through the estimates $\hat{\theta}(k)$ derived from the S.O.C. algorithm.
Combining (9.32) and (9.34) or (9.33) and (9.35) one obtains the
following 2n-dimensional dynamic systems, respectively:

$$\dot{x}(t) = f(x,c(\hat{x},\alpha,t),\theta,w,t) = f'(x,\hat{x},\theta,\alpha,w,t)$$

$$\hat{x}(t) = a(\hat{x},h(x,\theta,v,t),\alpha,t) = a'(x,\hat{x},\theta,\alpha,v,t) \tag{9.36}$$

or

$$x(k + 1) = f(x,c(\hat{x},\alpha,k),\theta,w,k) = f'(x,\hat{x},\theta,\alpha,w,k)$$

$$\hat{x}(k + 1) = a(\hat{x},h(x,\theta,v,k),\alpha,k) = a'(x,\hat{x},\theta,\alpha,v,k) \tag{9.37}$$

The above equations describe two sets of dynamic systems with forcing
functions, the environmental noise inputs $w(\cdot)$ and $v(\cdot)$, the unknown
parameters $\theta \in \Omega_\theta$, and the adjustable parameters $\alpha(\theta)$. Assuming
that the inputs $w(\cdot)$ and $v(\cdot)$ are bounded for the duration of the
process, one may find conditions for BIBO stability, sufficient
whenever they are available, for which either of the two systems are
bounded. Such sufficient conditions are usually in the form of a
set of inequalities involving the parameters θ and α, and are independent of $x,\hat{x},w,v$; e.g.,

$$S(\alpha(\theta),\theta) \geq 0 \tag{9.38}$$

Knowing that $\theta \in \Omega_\theta$, one may find a closed and bounded set Ω_α so that if $\alpha \in \Omega_\alpha$, $\theta \in \Omega_\theta$ conditions (9.38) are always satisfied. Constraining the parameter identification algorithm which estimates θ to yield $\alpha(\hat{\theta}) \in \Omega_\alpha$, one may guarantee BIBO stability for the S.O.C. system under consideration. In order to obtain this result, it was assumed that the disturbances are bounded

$$|w(\cdot)| \leq W; \quad |v(\cdot)| \leq V \tag{9.39}$$

Such an assumption includes certain random processes that are not bounded like the Gaussian noise, which considerably limits the applicability of the method. However, for most engineering this difficulty may be overcome by considering truncated approximations of these random variables, such as a truncated and renormalized Gaussian noise with a probability density function defined by

$$p_T(s) = \begin{cases} 0 & s > M \\ \dfrac{p(s)}{1 - 2P(M)} & |s| \leq M \\ 0 & s < -M \end{cases} \quad p(M) = \int_M^\infty p(s)\ ds \tag{9.40}$$

where M is some large number for which $P(M) \leq \varepsilon$, for some $\varepsilon > 0$.

This approach assumes the structure of the plant and the control known for most applications and therefore is more suitable for parameter-adaptive S.O.C. systems.

The method will be demonstrated for a linear continuous-time dynamic system for which explicit conditions of BIBO stability are available in the literature [9.1], [9.2].

A similar investigation for a discrete-time dynamic system may be found in Sec. 7.5.3.

A linear continuous-time dynamic stochastic system with constant but unknown coefficients θ and a feedback dynamic control with adjustable parameter α, would be described by a system of differential equations equivalent to (9.36)

$$\begin{bmatrix} \dot{x}(t) \\ \hline \dot{\hat{x}}(t) \end{bmatrix} = \begin{bmatrix} F(\theta) & \vdots & B(\theta)H \\ \hline c(\alpha)H & \vdots & A(\alpha) \end{bmatrix} \begin{bmatrix} x(t) \\ \hline \hat{x}(t) \end{bmatrix} + \begin{bmatrix} G(\theta) & \vdots & 0 \\ \hline 0 & \vdots & I \end{bmatrix} \begin{bmatrix} w(t) \\ \hline v(t) \end{bmatrix} \qquad (9.41)$$

are equivalent to

$$\dot{y}(t) = F(\theta,\alpha)y(t) + G(\theta)v(t)$$

where the symbols used are the matrices and vectors corresponding to the ones used throughout this book for a linear feedback control system, and may correspond to the optimal combination of coefficients with respect to some performance criterion for *a priori* known parameters θ. Estimates of the unknown $\theta \in \Omega_\theta = \{\theta_{0i} \leq \theta_i \leq \theta_{1i};$ $i = 1,2,\ldots,M\}$ are obtained through an identification algorithm, and are substituted in an appropriate expression to yield the appropriate adjustable parameter α

$$\hat{\theta}(t_{j+1}) = T(t_j)\,\hat{\theta}(t_j) + Q(t_j) \quad \theta \in \Omega_\theta \qquad (9.42)$$

$$\alpha(\hat{\theta}) = R(\hat{\theta})$$

where $t_{j+1} - t_j$ is the identification interval. Assuming that $w(t)$ and $v(t)$ satisfy (9.39) and (9.40), the BIBO stability conditions for the new time varying system (9.41) are given by

$$\int_0^{10} \mid \zeta_{kj}(t,\tau) \mid d\tau \leq \sum_{i=0}^{\infty} a\int_{t_i}^{t_{i+1}} \mid \zeta_{kj}(t-\tau,\theta_i) \mid d\tau$$

$$\leq \sum_{\ell=1}^{2n} A_\ell \int_0^{\infty} \ell^{\lambda_\ell(max)\tau} d\tau \quad k = 1,\ldots,2n$$

$$j = 1,\ldots,2m \qquad (9.43)$$

where $\zeta_{kj}(t,\tau)$ is the impulse response of the kth state $y_k(t)$ resulting from the jth input $v_j(\tau)$ applied at τ, $\zeta_{kj}(\tau,\theta_i)$ is the time invariant impulse response obtained during the ith identification interval where θ_i was used, a_i and A_ℓ are appropriate constants, and $\lambda_\ell(max)$ is the maximum of all the ℓth eigenvalues of $F(\theta)$ over $\theta \in \Omega_\theta$

$$\lambda_\ell(max) = \underset{\theta}{Max} \;\; \lambda_\ell[F(\theta)] \qquad (9.44)$$

Conditions (9.43) are always satisfied if

$$\text{Re}\{\lambda_\ell (\max)[F(\theta,\alpha)]\} < 0 \quad \ell = 1,\ldots,2n \quad\quad (9.45)$$

In view of Ω_θ, condition (9.45) derived for distinct eigenvalues λ_ℓ, may be solved simultaneously to yield the set of admissible α, Ω_α. Similar expressions may be obtained for multiple eigenvalues.

As an example consider the first-order dynamic system with its first-order dynamic feedback control

$$\dot{x}(t) = x + \theta\hat{x} + g_1 w$$

$$\dot{\hat{x}}(t) = \alpha_1(\theta)x + \alpha_2(\theta)\hat{x} + v$$

where θ may be selected from $\Omega_\theta = \{-1 \le \theta \le 1\}$
The eigenvalues of the F matrix are

$$\lambda_{1,2} = -\frac{1 + \alpha_2}{2} \pm \sqrt{\frac{(1 - \alpha_2)^2}{2} - \alpha_1\theta} \quad \theta \in \Omega_\theta = \{-1 \le \theta \le 1\}$$

and

1. $\alpha_1 > 0$, $\theta = -1$; $\alpha_1 < 0$, $\theta = +1$

$$\lambda_1(\max) = \frac{1 + \alpha_2}{2} + \sqrt{\frac{(1 - \alpha_2)^2}{2} + |\alpha_1|}$$

2. $\alpha_1 > 0$, $\theta = 1$; $\alpha_1 < 0$, $\theta = -1$

$$\lambda_2(\max) = \frac{1 + \alpha_2}{2} - \sqrt{\frac{(1 - \alpha_2)^2}{2} - |\alpha_1|}$$

In order to satisfy condition (9.45) one considers

1. $\alpha_1 < \alpha_2$, $\alpha_2 < -1$

2. $\alpha_1 > (\frac{1 - \alpha_2}{2})^2 \ge 0$, $\alpha_2 < -1$

3. $\alpha_1 < 0 \quad \alpha_2 < -\alpha_1$

or $\alpha_1 < -(\frac{1 - \alpha_2}{2})^2$, $\alpha_2 < -1$

$\alpha_1 < 0$, $\alpha_2 < \alpha_1$

Combining conditions 1. and 2. one ends up with the admissible set

$$\Omega_\alpha = \left\{ \alpha_1 < - \left[\frac{1 - \alpha_2}{2}\right]^2 , \alpha_2 < -1 \right\}$$

which defines a region of constraints for which the parameter identification algorithm for θ and consequently $\alpha(\hat{\theta})$ must be confined to
the stability of the system.   It is obvious that this method yields
only sufficient conditions and therefore a smaller constraint area
than the true one.   On the other hand, an alternative way to obtain
Ω_α is to find the α's for which $\lambda_{1,2}$ have always negative real parts,
no matter what $\theta \in \Omega_\theta$ is.

For higher-order systems the problem of finding Ω_α is tedious
but one may always find a subregion which will be sufficient for the
stability of the algorithm.

9.6.2 Stochastic Stability Investigation

The method discussed above has the disadvantage of assuming that
the disturbances are bounded and that the structure of the system is
a priori known, preferably to be linear.

In order to treat the stochastic processes properly and consider
the stability of systems controlled by a performance-adaptive S.O.C.
algorithm, one must investigate the system as a stochastic process.

There are several definitions and many ways to treat stochastic
stability [9.14], [9.9], [9.13]. Most of the methods require mathematics
beyond the level of this treatise.   However, since only the finiteness of the response of the system is required as the equivalent of
BIBO stability in the stochastic processes treated here, one may
define a concept of stability and prove certain necessary theorems,
based on the finiteness of the process in the identification subinterval $T_i = t_{i+1} - t_i$, $i = 1, 2,\ldots$, and in the overall process.

With regard to this development, definitions have been made to
provide a framework to relate behavior on the identification sub-
intervals T_i to behavior on the overall interval of the process
(T_0, ∞).   Let $x(t)$ represent the process governed by

$$\dot{x}(t) = f(x(t), u(t), w(t), t) \quad x(t_0) = x_0 \quad u(t) \in \Omega_u \qquad (9.46)$$

$$z(t) = h(x(t), v(t), t)$$

where $x(t)$ is an n-dimensional state vector in the state space Ω_x; $u(t)$ is an m-dimensional control vector in a closed and bounded set Ω_u of admissible controls; $x_0, w(t)$, and $v(t)$ are three vector-valued random variables of dimension n, q, and r, belonging, respectively, to independent stationary processes with unknown statistics; $z(t)$ is an s-dimensional output vector, $s \leq n$; and $f(\cdot)$ and $h(\cdot)$ are unknown nonlinear continuous bounded functions of their arguments of appropriate dimensions. The vectors $w(t)$ and $v(t)$ represent the plant-environment relationship. Then the following definitions are appropriate [9.22]:

Definition 9.1

x(t) is called "subinterval finite," if the probability
measure, which defined the performance functional of the extent
$x(t) \to \infty$ for $t \in T_i$ is zero.

Although Definition 9.1 looks similar to the definition of finite escape time, it is not, since any τ_i may be of infinite length.

Definition 9.2

x(t) is called "overall interval finite," if the probability
measure, which defined the performance functional of the extent
$x(t) \to \infty$ for $t \in (t_0, \infty)$ is zero.

Since the implementation of the expanding subinterval algorithm is contingent upon the performance index being integrable, "weakened" definitions may be stated to take care of this condition.

Definition 9.3

x(t) is called "subinterval performance integrable," if the
subgoal (8.100), for any τ_i is finite, i.e., the integral
exists.

Definition 9.4

x(t) is called "overall interval performance integrable," if the value of the overall performance index, is finite.

The performance functional is usually defined by

$$I(u) = E\{J(u)\} = E\left\{ \lim_{t_k \to \infty} \frac{1}{t_k - t_0} \int_{t_0}^{t_k} L(z(t),u(t),t) \ dt \right\} (9.47)$$

where L is a known nonlinear continuous bounded nonnegative operator. The following theorem demonstrates the connection between subinterval and overall interval stability properties:

Theorem 9.1

x(t) is subinterval finite for all i, if and only if x(t) is overall interval finite.

The proof is given in Ref. [9.22].

The necessity of relating subinterval stability properties to overall interval stability properties arises because of the assumption of a stationary process.   It will be possible to apply some existing stability concepts to a process which is described as a homogeneous stationary Markov process.   But since the controller may change some parameter from subinterval to subinterval, the connection had to be made.

Theorem 9.2

If x(t) is subinterval finite for all i, then x(t) is overall interval finite.

It is clear that the above results have been developed for the expanding subinterval algorithm of Sec. 8.6 where the identification subinterval T_i is the expanding subinterval [9.24].   However, it is reformulated here to hold for other performance-adaptive S.O.C. algorithms.   These results are also applicable to the discrete-time case [9.22].

Utilizing the results stated above and the existing stochastic Lyapunov theory [9.14], without getting into mathematical details, it is possible to restrict the space of allowable vector parameters

c which characterize the performance-adaptive feedback controller
so that the overall system is stable in the sense discussed previously.
Since it is required to evaluate a Lyapunov function candidate along
the trajectories of the process, the completely general case of
unknown $f(\cdot)$ and $g(\cdot)$ functions given in (9.46) has not been solved.
However, present efforts of system identification employing pattern-
recognition techniques indicates that systems may be classified
successfully according to some broad classes, e.g., linear or nonlinear, order of nonlinearity, etc. [9.26]. This makes the development of stability analysis to be performed within different classes
of systems a feasible method for systems with *a priori* unknown
dynamics.

The pattern-classification algorithm of Saridis and Hofstadter
provides a method to classify nonlinear systems according to the
type and location of the nonlinearity in the system [9.26]. Once
this is performed and the class of the nonlinear system is known,
a stability investigation is possible within that class to define
the admissible control parameter set. Such a procedure is applied
in the sequel as a feasibility study for three classes of nonlinearities, namely state-dependent noise, additive noise, and Luré type
nonlinearities in the feed-forward path. The method should be
extended to cover a larger class of nonlinear systems as described
in Ref. 9.22, to cccount for the systems encountered in practice.

9.6.2.1. *Linear Plant with State-Dependent Noise, State-Dependent Measurement Noise, and Linear Dynamic Controller.*

Although this type of system may be considered to be a subclass
of a more general classification, it is being considered separately
because of some existing NASC (necessary and sufficient conditions),
due to Kleinman [9.12], which facilitate the design of a stable
overall system.

Let the process be modeled by the following Ito-type equation

$$dx = Fx \, dt + \sum_{i=1}^{N} G_i x \, dw_i + Bu \, dt \qquad (9.48)$$

where F is a constant n x n matrix for which only bounds may be
known on its entries a_{ij}; G_i, i = 1,...,N, are constant n x n matrices
again for which only bounds may be known on their entries; B is a
known constant n x m matrix; and w_i, i = 1,...,N, represents a
Wiener process with

$$E\{[w_j(t) - w_i(\tau)][w_j(t) - w_j(\tau)]\} = \tau_{ij}|t - \tau|$$

The plant output is given by

$$z(t) = Hx(t) + \sum_{i=1}^{M} D_i v_i(t) \tag{9.49}$$

where H is a p x n known constant matrix; D_i, i = 1,...,M, are p x n
constant matrices with given bounds on their entries, and v_i,
i = 1,...,M, represents a Wiener process with

$$E\{[v_i(t) - v_i(\tau)][v_j(t) - v_j(\tau)]\} = \bar{\sigma}_{ij}|t - \tau|$$

The feedback control will be selected as the output of a linear
filter M, operating on the available output from the plant

$$u(t) = Ky(t)$$
$$\dot{y}(t) = Ay(t) + Ez(t) \tag{9.50}$$

where K is a known m x n constant matrix, y(t) is an n-vector, A is
an n x n matrix, and E is an n x p matrix. The elements of A and E
are the parameters on which the random search is performed, and they
may be written as components of an n(n + p)x1 random vector c, i.e.,

$$c_1 = a_{11}; \; c_2 = a_{12}; \; ...; \; c_n = a_{in}; \; ...; \; c_n^2 = a_{nn}; \; c_n^2{}_{+1} = e_{11};$$
$$...; \; c_{n(n+p)} = e_{np}.$$

The overall closed-loop system may be written as

$$d\hat{x} = \hat{F}\hat{x} \, dt + \sum_{i=1}^{N+M} G_i \hat{x} \, d\hat{w} \tag{9.51}$$

where $\hat{x}$ is a 2n-vector and $\hat{F}$ is a 2n x 2n matrix given by

$$\hat{x} = \begin{bmatrix} x \\ y \end{bmatrix}$$

$$\hat{F} = \begin{bmatrix} F & BK \\ EH & A \end{bmatrix} \tag{9.52}$$

or $i = 1, \ldots, N$

$$\hat{G}_i = \left[\begin{array}{c|c} G_i & 0 \\ \hline 0 & 0 \end{array} \right]_{(2n \times 2n)} \qquad \hat{w}_i = w_i$$

and for $i = N + 1, \ldots, N + M$

$$\hat{G}_i = \left[\begin{array}{c|c} 0 & 0 \\ \hline 0 & ED_i \end{array} \right]_{(2n \times 2n)} \qquad \hat{w}_i = v_i \qquad (9.53)$$

The noise is still assumed to be of the form

$$E\{ [\hat{w}_i(t) - \hat{w}_i(\tau)][\hat{w}_j(\tau) - \hat{w}_j(\tau)] \} = \hat{\sigma}_{ij} |t - \tau|$$

This indicates that w_i and v_i are together assumed to be white.

From the results obtained by Kleinman [9.12] we select the following lemma:

Lemma 9.1

The stochastic process defined in (9.53) approaches zero with probability one, if and only if for any $Q > 0$, $V(\hat{x}) = \hat{x}'P\hat{x}$ is a stochastic Lyapunov function where $P > 0$ satisfies

$$0 = Q + \hat{F}^T P + P\hat{F} + \sum_{i=1}^{N+M} \sum_{j=1}^{N+M} \hat{\sigma}_{ij} \hat{G}_i P \hat{G}_j \qquad (9.54)$$

Thus, there is a linear matrix equation which P must satisfy to insure the stability of the stochastic process, $x(t)$. The entries of $\hat{F}$ and $\hat{G}_i$, $i = 1, \ldots, N + M$, may be restricted so that the equation is satisfied on a subinterval basis. Applying then the theorems of the previous section yields the overall stability result.

The following example illustrates the approach:

Consider a scalar process described as

$$dx = x \, dt + u \, dt + x \, dw$$

whose output is given by

$$z = xv$$

w and v are uncorrelated Gaussian white noise, and the arbitrary bounds on "f," the coefficient scalar of the state x, are $-1/2 < f < 0$.

Let the feedback controller be represented by a dynamic scalar process

$$\dot{y} = ay + cz$$

and let the input to the original plant be the output of this controller

$$u = cy$$

The above closed-loop system described may now be written as

$$\begin{bmatrix} d\,x \\ d\,y \end{bmatrix} = \begin{bmatrix} f & 1 \\ 0 & a \end{bmatrix} \begin{bmatrix} x \\ y \end{bmatrix} dt + \begin{bmatrix} 1 & 0 \\ 0 & 0 \end{bmatrix} dw + \begin{bmatrix} 0 & 0 \\ ae & 0 \end{bmatrix} dv$$

In completing the algebra to insure that condition (9.54) is satisfied, it is found that the controller parameter values must be restricted by

$$|e| < 1 \text{ and } a < 0$$

The interesting result accompanying this class of systems is that both necessary and sufficient conditions have been satisfied. It should be noted also that the work required to obtain the last conditions reflects the added information.

9.6.2.2 *Linear Plant with Additive Noise, Additive Measurement Noise, and Linear Dynamic Controller*

The main stability theorem to be used here is due to Wonham [9.33], and it is not restricted to this class of systems.

Let the process be modeled by the following Ito-type equation

$$dx = Fx\,dt + Bu\,dt + G\,dw \qquad\qquad (9.55)$$

where x is an n-vector, u an m-vector, and w an r-vector Gaussian white noise process; F, B, and G are constant matrices of appropriate dimension.

The output is also noise corrupted as follows:

$$z(t) = Hx(t) + E\,v(t) \qquad\qquad (9.56)$$

where H is a known p x n matrix and E is a constant p x n matrix.

The feedback controller is chosen as the output of a linear dynamic
system

$$\dot{y} = Ay + Dz$$
$$u = cy \tag{9.57}$$

With this controller the overall system may be written as follows:

$$d\hat{x} = \hat{F}\hat{x}\ dt + \hat{G}\ dw \tag{9.58}$$

where

$$\hat{x} = \begin{bmatrix} x \\ - \\ y \end{bmatrix} \quad \hat{F} = \begin{bmatrix} F & \vdots & Bc \\ - & - & - \\ DH & \vdots & A \end{bmatrix}$$

$$\hat{w} = \begin{bmatrix} w \\ - \\ v \end{bmatrix} \quad \hat{G} = \begin{bmatrix} G & \vdots & 0 \\ - & - & - \\ 0 & \vdots & DE \end{bmatrix}$$

The stability criteria given by Wonham is a Lyapunov function
approach. Let $V(x)$ be a class of real-valued functions with the
following properties:

P1.   V is defined for $x \in \bar{D}_v$ where $\bar{D}_v = \{x: |x| > R\}$ ($R < \infty$ is arbitrary).

P2.   V is continuous in $\bar{D}_v$ and is twice continuously differentiable
      in $\bar{D}_v$.

P3. $V(x) \geq 0$, $x \in \bar{D}_v$, and $V(x) \to +\infty$ as $x \to \infty$.

Theorem 9.3

    If there exists a function V with properties P1-P3 and if
    $L[V(x)] \leq 0 \quad x \in \bar{D}_v$,
    then the process is recurrent [9.33].

    Here L is the differential generator of the process, defined
in Ref. 9.3.

Theorem 9.4

    The process x defined by (9.50) is recurrent if $\text{Re}\lambda_i < 0$, i =
    1,...,2n, where λ_i are the eigenvalues of A.

    (This theorem is a variation of Theorem 5 in Ref. 9.13).

    The proofs of Theorems 9.3 and 9.4 are given in Ref.[9.22].

A few remarks are in order concerning this example. The
theorem gives conditions for recurrence of a Markov process;
Khas'minskii [9.11] has shown the equivalence of recurrence and the
existence of an invariant probability measure to describe the transition probabilities diffusion processes. In essence, what was
shown here is that any hypersphere in the state space is hit
eventually with probability 1. This and the conditions developed in
the sequel are sufficient to satisfy the integrability of performance
index (9.47) which is necessary for the expanding subinterval performance-adaptive algorithm.

The following example illustrates the procedure:

Let the plant and output be represented by

$$dx = \begin{bmatrix} 0 & 1 \\ f_1 & f_2 \end{bmatrix} x\ dt + \begin{bmatrix} b_1 \\ b_2 \end{bmatrix} u\ dt + \begin{bmatrix} 1 \\ 1 \end{bmatrix} dw$$

and

$$z(t) = [1,0]x(t) + v(t)$$

where the noises w and v and the initial state $x(0)$ are modeled by
Gaussian white-noise processes. Assume that for design purposes the
only knowledge of the plant parameters is $f_1 < -1$, $f_2 < -1$, $b_1 = 0$,
and $0 < b_2 < 3$.

The controller is described by the second-order system

$$\dot{y} = \begin{bmatrix} 0 & 1 \\ a_1 & a_2 \end{bmatrix} y + \begin{bmatrix} e_1 \\ e_2 \end{bmatrix} z(t)$$

and the input to the system is

$$u(t) = [-1,0]y(t)$$

The above closed-loop system will satisfy the sufficiency
Theorem 9.4, if $a_1 < -1$, $a_2 < -1$, $0 < e_1 < 1$, and $e_2 < -1$. Since
only the eigenvalues of the closed-loop system are of concern, the
effort in calculations is minimal.

9.6.2.3 *Linear Plant with Luré-Type Nonlinearity, Additive Noise, and Linear Dynamic Controller*

Wonham's stability theorem [9.33] is again to be used to develop a criterion for recurrence.

Let the process be defined by

$$dx = Fx\ dt - bf(\gamma)\ dt + G(x)\ dw + Bu\ dt \qquad (9.59)$$

$$\gamma = c^T x$$

where x is an n-vector, u an m-vector, and w an n-vector Gaussian white-noise process; F, b, and B are constant matrices of appropriate dimension, G(x) is a bounded m x n nonsingular matrix, i.e., $||G(x)|| < K < \infty$ and

$$y^T G(x) G(x)^T y \geq c_2 y^T y \quad (x,y) \in R^n$$

and f($\cdot$) is a scalar-valued nonlinear function of its scalar argument.

The available output is given by

$$z(t) = Hx(t) + Ev(t) \qquad (9.60)$$

where H is a known constant p x n matrix and E is a known constant p x n matrix. The feedback is again chosen to be a dynamic controller.

$$\dot{y}(t) = Ay(t) + Dz(t) \qquad (9.61)$$

$$u(t) = Cy(t)$$

Now the closed-loop system may be written as follows:

$$d\hat{x} = \hat{F}\hat{x}\ dt + \hat{G}(\hat{x})\ d\hat{w} - \hat{b}f(\gamma)\ dt \qquad (9.62)$$

where

$$\hat{G}(\hat{x}) = \begin{bmatrix} G(x) & | & 0 \\ \text{----} & | & \text{----} \\ 0 & | & DE \end{bmatrix}, \quad \hat{w} = \begin{bmatrix} w \\ v \end{bmatrix}$$

Theorem 9.5

Let the closed-loop process given in (9.62) satisfy the following conditions:

1. All of the eigenvalues of F have negative real parts.
2. $\gamma f(\gamma) > 0$ if $\gamma \neq 0$ and $f(0) = 0$; $f(\gamma)$ is continuously differentiable, and

$$\left| \frac{df(\gamma)}{d\gamma} \right| < K < \infty \quad -\infty < \gamma \propto$$

3. There exist two nonnegative constants α and β so that $\alpha + \beta > 0$ and

$$Re(\alpha + j\omega\beta)c^T(j\omega I - F)^{-1} b > 0$$

   for all real ω.
   The process (9.62) is recurrent.
   (This theorem is Theorem 6 in Ref. [9.13]).

The proof of this theorem is given in Ref. 9.22. The following example illustrates the method:

Let the plant be modeled by the scalar process

$$dx = fx \, dt - \tan^{-1}x + 2 \, dw + u \, dt$$

with the corrupted output

$$z = x + v$$

w and v are Gaussian white-noise processes.

The controller is as before

$$\dot{y} = ay + dz$$

with the input to the plant

$$u = y$$

The closed-loop system is described by the second-order process

$$\begin{bmatrix} dx \\ dy \end{bmatrix} = \begin{bmatrix} f & 1 \\ d & a \end{bmatrix} \begin{bmatrix} x \\ y \end{bmatrix} dt - \begin{bmatrix} 1 \\ 0 \end{bmatrix} \tan^{-1}x \, dt + \begin{bmatrix} 2 \\ d \end{bmatrix} dv$$

where for design consideration it was assumed $|f| < 1$.

In order to satisfy the two algebraic criteria of Theorem 9.5 the parameters of the controller must lie in the following regions:

$$f > 1 \quad \text{and} \quad d < -f(2 + f)$$

Because of the nature of this example the parameters are coupled; this is not a harmful result nor will it necessarily arise in any arbitrary plant and controller.

As mentioned before, this stability investigation was primarily developed for Saridis' expanding subinterval algorithm (Sec. 8.6) which requires the integrability of the performance criterion $I(u_c)$ used.

Theorems 9.1 to 9.5 guarantee that the actual trajectories (regardless of which sample path is chosen) remain finite. This then insures that the performance index is integrable; however, the proofs of Theorems 1 and 2 require existence conditions on the performance index $I(u_c)$.

The following theorem, proven by Kushner [9.14], gives conditions for boundedness of the performance index under the given choice of a stochastic Lyapunov function.

Theorem 9.6

Suppose that $V(x)$ is bounded for finite x, and that $x(t)$ does not have a finite escape time (with probability one) to ∞. Let

$$D[V(x)] \geq -L(x) + k \tag{9.63}$$

where $L(x) \geq 0$ and $0 < k < \infty$. Then

$$\lim_{T \to \infty} \int_{t_o}^{T} \frac{E_x L(x(s))}{k} \, ds \leq 1 \tag{9.64}$$

(This theorem is Theorem 6 in Ref. [9.14]).

In order to complete the conditions for the existence of the performance index, the interchange of the order of integration and expectation in (9.64) is necessary. However, the class of physical systems considered here is so that Fubini's theorem is satisfied, and this is accomplished.

Theorems similar to Theorem 9.6 have been proven also by Wonham [9.35] and Zakai [9.36].

This completes the presentation of the stochastic stability investigation of the S.O.C. systems. It has generated a method which allows the designer to insure the necessary stable behavior of the closed-loop plant, provided that certain information of the non-linearity and its location in the system is available. Such information is either roughly available or it is obtainable through a non-linear classification method. The stability within each class can then be studied independently, a project which is complex by itself. Since proof involving advanced mathematical procedure has been omitted, the interested reader is referred to Ref. 9.22 and the associated references.

9.7 DISCUSSION

As with all new methodologies proposed to solve a class of problems, a qualitative and quantitative evaluation has been presented for the S.O.C. systems. Their feasibility has been proven and even a relative ranking has been proposed, based on the simulation results presented and the experience acquired by the author and his colleagues in the past ten years.

However, the most general conclusion from this study can be summarized by stating that S.O.C.s are still at a very early developmental stage and a considerable amount of research work remains to be done to strengthen, unify, and improve the discipline to maturity. As a result, each algorithm discussed in this book may look more suitable to the particular problem it was originally invented for, and there are considerable difficulties to adapt it for more general problems.

Designers asked to use these algorithms voiced basic criticisms pertaining to the stability problem. Such questions were answered to a certain extent in this chapter but by no means exhausted the problem of an efficient stable operation of such a system. More work is needed to have satisfactory answers to the practicing engineer, who requests a well-tested, functional methodology with built-in safety tests to apply to his process. Such applications,

along with proposed investigations, are discussed in the next
chapter.

However, from the evaluation performed in this chapter one may
move with confidence into the formulation of more advanced intelligent systems, in the same way control scientists have moved from
classical control to parameter-adaptive to performance-adaptive S.O.C.
The next chapter will provide some of those excursions of the authors
imagination in the future of automatic control.

9.8 REFERENCES

[9.1] Bower, J. L., and Schultheiss, P. M., *Introduction to the Design of Servomechanisms*, Wiley, New York, 1958.

[9.2] D'Azzo, J. J., and Houpis, C. H., *Feedback Control Analysis and Synthesis*, 2d ed., McGraw-Hill, New York, 1966.

[9.3] Dynkin, E. B., *Markov Processes*, Springer-Verlag, New York, 1965.

[9.4] Fu, K. S., and Tou, J., Proceedings of Research Workshop on Learning System Theory and Its Practical Applications, sponsored by NSF, University of Florida, Gainesville, Fla., Oct. 18-20, 1973.

[9.5] Gibson, J. E., *Non-Linear Automatic Control*, McGraw-Hill, New York, 1962.

[9.6] Hahn, W., *Theory and Applications of Lyapunov's Direct Method*, Prentice-Hall, Englewood Cliffs, N. J., 1963.

[9.7] Horowitz, I. C., and Shaked, U., Superiority of Transfer Function over State-Variable Methods in Linear Time-Invariant Feedback System Design," *IEEE Trans. Automatic Control*, <u>AC-20</u> (1), 84-97 (1975).

[9.8] Jury, E., *Theory and Applications of z-Transform Method*, Wiley, New York, 1964.

[9.9] Kailath, T., Kozin, F., Bucy, R., and Wonham, W. M., *Stochastic Problems in Control*, ASME Publ., New York, June 1968.

[9.10] Kitahara, R. T., and Saridis, G. N., "Computational Aspects of Performance-Adaptive Self-Organizing Control Algorithms," *Proc. 1972 Conference on Decision and Control*, New Orleans, Dec. 13-15, 1972.

[9.11] Khas'minskii, R. Z., "Ergodic Properties of Recurrent Diffusion Processes and Stabilization of the Solution to the Cauchy Problem for Parabolic Equations," *Theory of Probability and Its Applications*, <u>5</u>, (2), 179-196, (1960).

[9.12] Kleinman, D. L., "On the Stability of Linear Stochastic Systems," *IEEE Trans. Automatic Control*, AC-14, (4), 429-430 (1969).

[9.13] Kushner, H. J., "The Concept of Invariant Set for Stochastic Dynamical Systems and Application to Stochastic Stability," in *Stochastic Optimization and Control* (H. F. Karrmen, ed.) Wiley, New York, 1968.

[9.14]   Kushner, H. J., *Stochastic Stability and Control*, Academic
         Press, New York, 1967.

[9.15]   Kushner, H. J., *Introduction to Stochastic Control*, Holt,
         Reinhart & Winston, New York, 1971.

[9.16]   Meyer, K. R., "Lyapunov Functions for the Problem of Luré,"
         Proc. Natl. Acad. Sci. US, (March 1965).

[9.17]   Sage, A. P., and Melsa, J. L., *Estimation Theory with
         Applications to Communications and Control*, McGraw-Hill,
         New York, 1971.

[9.18]   Riordon, J. S., "An Adaptive Automaton Controller for
         Discrete-Time Markov Processes," *Automatica - IFAC J.*, $\underline{5}$,
         721-730 (1959).

[9.19]   Saridis, G. N., "On a Class of Performance Adaptive SOC
         Systems," in *Pattern Recognition and Machine Learning*,
         (K. S. Fu, ed.), Plenum Press, New York, 1971.

[9.20]   Saridis, G. N., "Stochastic Approximation Methods for
         Identification and Control-A Survey," *IEEE Trans. Automatic
         Control*, $\underline{AC-19}$, (6), 798-809, (6), (1974).

[9.21]   Saridis, G. N., "Learning System Theory for General System
         Studies," Proc. 3d Annual Meeting of the Society of General
         System Research, University of Maryland, College Park, Md.,
         Sept. 1974.

[9.22]   Saridis, G. N., and Fensel, P. A., "Stability and Performance
         of Self-Organizing Control Systems," Tech. Rept. TR-EE-73-27,
         School of Electrical Engineering, Purdue University, Lafayette,
         Ind., Aug. 1973.

[9.23]   Saridis, G. N., and Fensel, P. A., "On the Stability of a
         Performance-Adaptive S.O. Algorithm Applied to Linear
         Stochastic Systems," Proc. 1972 JACC, Paper, No. TR-4,
         Stanford University, Stanford, Calif., June 1972.

[9.24]   Saridis, G. N., and Fensel, P. A., "Stability Consideration
         of the Expanding Subinterval Performance Adaptive S.O.C.
         Algorithm," *J. Cybern.*, $\underline{3}$, (2), 26-39 (1973).

[9.25]   Saridis, G. N., and Hofstadter, R. F., "A Pattern Recognition
         Approach to the Classification of Nonlinear Systems," *IEEE
         Trans. Systems Man Cybern.*, $\underline{SMC-4}$, (4), 362-371 (1974).

[9.26] Saridis, G. N., and Hofstadter, R. F., "A Pattern Recognition Approach to the Classification of Stochastic Non-Linear Systems," Tech. Rept. TR-EE-74-13, School of Electrical Engineering, Purdue University, Lafayette, Ind., June 1974.

[9.27] Saridis, G. N., and Kitahara, R. T., "Computation Aspects of Performance-Adaptive SOC Algorithms," Tech. Rept. TR-EE-71-41, School of Electrical Engineering, Purdue University, Lafayette, Ind., Jan. 1972.

[9.28] Saridis, G. N., and Lobbia, R. N., "Parameter-Adaptive Self-Organizing Control of Stochastic Linear Discrete-Time System," Tech. Rept. TR-EE-71-43, School of Electrical Engineering, Purdue University, Lafayette, Ind., April 1972.

[9.29] Saridis, G. N., and Lobbia, R. N., "Parameter Identification and Control of Linear Discrete-Time Systems," *IEEE Trans. Automatic Control*, 17, (1), 52-60 (1972).

[9.30] Saridis, G. N., and Lobbia, R. N., "A Note on Parameter Identification of Linear Discrete-Time Systems," *IEEE Trans. Automatic Control*, AC-20, (3), 40-59 (1975).

[9.31] Tsypkin, Ya. Z., *Adaptation and Learning in Automatic Systems*, translated by Z. J. Nikolic, Academic Press, New York, 1971.

[9.32] Tsypkin, Ya. Z., *Foundation of the Theory of Learning Systems*, translated by Z. J. Nikolic, Academic Press, New York, 1973.

[9.33] Wonham, W. M., "Lyapunov Criteria for Weak Stochastic Stability," *J. Differential Equations*, 2, 195-207 (1966).

[9.34] Wonham, W. M., "On Weak Stochastic Stability of Systems Perturbed by Noise," Tech. Rept. TR-65-4, Center for Dynamical Systems, Brown University, Providence, R. I., March 1965.

[9.35] Wonham, W. M., "A Lyapunov Method for the Estimation of Statistical Averages," *J. Differential Equations*, 2, 365-377 (1966).

[9.36] Zakai, M., "A Lyapunov Criterion for the Existence of Stationary Probability Distributions for Systems Perturbed by Noise," Society of Industrial and Applied Mathematics, *J. Control*, 7, (3) 390-397 (1969).

Chapter 10

THE FUTURE OF SELF-ORGANIZING CONTROL:
LEARNING AND HIERARCHICAL INTELLIGENT CONTROL SYSTEMS

10.1 INTRODUCTION

The S.O.C. algorithms analyzed and evaluated in the last three
chapters represent methodologies ranking very high in sophistication
in the hierarchy of control techniques. They have been designed to
treat only systems that require a high degree of sophistication in
the control, due mainly to the existing uncertainties of their
dynamics and their relation with the environment. As a matter of
fact, these algorithms have been developed as the ultimate in control
systems design, endowed with capabilities of conditional decision
making under unknown environmental circumstances. However, it must
be emphasized that this is by no means the final accomplishment,
but may be the beginning of a new generation of advanced controls.
It is a fact that the systems designed today cover a larger spectrum
than ever before and treat problems that require "intelligent"
capabilities and complex decision making in uncertain changing environments. Such areas of system research are in systems with a human
controller, man-machine interactive systems, artificial systems,
artificial intelligence, robotics, remote manipulation, industrial
automation, exploration in hazardous environments, learning prostheses,
and bionic systems. Furthermore, since the applications are not
limited to electromechanical powered systems, *intelligent* complex
decision making has been proposed for highly complicated societal,
economic, and other *large-scale systems* [10.62], [10.84], [10.100],
[10.128], [10.122].

In order to solve those problems that require control systems
with intelligent functions such as simultaneous utilization of a

446

memory, learning, or multilevel decision making in response to
"fuzzy" or qualitative commands, a new generation of controls is
needed with capabilities much more advanced than the existing ones.

How does one go about designing such "intelligent" controls?
The best way is to use the human brain as a guide, since it is still
the far most "intelligent" control available for analysis and investigation. Such an approach was taken in developing certain
"behavioral" S.O.C.s in Chap. 8, which represent a subclass of
systems defined as "learning systems" at the beginning of this book.
A further discussion of the state of the art of the "learning systems" will be given in the next section and a detailed bibliography
at the end of the chapter [10.30], [10.43], [10.45], [10.46], [10.141].

Learning systems may be considered as a class of systems
designed to reproduce lower-level "behavioral" functions of living
systems. As such they may be considered a special class of systems
belonging to a field characterized by researchers as "artificial
intelligence" [10.37], [10.101]. This field is very broad and covers
the intelligent behavior of a modern computer as paralleling the
intelligent behavior of human artificial systems that imitate the
intelligent function of biological systems called *bionic systems*,
and they may also be considered as belonging to the field of artificial intelligence covering the upper level in the hierarchy.
Bionic systems as control systems have been called *intelligent controls* [10.45] and have been applied to systems with human controllers,
man-machine interactive control systems such as remote manipulators,
etc., and autonomous robot systems [10.12], [10.40], [10.42], [10.501,
[10.92], [10.155]. These systems will be discussed as part of a hierarchical intelligent control. Fuzzy automata and linguistic methods
belonging to the highest level of intelligent systems are also of
major importance.

The fuzzy automaton [10.7], [10.170], [10.171], is a learning device
with decision-making capabilities that may function in uncertainly
defined situations and therefore qualifies as a multilevel control
organizer. A brief discussion of the fuzzy automaton will be given
in Sec. 10.4.

Finally, the linguistic approach that has been applied successfully to pattern-recognition problems [10.51], [10.60] will be discussed
as a possible decision maker for qualitative commands. Because of
their relation to stochastic and fuzzy automata, stochastic languages
may be related to other learning and intelligent controls and therefore serve in a hierarchical structure of intelligent decision making.

Since an intelligent multilevel decision maker with possible
applications to bionic systems, etc., is sought as the next step in
the control hierarchy, such a hierarchical control will be proposed.
It is based on the principle that hierarchical controls should use
decision makers of increasing degrees of sophistication which could
handle commands of a lower degree of certainty and precision as one
goes up the levels of the hierarchy. Such a control for bionic
prosthetic devices is proposed and discussed, comprising three levels
of coordinated hierarchical control. The lowest level should contain a self-organizing controller for the mechanical motion, which
should be coordinated by a fuzzy automaton at the next level, directed
by a linguistic decision maker at the top. Other applications are
also discussed.

Other developments, improvements, and applications are left to
the creativity of the reader.

10.2 LEARNING SYSTEMS

Learning systems have been defined in Chap. 1 and have been
discussed both in Chaps. 1 and 8 from the point of view of control
systems. However, learning systems have not been developed only to
be used as simple controllers in a system [10.46]. Such systems may
be designed to perform several functions as imitations of lower-
level functions of an intelligent being, such as pattern recognition,
information storage and retrieval, basic decision making in the
presence of incomplete information of a process operating in a random
environment, etc. (see Fig. 10.1). Since most of the methods are
intended to duplicate intelligent functions of the lowest level, e.g.,

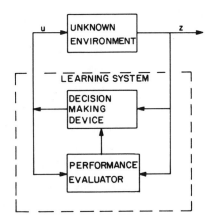

Fig. 10.1.   Block Diagram of a Learning System

the reflexes of an intelligent being, simple statistical successive
approximation or probabilistic sequential estimation methods,
studied in Chap. 6, are sufficient for their implementation [10.46].
In general, the theory of approximations has found application in
the study of learning systems, where various assumptions are made
about the knowledge of the process, and the appropriate information
regarding the performance of the system is learned via various analytic methods.   The most popular mathematical schemes used for the
study of learning process are summarized as follows [10.44], [10.45]:
1.   Trainable systems using pattern classifiers.
2.   Reinforcement learning systems.
3.   Bayesian estimation.
4.   Stochastic approximation.
5.   Stochastic automata models.
6.   Fuzzy automata models.
7.   Linguistic methods.
   Most of these methods have been discussed in a different form
in various parts of this book and should be familiar to the reader.
The last two, which will be discussed later, are exceptions.

The fairly detailed bibliography at the end of this chapter
provides sufficient information regarding the analytic description
and application of such algorithms. Since the purpose of this
chapter is only to suggest future developments, a detailed description of each of these schemes is omitted. However, the first five
methods have been used in Chap. 8 to derive performance-adaptive
S.O.C. systems. Therefore, it may be recommended that such schemes
be used in the lowest level of a hierarchically "intelligent" system,
because of their capabilities of easily handling basic forms of
learning. The last two methods, namely, the fuzzy automaton and
the linguistic methods, provide a higher level of intelligent capabilities in the processing of commands and decision making, and are
therefore recommended for higher levels of hierarchical controls.
For this reason, their potential as higher-level controllers is
discussed separately in the sequel.

Finally, learning system theory has been proposed to serve as
the missing dimension in the analytic treatment of problems arising
in the so-called "soft sciences," treating societal, environmental,
political and other systems [10.122]. In spite of the controversial
aspect of such an approach and the marginal success of analytical
solutions, a considerable research effort is still in progress in
this field [10.84], [10.100], [10.121]. Learning system theory may
serve to represent and evaluate the social, economic and other values
in the system which are not easily described by common analytic
methods.

10.3 BIONICS AND THE INTELLIGENT SYSTEM

The name "cybernetics" was first used by Wiener [10.159] to
characterize the field dealing with the interaction of man and
machine. Even though the name is still quite popular with some researchers, the area has been divided into various subareas with
emphasis on the specific use of the system under consideration.
"Artificial intelligence" has been developed to investigate the

intelligent behavior of a digital computer while applied to recognize objects, prove theorems, play games, or drive robots.   The
emphasis here is on the digital computer and its enormous programming
capabilities which may be trained to perform intelligent functions
almost like a human being.   For more details, see references [10.37],
[10.70], [10.71], [10.101].

10.3.1   Bionic Systems

   Bionic systems are defined as those which duplicate the functions of biological systems at all levels of intelligence, and therefore present a significant overlap with artificial intelligent systems [10.101].   However, bionic systems are usually designed to
perform a complete function such as artificial limb, and therefore
require several levels of controls from the actual mechanical hardware to the computer control interfaced with the amputee.   Such
systems are highly complex and require a hierarchical control
design, and one is proposed in the next section.   One of the first
existing bionic systems are the probability state variable (P.S.V.)
devices which have been investigated by Adaptronics, Inc. and have
been reported in Ref. [10.10].   The controller is composed of a
large network of components called *neurotrons* systematically or
randomly connected which can adjust their function during the operation as a result of a search for the improvement of the performance
of the system.   The probability state variable devices, and in
particular the *neurotron*, belong to the class of components that can
be trained to learn certain functions.   The *neurotron* is a two-
input/one-output digital device which can accomplish both digital
and analog operations on pulse-encoded signals.   The specific function of each of the P.S.V. devices is controlled by a random sequence
properly biased by a reward or penalty input guided by the performance evaluator.   A typical *neurotron* is depicted in Fig. 10.2.   The
interaction between P.S.V. devices and the function that they accomplish is thus governed by the improvement of the performance of the
system.   Although the idea is certainly along the lines of the

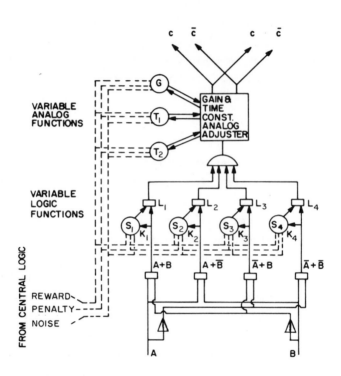

Fig. 10.2.   An OR Neurotron, Element of Probability State
             Variable Device

definition of bionic controls, the method of implementing it was
never studied sufficiently to result in an acceptable technique.
Adaptronics concentrated more on the construction of very fast P.S.V.
devices coordinated in a rather heuristic manner. This fact and the
problem of overdesigning P.S.V. controllers for relatively simple
systems not requiring a bionic approach made the method rather
unpopular.

Other bionic devices designed mainly for medical purposes comprise artificial arms and legs, artificial tactile vision as well as
wheelchairs etc., for the disabled [10.14], [10.65], [10.81], [10.92],
[10.127]. Many of those are controlled through electromyographic
signals from muscles located on the shoulder or other intact muscles
of the body [10.62], [10.127].

Together with bionic systems one may classify the simulation of the human operator, performed to analyze and imitate the function of a human when it performs certain simple tasks. The models obtained up to now are simplistic and limited to only very specific easy tasks. One of the problems in their concept is that they use rather low-order models to simulate the function of a highly sophisticated and high-dimensional system, changing the performance criterion whenever the task is changed. This has not been found very effective for modeling.

The most advanced simulation of a human subject who operates a joy stick to locate a target, has been developed by Gilstad and Fu and is described in detail in Ref. [10.64]. A block diagram of this model is given in Fig. 10.3

10.3.2 Man-Machine Control Systems

Man-machine control systems have been mainly developed to initiate human functions such as handling and manipulative operation in remote or hazardous environments, e.g., the Mars-rover and the

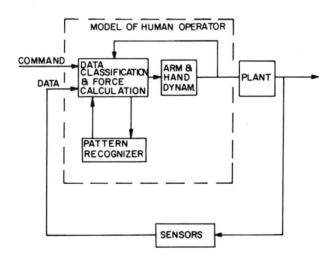

Fig. 10.3. Block Diagram of Gilstad-Fu Model for Manual Control System

radioactive materials operator, respectively [10.12].   The machine
part of the controller is relatively autonomous and therefore accepts
trainability, some learning, and some autonomous decision making for
elementary situations.   In other words, simple control tasks of an
adaptive or learning nature may be performed by the machine represented by a computer, while more complicated tasks such as advanced
decision making, coordination of motion, etc., are performed by the
human operator.   Through its learning capabilities the machine will
gradually replace some of the decision making performed by man.   The
function of such a system is given in the block diagram of Fig. 10.4.
The on-board remote computer serves as the short-range goal feedback
controller for the plant-environment combinations composed by the
manipulator and its site of operation.   The computer located in the
control headquarters is larger and has learning capabilities for
long-range decision making under the supervision and the training of
the human operator in the loop.

Various formulations of the remote manipulation have been proposed; the following are the most successful: Whitney [10.55] has
recommended a state space approach to control the plant-environment
combination of Fig. 10.4.   A state vector is used that contains the

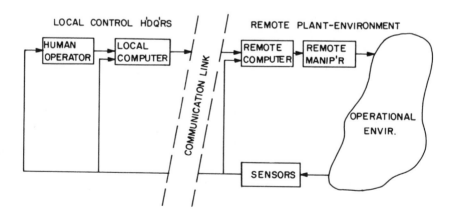

Fig. 10.4.   Block Diagram of a Man-Machine Controller Remote
            Manipulator

manipulator variables and important environmental parameters defined
over a quantized state space, representing different control situations in the same sense as in Sec. 8.4.2. An optimal control policy
is sought to move the manipulator through various control situations
until it reaches a predetermined state at minimum cost. A dynamic
programming type of procedure is employed for this purpose.

Freedy, Lyman and their colleagues [10.39], [10.40], [10.41],
used learning techniques to implement a man-machine controller for
the remote manipulation system of Fig. 10.4. They utilized the concept features of a trainable control network to observe the operators
control actions, learn the task at hand, and gradually take over a
part of the control action originally applied by the operator. The
computers in the loops serve as the interface between the operator
and manipulator as well as the intelligent control which learns the
various tasks and participates actively in the organization and
decision making. A discretization of the state space to define the
control situations is also performed and the transition from one
situation to another is performed through a trainable conditional
probability transition matrix which coordinates the elementary sub-
tasks involved. Training takes place when the operator makes the
original moves by reinforcing the appropriate elements of the matrix.
The total control is subdivided into two separate functions, the
trajectory control with its learning control network located on the
on board computer, and the sensor control with its learning control
network located in the local computer. The control commands for
short-range actions are generated by the two networks and supervised
and trained by the operator. Overall performance improvement is
heuristically evaluated by the operator's experience while the
short-range performance is improved by the reduction of the probability of errors. This manipulator has been built with a considerable
degree of success.

Saridis [10.123] has proposed an optimal control formulation of
the motion of the manipulator with trainable performance criterion
to accomplish nearly anthropomorphic movements. A nonlinear controller is structurally designed on the on-board computer as an

approximation to control the nonlinear plant with adjustable gains
controlled from the local computer of Fig. 10.4. This latter computer
produces, along with the learned gains, the terminal position of the
manipulator and its final task. The conditions of level coordination
and of avoiding various objects are imposed as constraints on the
state variables of the control system. Only simulations are at
present available on this study.

Several more studies on manipulator systems will be found in the
bibliography at the end of this chapter.

10.3.3 Autonomous Robots

Autonomous robots are considered to be all-purpose intelligent
systems endowed with anthropomorphic functions which operate in an
environment without the assistance of an on-line human operator, and
improve their performance by interval learning without external
supervision.

It is clear that this is the most ambitious project to be
undertaken by modern engineering, since it involves the construction
of an efficient structure capable of performing anthropomorphic
functions. Such a device will comprise the design of lower-level
intelligence controls to collect data from sensors and perform
routine decisions and on-line computation for a reflex type of
tasks. Then the realization of higher-level intelligence controls
will replace the human operator's intelligence of the previous systems to guide and train the robot to perform tasks that it is asked
to perform through some kind of qualitative set of commands [10.61],
[10.102], [10.103], [10.111], [10.119], [10.45], [10.132], [10.154].

A considerable amount of research has been performed in developing the individual components of such a system such as (1) perception,
(2) modeling of the environment, (3) advanced decision making (problem solving), and (4) efficient locomotion [10.42], [10.70], [10.71],
[10.121], [10.149]. However, the assembled robots that have been produced up to now are rather primitive, naive, bulky, slow, and clumsy,
to say the least [10.42], [10.61], [10.119]. Scientists at the

Stanford Research Institute, where in addition to MIT and Stanford University major research efforts have been made, indicate that it will take another 50 years before an efficient autonomous robot is constructed.

For information purposes, robot systems built by the Stanford Research Institute laboratories (affectionately called "Shaky" for obvious reasons) will be briefly discussed here. A block diagram representing its major functions is given in Fig. 10.5. The robot vehicle is propelled by two wheels, driven independently by stepping motors with local control loops and protection devices. It carries one vidicon TV camera with an optical range finder and a control computer which channels the commands from the major computer down to the appropriate actuators for locomotion and sends back telemetry and video information through two radio communication links. The major computer performs the intelligent control functions on two levels: the lower one handles the modeling of the new environment

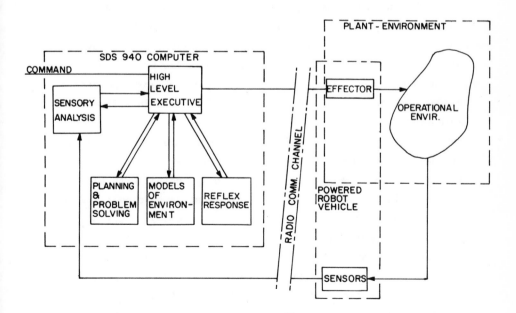

Fig. 10.5.   Block Diagram of the Stanford Research Institute
            Robot System

from sensory information acquired, and plans and solves the problem
of routing the vehicle through the new environment, while the reflex
response box assigns the lower-level control commands for the vehicle
in parallel with the other functions. The high-level executive in-
terprets the outside commands received into a task, surveys the scene
by performing an appropriate scene analysis and object recognition
through the process of the sensory data acquired, and assigns proper
investigations to the lower levels. Then, after a plan of action is
worked out by the lower-level, it assigns a sequence of elementary
actions to the robot with the purpose of carrying out the prespecified command. The executive also evaluates performance of the system through the sensory feedback, in order to improve the future
operation in cooperation with the modeling function. As the environment changes, the model is revised accordingly. A typical example of
the robots functions is to enter a room and rearrange simple-shaped
objects lying on the floor. The robot should locate the proper
door, enter without bumping the wall or other objects, identify
the objects to be rearranged by their shape, approach them from the
right angle, and push them to the right location. It is obvious
that such a machine is far from resembling a living system in performing simple functions. It does not even look anthropomorphic,
being a cart on wheels. In order to perform more intelligent tasks,
improvements in the hardware and the software design are necessary.
Sophisticated pattern-recognition techniques are required to improve
the analysis of complex environment, and faster and larger computers
are required for fast modeling and action planning, with additional
links which increase the degree of complication. However, "Shaky"
is a step in the direction of designing autonomous robots.

Other attempts to design such robots in the United States and
in other countries have been reported in the literature and can be
found in the bibliography at the end of this chapter. Most of them
suffer from hardware shortcomings, like sluggishness and clumsiness
or limitation in the functions, but all of them suffer from the
limitation of the intelligent higher-level decision maker and

coordinator.   This latter shortcoming can be improved only by progress
in computer technology.

10.4 A HIERARCHICALLY INTELLIGENT CONTROL

The systems discussed in the preceding section represent
attempts to design highly intelligent control systems, endowed with
problem solving, planning, cognitive and other capabilities characteristic only of the human operator.   However, even with the current
advancements of technology, such intelligent controls are not yet
capable of duplicating the human brain in capacity, versatility, and
intelligent functions.   Therefore, it is the author's feeling that
systems with a human brain as the supreme commander would stand a
better chance of representing a good design as shown by increased
capabilities, versatility, and intelligent operation.

A design for a p-degree of freedom bionic arm with a hierarchically intelligent controller, supervised and commanded by the brain
has been the work of the author and his colleagues to verify this
claim and establish the feasibility of such a control system [10.123]
[10.127].   The purpose is to construct a prosthetic upper limb for
amputated persons or hazardous environment use with anthropomorphic
characteristics of motion which would receive commands directly from
the brain of the operator either through a direct tapping of the
nervous system or through electromyographic signals or any other
source of qualitative commands, resulting from the operator's survey of the environment.   It is conjectured that such an anthropomorphic function of the arm, in response to crude qualitative commands, can be obtained only by a hierarchical, say, three-level control, with hierarchically increasing order of intelligence and
hierarchically decreasing order of precision of control signals as
one moves from the electromechanical actuators of the arm to the
interface with brain.   A block diagram in Fig. 10.6 illustrates the
concept of such a hierarchically intelligent control system.   The
three levels of control are (1) a linguistic organizer, (2) a fuzzy

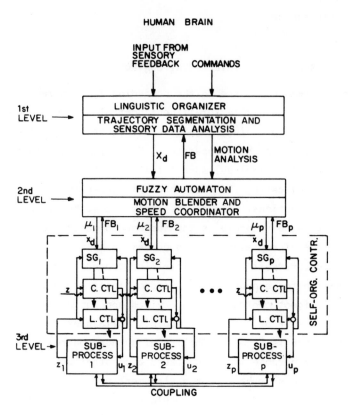

Fig. 10.6. A Hierarchically Intelligent Control for a Bionic
p-Degree-of-Freedom Arm

automaton, and (3) a bank of self-organizing controls. They are
described briefly below along with their function in the control
system.

10.4.1 The Self-Organizing Control Level

The arm as a whole process may in general be subdivided into
p subprocesses, one per degree of freedom, described by the following set of generalized differential equations:

$$\dot{x}_i = F_i x_i + B_i(\bar{x}_i)u_i + f_i(x,\dot{x},w_i)$$

$$z_i = H_i x_i + v_i \qquad\qquad i = 1, 2,\ldots,p \qquad (10.1)$$

where x_i is the n_i-dimensional state vector, z_i is the r_i-dimensional output vector, and u_i is the m_i-dimensional control vector of the ith subprocess; F_i, $B_i(\bar{x}_i)$, and H_i are matrices of appropriate dimension, w_i and v_i appropriate noise vectors, and $f_i(\cdot)$ nonlinear functions representing the gravity influence and the coupling terms from other subsystems through various reaction forces [10.127]. If the subsystem was isolated from those force fields, these terms would be zero and the system would be linear in x_i. Furthermore, the state of the overall system would be

$$x^T = [x_1^T,\ldots,x_p^T] \quad \bar{x}_i^T = [x_1^T,\ldots,x_{i-1}^T, x_{i+1}^T,\ldots,x_p^T]$$

$$z^T = [z_1^T,\ldots,z_p^T] \qquad\qquad (10.2)$$

A performance criterion for the proper mechanical function of the system may be defined as

$$J = \sum_{i=1}^{p} \mu_i J_i(\alpha_i) = \sum_{i=1}^{p} \mu_i \lim_{T\to\infty} \frac{1}{T}\left\{\int_{t_0}^{T} (z_i(t) - x_i^d(t))^T M_i(\alpha_i)\right.$$

$$\left.\cdot (z_i(t) - x_i^d(t))\ dt + \int_{t_0}^{T} u_i^T(t)N_i(\bar{x}_i)u_i(t)\ dt\right\} \qquad (10.3)$$

where $x_i^d(T)$ is the vector of desired final states, $M_i(\alpha_i)$ and $N_i(\bar{x}_i)$ are diagonal matrices, α_i are adjustable coefficients relative to the speed of response of the subprocesses, and μ_i are adjustable coefficients $0 \leq \mu_i \leq 1$, $i = 1,\ldots,p$, relative to the appropriate blending of primitive motions defined by the individual subprocesses to generate an appropriate compound motion for the arm. Obviously $J_i(\alpha_i)$ is the performance criterion for each subprocess, assumed to be of infinite duration for simplicity of implementation, a conjecture verified experimentally. A feedback control may be structured per subsystem as

$$u_i(z) = K_i \phi(z_i) + C_i \psi(z) \tag{10.4}$$

where the first part is the optimal control (O.C.) for the uncoupled
subsystem, while the second term represents a nonlinear term depending on the nonlinear coupling with the other subsystems. The expanding subinterval algorithm of Sec. 8.6 may be applied to yield the
asymptotically optimal coefficients K_i and C_i for each subprocess,
thus creating performance-adaptive S.O.C.s for the lowest level of
the hierarchically intelligent control system. Self-organization is
necessary because of the complexity of the process involved.

The higher levels are interfaced with the individual self-
organizing controls through the adjustable relative speed coefficients
α_i, the blending coefficients μ_i, and the desired final states $x_i^d(T)$.

From the preceding discussion, it is obvious that the third-
level S.O.C.s are designed for precision control of the mechanical
subprocesses at hand but do not exhibit higher-quality intelligent
functions, such as intelligent decision making for motion coordination, direction, and goal accomplishment. With their limited capabilities they resemble more the reflexes of a biological system.
The higher intelligent functions are hierarchically distributed to
the higher levels of the controls which are described next.

10.4.2 The Fuzzy Automaton as a Control Coordinator

A *fuzzy automaton* has been presented [10.52], [10.171] as an ex-
tension of the variable structure stochastic automata discussed in
Sec. 8.4.3 to accommodate inputs of a "fuzzy" nature defined by
Zadeh [10.168], [10.170].

A fuzzy quantity, belongs to a set of values describing an
"object" with undefined boundaries that are characterized individually by a membership coefficient to the set [10.170].

A fuzzy automaton is therefore defined as a sixtuple $(Z,Q,U.F,$
$H,\zeta)$, where, in addition to the finite set of "fuzzy" inputs $Z = \{z\}$,
finite set of states $Q = \{q\}$, finite set of outputs $U = \{u\}$, state
transition function F, and output function H, a fuzzy membership

vector ζ is assigned to the states of the automaton [10.7], [10.153], [10.171]. A membership transition matrix $\Xi^k(n) = \left\{ \xi_n^k(n); \ \sum_{i=1}^{r} \xi_{ij}(n) = 1 \right\}$ for each fuzzy input z^k may be assigned to update the membership functions ζ in the same way as the state transition probability matrix $T^k(n)$ in Sec. 8.4.3.

Such a fuzzy automaton may be used to ccordinate the p-primitive motions of the subprocess to form a desired compound movement of the arm from an initial state $x_0(t_0)$ to a final predefined state $x^d(T)$. Such a coordination is needed to put together the right amount of primitive motions and their proper velocities in order to accomplish the proper compound motion in response to a fuzzy command of the higher-level intelligent controller which is not to be bothered with the details of coordination. A *fuzzy automaton* (C,Q,Q,F,I,ζ) has a structure natural for such a coordinator, if $C = \{c\}$ is the set of fuzzy command inputs transmitted from the organizer, $Q = \{q\}$ is the set of the states as well as outputs of the automaton representing the motions and relative velocities of the subprocesses with membership coefficients $\zeta^T = \{\mu_1,\ldots,\mu_p,\alpha_1,\ldots,\alpha_p\}$.

The coefficients ζ will be trained by, say a reinforcement algorithm, in the automaton to generate a proper combination of primitive motions and velocities for each fuzzy command input corresponding to compound motion, by producing different gains K and C in the S.O.C. (10.4) by affecting the expanding subinterval subgoals generated from the performance criterion (10.3). A memory file is provided to store the learned ζ^k for each command c^k for future use.

10.4.3   Linguistic Methods for Intelligent Organization

Based on the preceding discussion, given a command and a terminal state, the fuzzy automaton can be trained to produce the proper compound motion for the arm, which represents a level of control more intelligent than the self-organizing. However, the commands generated from the brain of the human operator are more in the form of compound tasks, like picking up a glass of water to drink.

Therefore an intelligent control system is needed to interface the
brain with the fuzzy automaton and translate the above qualitative
command to a sequence of compound motions of the area that will
accomplish the task. Such a control will be required to produce a
segmentation of the task providing appropriate $x^d(T)$ and qualitative
information of the compound motion of the arm for each segment. It
should produce <u>on-line</u> information about the change of direction,
combination, or expansion of segments, evaluation of the accomplishment of the task and processing of sensory feedback information from
the brain, etc., without burdening the operator with unnecessary
details about the function of the arm.

A machine producing decisions and functions of such a high
level of intelligence must be an advanced digital computer, capable
of processing qualitative information of high content, but also of
fuzzy nature in the sense that high precision in execution is not
required. A natural system for this type of information processing
is the linguistic methods approach which has been developed in the
modern literature for artificial intelligence, pattern-recognition,
scene analysis, and other functions [10.51], [10.60], [10.83], [10.117].
Such methods process strings of words with logic instruction to
accomplish the task according to certain predetermined grammar and
syntax in manner similar to natural languages. In particular,
stochastic grammars developed by Fu and Swain [10.60] for syntactic
pattern recognition or fuzzy grammars proposed by Zadeh [10.170]
are most desirable for a generation of the command strings appropriate to organize the motion of the artificial arm.

Since a detailed discussion of this subject is beyond the scope
of this book, and only preliminary results have yet been obtained
at this level of design of the intelligent control of the bionic
arm, the reader is referred to Ref. [10.127] for additional information.

However, through the example of the control of the bionic arm,
the concept of a hierarchical intelligent man-machine interactive
control system has been proposed and its feasibility established.
Generalization to other man-machine interactive systems or even

autonomous robots should be straightforward and would be one of the areas of future research in control systems.

10.5 OTHER APPLICATIONS OF SELF-ORGANIZING CONTROLS

This chapter has been devoted mostly to the outlook for S.O.C. systems with emphasis on the future developments and their applications. However, before closing the chapter it is only appropriate to emphasize the attempted applications of S.O.C.s to technological systems to improve their credibility and change the minds of possible nonbelievers.

The applications of S.O.C.s to bionic, prosthetic, man-machine manipulator and robotic systems have been already discussed in the previous chapters. Other similar applications can be found in the bibliography at the end of this chapter.

It was communicated to the author by Stein [10.137] that the first fully parameter-adaptive S.O.C. aircraft autopilot has been designed and is being tested by Honeywell, Inc. In addition, feasibility studies of a parameter-adaptive S.O.C. for a booster-stage attitude control problem for a space vehicle designed by Boeing have been reported by Saridis and Lobbia [10.126]. A feasibility study of a performance-adaptive self-organizing attitude control of an orbiting satellite with random jet firing, using the expanding subinterval approach has also been reported by Saridis and Gilbert [10.125].

Studies of applications of reinforcement having control techniques to space-craft control have been reported by Mendel [10.97]. Other applications of self-organizing and learning control can be found in industrial processes and control [10.8], [10.63], [10.105].

In conclusion, self-organizing control is already in the process of being tested by the various technological sectors of industry with successful results. Therefore the whole discipline has moved out of the realm of pure research into research and development as more applications are reported.

10.6 DISCUSSION AND SUGGESTIONS FOR FUTURE RESEARCH

In concluding the presentation of the stochastic S.O.C. system
a summary and evaluation of the work is felt appropriate. Venturing
into an area of research for which only fragmentany articles were
available, an attempt was made to unify and consolidate the methods
that are treating the control problem of dynamic systems with completely or partially unknown dynamics operating in a stochastic
environment. For this purpose, the need to study of such systems
was discussed in Part I of this book and background material in
estimation theory, stochastic and dual optimal control, and
parameter identification necessary for the understanding of S.O.C.s
was presented in Part II. Part III was devoted to the development
of various self-organizing methods, their comparisons, applications,
and their natural future extensions into the areas of learning and
intelligent controls.

The presentation is not claimed to be complete and the reader
may find more questions in the text than answers.

Considerable research effort is still needed in the areas of
stability investigation, structural choice of a feedback self-
organizing controller, space discretization, acceleration of convergence, etc., of the existing algorithms. On the other hand, new
more efficient S.O.C. algorithms, which will reduce the amount of
computations and the size of the computer, are more than welcome.
Such new algorithms as well as improvements of the existing ones
would result in a miniaturization of the hardware and produce
feasible on-board controllers for among other applications the bionic
systems for which small size and weight are essential.

Looking into other areas outside of the hard sciences, self-
organizing, learning and hierarchically intelligent controls may
provide a vehicle of including the missing dimension of values in
the study by system methods of societal, environmental, judicial,
urban economic, and other such processes with the purpose of prediction, analysis, and control for the benefit of humanity. Self-
organizing controls have been developed as a first step in a ladder

of technological development utilizing a new scientific approach,
and have a long way to go.

10.7 REFERENCES AND BIBLIOGRAPHY

[10.1]   Ackerson, G. A., and Fu, K. S., "On State Estimation in Switching Environment," *IEEE Trans. Automatic Control,* AC-15, (1) 10-17 (1970).

[10.2]   Agrawala, A. K., "Learning with a Probabilistic Teacher," *IEEE Trans. Inform. Theory,* IT-16, (4) 373-378 (1970).

[10.3]   Aiserman, M. A., Braverman, E. H., and Rozonoer, L. I., , *Potential Function Method in the Problems of Machine Learning,* Nauka, Moscow, 1970, (in Russian).

[10.4]   Aiserman, M. A., Braverman, E. H., and Rozonoer, L. I., "Theoretical Foundations of the Potential Function Method in Pattern Recognition," *Avtomatika Telemekhanika,* 25, 917-936, (1964).

[10.5]   Aiserman, M. A., "The Method of Potential Functions for the Problemsof Restoring the Characteristic of a Function Converter from Randomly Observed Points," *Avtomatika Telemekhanika,* 25, 1705-2213 (1964).

[10.6]   Aiserman, M. A., "The Probability Problem of Pattern Recognition Learning and the Method of Potential Functions," *Avtomatika Telemekhanika,* 25, 1307-1323 (1964).

[10.7]   Asai, K., and Kitajima, S., "A Method for Optimizing Control of Multimodal Systems Using Fuzzy Automata," *Inform. Sci.,* 39, (4), 343-353 (1971).

[10.8]   Åström, K. J., and Wittenmark, B., "Control of Constant But Unknown Systems," Fifth World Congress of IFAC, Paris, June 12-17, 1972.

[10.9]   Atkinson, R. C., Bower, G. H., and Crothers, E. J., *An Introduction to Mathematical Learning Theory,* Wiley, New York, 1965.

[10.10] Barron, R., "Self-Organizing Control," *Control Engineering,* Part I, 15, (2), 70-74, Part II 15, (3) 69-74, (1968).

[10.11] Bar-Shalom, Y., Larson, R. E., and Grossbery, M. A., "Application of Stochastic Control Theory to Resource Allocation Under Uncertainty," *IEEE Trans. Automatic Control,* AC-19, 1-7 (1974).

[10.12] Bejczy, A. K., "Remote Manipulator System Technology Review," Tech. Rep. No. 760-77, Jet Propulsion Lab., Pasadena, Calif., July 1972.

[10.13] Bejczy, A. K., "New Techniques for Terminal Phase Control of Manipulators," Techn. Rep. 760-98, Jet Propulsion Lab., Pasadena, Calif. Feb. 1974.

[10.14] Bejczy, A. K., "Robot Arm Dynamics and Control," Tech. Rep., 33-669, Jet Propulsion Lab., Pasadena, Calif., Feb. 1970.

[10.15] Braverman, D., and Abramson, N., "Learning to Recognize Patterns in a Random Environment," *IRE Trans. Inform. Theory,* IT-8, (5) 58-63, (1962).

[10.16] Bruce, G. D., and Fu, K. S., "A Model for Finite-State Probabilistic Systems," *Proc. First Allerton Conference on Circuit and System Theory,* University of Michigan Press, Ann Arbor, Michigan, 1963.

[10.17] Bush, R. R., and Mosteller, F., *Stochastic Models for Learning Theory,* Wiley, New York, 1955.

[10.18] Bush, R. R., and Estes, W. K. (eds.), *Studies in Mathematical Learning Theory,* Stanford University Press, Stanford, Calif., 1959.

[10.19] Butz, A. R., "Learning Bang-Bang Regulators," Proc. Hawaii International Conference on System Sciences, Jan. 1968.

[10.20] Butz, A. R., "Perceptron Type Learning Algorithms in Nonseparable Situations," *J. Math. Anal. Appl.,* 17, 560-576, (1967).

[10.21] Chandrasekaran, B., and Shen, D. W. C., "Adaptation of Stochastic Automata in Non-Stantionary Environments," *Proc. Natl. Electron. Conf.,* 23, 39-45, (1967).

[10.22] Chandrasekaran, B., and Shen, D. W. C., "On Expediency and Convergence in Variable Structure Automata," *IEEE Trans. Systems Systems Sci. Cybern.,* SSC-4, (1) 52-60 (1968).

[10.23] Chang, S. S. L., and Zadeh, L. A., "Fuzzy Mapping and Control," *IEEE Trans. Systems Sci. Cybern.,* SMC-2, 30-42 (1972).

[10.24] Chaplin, W. G., and Levadi, V. S., "A Generalization of the Linear Threshold Decision Algorithm to Multiple Classes," in *Computer and Information Sciences,* II (J. T. Tou, ed.) Academic Press, New York, 1967.

[10.25] Chien, Y. T., and Fu, K. S., "Stochastic Learning of Time-Varying Parameters in Random Environment," *IEEE Trans. Systems Sci. Cybern.,* SSC-5, (3) 237-245 (1969).

[10.26] Chien, Y. T., and Fu, K. S., "On Bayesian Learning and Stochastic Approximation," *IEEE Trans. Systems Sci. Cybern.*, SSC-3, (1) 28-38 (1967).

[10.27] Cockrell, L. D., and Fu, K. S., "On Search Techniques in Switching Environment," *Proc. 9th Symposium on Adaptive Processes*, Austin, Texas, Dec. 7-10, 1970.

[10.28] Cooper, D. B., "Adaptive Pattern Recognition and Signal Detection Using Stochastic Approximation," *IEEE Trans. Electr. Computers*, EC-13, (3) 306-307, (1964).

[10.29] Cooper, P. W., "Some Topics on Non-supervised Adaptive Detection for Multivariate Normal Distributions," in *Computer and Information Sciences*, II, (J. T. Tou, ed.) Academic Press, New York, 1967.

[10.30] Cover, T. M., and Hellman, M. E., "The Two-Armed-Bandit Problem with Time-Invariant Finite Memory," *IEEE Trans. Inform. Theory*, IT-16, (1970).

[10.31] Cunningham, D. R., and Breipohl, A. M., "Empirical Bayesian Learning," *IEEE Trans. Systems Man Cybern.*, SMC-1, (1) 19-23 (1971).

[10.32] DeFigueiredo, R., "Convergent Algorithms for Pattern Recognition in Nonlinearly Evolving Nonstationary Environment," *Proc. IEEE,* 56 (2) 188-189 (1968).

[10.33] Dorofeyuk, A. A., "Algorithms of Teaching the Machine the Pattern Recognition Without Teacher Based on the Method of Potential Function," *Automatikai Telemekhanika,* 27, (10), (1966).

[10.34] Drimi, M., and Hans, O., "On Experience Theory Problems," Proc. Second Prague Conference on Information Theory, Statistical Decision Functions and Random Processes, 1959.

[10.35] Duda, R. O., and Fossum, H., "Pattern Classification by Iteratively Determined Linear and Piecewise Linear Discriminant Functions," *IEEE Trans. Electr. Computers,* EC-15, 220-232 (1966).

[10.36] Duda, R. O., and Singleton, R. C., "Training a Threshold Logic Unit with Imperfectly Classified Patterns," Presented at the WESCON (Western Electric Show and Convention), Los Angeles, Calif., Aug. 1964.

[10.37] Fergenbaum, E. A. and Feldman J. (eds.), *Computer and Thought,* McGraw-Hill, New York, 1963.

[10.38] Fralick, S. C., "Learning to Recognize a Pattern Without a Teacher," *IEEE Trans. Inform. Theory,* <u>IT-13</u>, (1) 57-65 (1967).

[10.39] Freedy, A., Hull, F. C., Lucaccini, L. F., and Lyman, J., "A Computer-Based Learning System for Remote Manipulator Control," *IEEE Trans. Systems Man Cybern.,* <u>SMC-1</u> (4), 356-364, (1971).

[10.40] Freedy, A., Welthman, G., and Lyman, J., "Learning Control Systems Using Computers with Application to Remote Manipulation," Proc. 1972 IEEE Conference on Decision and Control, New Orleans, Dec. 13-15, 1972.

[10.41] Freedy, A., and Welthman, G., "A Prototype Learning System as a Potential Controller for Industrial Robot Arms," Proc. Second International Symposium on Industrial Robots, Chicago, Ill., May 16-18, 1972.

[10.42] Firschein, O., Fischier, M., Coles, S. L., and Tenenbaum, J. H., "Forecasting and Assessing the Impact of Artificial Intelligence on Society," Techn. Rep., Stanford Research Institute, Menlo Park, Calif., Feb. 1973.

[10.43] Fu, K. S., "Learning Control Systems," in *Advances in Information Systems Science,* (J. T. Tou, ed.) 1969, New York, Plenum Press.

[10.44] Fu, K. S., "Learning Control Systems - Review and Outlook," *IEEE Trans. Automatic Control,* <u>AC-15</u>, (2), 210-221 (1970).

[10.45] Fu, K. S., "Learning Control Systems and Intelligent Control Systems: An Intersection of Artificial Intelligence and Automatic Control," *IEEE Trans. Automatic Control,* <u>AC-16</u>, (1) 70-72 (1971).

[10.46] Fu, K. S., "Learning System Theory," Chap. 11, in *Systems Theory,* (L. A. Zadeh and E. Polak, eds.) McGraw-Hill, New York, 1969.

[10.47] Fu, K. S. (ed.), *Pattern Recognition and Machine Learning,* Plenum Press, New York, 1971.

[10.48] Fu, K. S., *Sequential Methods in Pattern Recognition and Machine Learning,* Academic Press, New York, 1968.

[10.49] Fu, K. S., *Syntactic Methods in Pattern Recognition,* Academic Press, New York, 1974.

[10.50] Fu, K. S. (ed.), *Learning Systems*, Chap. 1: J. M. Mendel,
        "Reinforcement Learning Models and Their Application to
        Control Problems"; Chap. 2: G. N. Saridis, "On-Line Learning
        Control Algorithms"; Chap. 3: A. Freedy, "Learning Control
        in Remote Manipulator and Robot Systems"; K. S. Fu,
        "Bibliography," ASME Publ., New York, 1973.

[10.51] Fu, K. S., "Stochastic Automata, Stochastic Languages and
        Pattern Recognition," *J. Cybernetics*, 1, (33), 31-49 (1971).

[10.52] Fu, K. S., "Learning Control Systems," in *Computer and
        Information Sciences* (J. T. Tou and R. H. Wilcox, (eds.),
        Spartan Books, New York, 1964.

[10.53] Fu, K. S., "On Syntactic Pattern Recognition and Stochastic
        Languages", in *Frontiers of Pattern Recognition* (S. Watanabe,
        ed.), Academic Press, New York, 1972.

[10.54] Fu, K. S., "Learning Techniques in System Design - A Brief
        Review," Fifth World Congress of IFAC, Paris, June 12-16,
        1972.

[10.55] Fu, K. S., Chien, Y. T., and Cardillo, G. P., "A Dynamic
        Programming Approach to Sequential Pattern Recognition,"
        IEEE Trans. Electr. Computers, <u>EC-16</u>, (6) 790-803 (1967).

[10.56] Fu, K. S., and Fung, L. W., "Decision Making in a Fuzzy
        Environment," Tech. Rept. TR-EE-73-22, Purdue University,
        Lafayette, Ind., May 1973.

[10.57] Fu, K. S., Gibson, J. E., Hill, J. D., Luisi, J. A.,
        Raible, R. H., and Waltz, M. D., "Philosophy and State of the
        Art of Learning Control Systems," Tech. Rept. TR-EE-63-7,
        Purdue University, Lafayette, Ind., Nov. 1963.

[10.58] Fu, K. S., and Mc Laren, R. W., "An Application of Stochastic
        Automata to the Synthesis of Learning Systems," Tech. Rept.
        TR-EE-65-17, Purdue University, Lafayette, Ind., Sept. 1965.

[10.59] Fu, K. S., and Nikolic, Z. J., "A Study of Learning Systems
        Operating in Unknown Stationary Environments," Tech. Rept.
        TR-EE-66-20, School of Electrical Engineering, Purdue
        University, Lafayette, Ind., Nov. 1966.

[10.60] Fu, K. S., and Swain, P. A., "On Syntactic Pattern
        Recognition," COINS-69 Symposium, *Software Engineering*, Vol.
        II, (J. T. Tou, ed.), Academic Press, November 1971.

[10.61] Fu, K. S., and Tou, J., *Learning Systems and Intelligent
        Robots*, Plenum Press, New York, 1974.

[10.62] Fu, K. S., and Tou, J., Proceedings of Research Workshop on Learning System Theory and its Practical Applications, University of Florida, Gainsville, Fa., Oct. 18-20, 1973.

[10.63] Garden, M., "Learning Control of Valve Actuators in Direct Digital Control Systems," Preprints JACC, 1967.

[10.64] Gilstad, D. W., and Fu, K. S., "A Two-Dimensional Pattern-Recognizing Adaptive Model of a Human Controller," *IEEE Trans. Systems Man Cybern.*, SMC-1 (3) 261-266 (1971).

[10.65] Graupe, D., "Control of Artificial Upper Extremity Prostheses in Several Degrees of Freedom," Progress Report VA Grant V10(134) Dept. of Electrical Engineering, Colorado State University, Fort Collins, Colo., Feb. 1974.

[10.66] Hammer, A., Lynn, J. W., and Graupe, D., "Investigation of a Learning Control Systems with Interpolation," *IEEE Trans. Systems Man Cybern.*, SMC-2, (4) 288-296 (1972).

[10.67] Henrichon, E. G., Jr., and Fu, K. S., "Calamity Detection Using Non-Parametric Statistics," *IEEE Trans. Systems Sci. Cybern.*, SSC-5, (2), 150-156 (1969).

[10.68] Hilborn, C. G., Jr., and Lainiotis, D. G., "Optimal Unsupervised Learning Multicategory Dependent Hypothesis Pattern Recognition," *IEEE Trans. Inform. Theory*, IT-14, (3) 468-470 (1968).

[10.69] Hill, J. D., and Fu, K. S., "A Control System Using Stochastic Approximation for Hill-Climbing," Preprints JACC, 1965, p. 315.

[10.70] Hill, J. W., and Bliss, J. C., "A Tactile Perception Studies Related to Teleoperator Systems," Memorandum Report, Contract NAS2-5409, Final Report 2, Stanford Research Institute, Menlo Park, Calif., June 1971.

[10.71] Hill, J. W., and Sword, A. J., "Studies to Design and Develop Improved Remote Manipulator Systems," SRI Tech. Rept. No. 1587, Stanford Research Institute, Menlo Park, Calif., Nov. 1972.

[10.72] Ho, Y. C., and Agrawala, A. K., "On Pattern Classification Algorithms," *IEEE Trans. Automatic Control*, AC-13, (6) 676-690 (1968).

[10.73] Ivakhnenko, A. G., "Heuristic Self-Organization in Problems of Engineering Cybernetics," 4th World Congress of IFAC, Warsaw, Poland, June 16-21, 1969.

[10.74] Ivakhnenko, A. G., "Method of Data Handling by Groups as Competitor to Stochastic Approximation," *Automatika,* (3) (1968).

[10.75] Ivakhnenko, A. G., "Polynomial Theory of Complex Systems," *IEEE Trans. Systems Man Cybern.,* SMC-1 (4) 364-378 (1971).

[10.76] Ivakhnenko, A. G., and Dimitrov, V. D., "Stochastic Algorithms of The Method of Data Handling by Groups in the Problem of Predicting the Random Events," *Automatika,* (3) (1969).

[10.77] Ivanov, A. Z., Krug, G. K, Kushelev, Yu. N., et.al., "Learning-Type Control Systems," *Automatica Telemekhanika,* Tr. MEI 44 (in Russian) (1962).

[10.78] Jones, L. E., "On the Choice of Subgoals for Learning Control Systems," *IEEE Trans. Automatic Control,* AC-13, (6) 613-621 (1968).

[10.79] Kanal, L. N. (ed.), *Pattern Recognition,* Thompson Book Co., Washington, D.C., 1968.

[10.80] Kahne, S. J., and Fu, K. S., "Learning System Heuristics, Correspondence and Response," *IEEE Trans. Automatic Control,* AC-11, (4) 611-612 (1966).

[10.81] Karchak, A. Jr., and Allen, J. B., "Investigation of Powered Orthotic Devices," Final Report VRA Grant RD-1461-M-67, Rancho Los Amigos Hospital, University of Los Angeles, Calif., 1967.

[10.82] Kashyap, R. L., and Blaydon, C. C., "Recovery of Functions from Noisy Measurements taken at Randomly Selected Points and Its Applications to Pattern Classifications," *Proc. IEEE,* (54), (8) 1127-1129 (1966).

[10.83] Klinger, A., "Natural Language, Linguistic Processing, and Speed Understanding: Recent Research and Future Goals," UCLA Tech. Report R-1377ARPA Rept., Los Angeles, Calif., Dec. 1973.

[10.84] Klir, G. (ed.), *Trends in General System Theory,* Wiley - Interscience, New York, 1972.

[10.85] Koford, J. S., and Groner, G. F., "The Use of an Adaptive Threshold Element to Design a Linear Optimal Pattern Classifier," *IEEE Trans. Inform. Theory,* IT-12, (1) 42-50 (1966).

[10.86] Krug, G. K., and Netushil, A. V., "Automatic Systems with Learning Elements," Second IFAC Congress, Basel, Switzerland, 1963.

[10.87] Lambert, J. D., and Levine, M. D., "A Two-Stage Learning Control System," *IEEE Trans. Automatic Control*, <u>AC-15</u>, (3) 351-354 (1970).

[10.88] Lambert, J. D., and Levine, M. D., "Learning Control Heuristics," *IEEE Trans. Automatic Control*, <u>AC-13</u> 741-742 (1968).

[10.89] Lemke, R. R., "On the Application of the Potential Function Method to Pattern Recognition and System Identification," Tech. Rept. EE-68-8, School of Electrical Engineering, Purdue University, Lafayette, Ind., April 1968.

[10.90] Leondes, C. T., and Mendel, J. M., "Artificial Intelligence Control," Survey of Cybernetics, (R. Rose, ed.) ILLIFE Press, London, 1969.

[10.91] Luce, R. D., Bush, R. R., and Galanter, E. (eds.), *Handbook of Mathematical Psychology*, Vols. I and II, Wiley, New York, 1963.

[10.92] Lyman, J., and Freedy, A., "Summary of Research Activities on Upper Prosthesis Control," *Bull. Prosthetic Res.*, <u>BPR 19</u>, Winter (1973).

[10.93] Mangasarian, O. L., "Linear and Nonlinear Separation of Patterns by Linear Programming," *Operations Res.*, <u>13</u>, 444-452 (1965).

[10.94] Mc Laren R. W., "A Stochastic Automation Model for the Synthesis of Learning Systems," *IEEE Trans. Systems Sci. Cybern.*, <u>SSC-2</u>, 109-114 (1966).

[10.95] McMurtry, G. J., and Fu, K. S., "A Variable Structure Automaton Used as a Multi-Modal Searching Technique," *IEEE Trans. Automatic Control*, <u>AC-11</u>, (3) 379-388 (1966).

[10.96] Meisel, W. S., "Potential Functions in Mathematical Pattern Recognition," *IEEE Trans. Computers*, <u>C-18</u>, 911-917, (1969).

[10.97] Mendel, J. M., "Applications of Artificial Intelligence Techniques to a Spacecraft Control Problem," Douglas Rept. DAC-59328, Douglas Aircraft Corp., Santa Monica, Calif., 1966.

[10.98] Mendel, J. M., "Survey of Learning Control Systems for Space Vehicle Applications," Preprints JACC, Aug. 1966.

[10.99] Mendel, J. M., and Fu, K. S. (eds.), *Adaptive, Learning and Pattern Recognition Systems Theory and Applications*, Academic Press, New York, 1970.

[10.100] Mesarovic, M., Macko, D., and Takahara, Y., *Theory of Hierarchical Multilevel Systems*, Academic Press, New York, 1970.

[10.101] Minsky, M. L., *Artificial Intelligence*, McGraw-Hill, New York, 1972.

[10.102] Minsky, M. L., "Artificial Intelligence and Intelligent Automata," Project MAC Progress Report, 1968-1969, MIT, Cambridge, Mass., 1969.

[10.103] Minsky, M. L., and Papert, S. A., "Research on Intelligent Automata," Project MAC, Status Report II, MIT, Cambridge, Mass., Sept. 1967.

[10.104] Mowrer, O. H., *Learning Theory and Behavior*, Wiley, New York, 1960.

[10.105] Micciardi, A. N., "Elements of Learning Control Systems with Applications to Industrical Processes," Proc. 1972 IEEE Conference on Decision and Control, New Orleans, Dec. 13-15, 1972.

[10.106] Nakamura, K., and Oda, M., "Fundamental Principle and Behavior of Learntrols," *Computer and Information Sciences*, Vol. II, (J. T. Tou, ed.) Academic Press, New York, 1967.

[10.107] Narendra, K. S., and Streeter, D. N., "A Self-Organizing Control System Based on Correlation Techniques and Selective Reinforcement," Tech. Rept. No. 359, Cruft Lab., Havard University, Cambridge, Mass., July 1962.

[10.108] Netushil, A. V., Krug, G. K., and Letshkii, E. K., "Use of Learning Systems for the Automation of Complex Production Processes," *Izv. VUZOV SSR, Mashinostroyenie*, 12, (1961).

[10.109] Nilsson, N. J., "A Mobile Automaton: An Application of Artificial Intelligence Techniques," Proc. International Joint Conference on Artificial Intelligence, Washington D.C., May 1969.

[10.110] Nilsson, N. J., "Current Artificial Intelligence Research at SRI," Oral Presentation, University of California, Los Angeles, May 1971.

[10.111] Nilsson, N. J., *Learning Machines*, McGraw-Hill, New York, 1965.

[10.112] Norman, M. F., "Mathematical Learning Theory," in *Mathematics of the Decision Sciences*, (G. B. Dantzig and A. F. Veinott, eds.), Am. Math. Soc. Publications, New York, 1968.

[10.113] Northouse, R. A., and Fu, K. S., "Dynamic Scheduling of Large Digital Computer Systems Using Adaptive Control and Clustering Techniques," *IEEE Trans. Systems Man Cybern.*, SMC-3, (3) 225-234 (1973).

[10.114] Novikoff, A. B. J., "On Convergence Proofs for Perceptrons," Proc. Symp. Math. Theory of Automata, Polytechnic Institute of Brooklyn, New York, 1962.

[10.115] Patrick, E. A., and Hancock, J. C., "Nonsupervised Sequential Classification and Recognition of Patterns," *IEEE Trans. Information Theory*, IT-12, (3) 362-372 (1966).

[10.116] Pavlidis, T., "Grammatical and Graph Theoretic Analysis of Pictures," in *Graphic Languages*, (F. Nake and A. Rosenfeld, eds.), N. Holland Publishing Co., London, 1972.

[10.117] Pavlidis, T., "Segmentation of Pictures and Maps Through Functional Approximation," *Computer Graphics Image Processing*, 1, 360-372 (1972).

[10.118] Pingle, K., Singer, J. A., and Wichman, W. M., "Computer Control of a Mechanical Arm Through Visual Input," Proc. IFIP Conference, Vol. II, Edinburgh, 1968.

[10.119] Rosen, C. A., and Nilsson, N. J., "An Intelligent Automaton," 1967 IEEE International Convention Record, Part 9, New York, March 1967.

[10.120] Ruspini, E. H., "Numerical Methods for Fuzzy Clustering," *Inform. Sci.*, 2, (3), 319-350 (1970).

[10.121] Sage, A. P., and Arafeh, S. A., "On the Multilevel Hierarchical Decomposition and Coordination Methods in Large Scale Systems," Proc. of Third Milwaukee Symposium on Automatic Computation and Control MSAC2-75, Milwaukee, Wis., April 1975.

[10.122] Saridis, G. N., "Learning System Theory for General System Studies," Proc. of Third Annual Meeting, Society of General Systems Research, University of Maryland, College Park, Md., Sept. 1974.

[10.123] Saridis, G. N., "Self-Organizing Control and Application to Trainable Manipulators and Learning Prostheses," Proc. Sixth IFAC Congress, Cambridge, Mass., Aug. 1975.

[10.124] Saridis, G. N., "Fuzzy Notions in Nonlinear System Classification," *J. Cybern.*, 3, (3) 67-82 (1975).

[10.125] Saridis, G. N., and Gilbert, H. D., "On the Stochastic
         Fuel-Regulator Problem," Tech. Rept., TR-EE-68-9, Purdue
         University, Lafayette, Ind., Aug. 1968.

[10.126] Saridis, G. N., and Lobbia, R. N., "Parameter Adaptive
         S.O. Control of Stochastic Linear Discrete-Time Systems,"
         Tech. Rept. TR-EE-71-43, School of Electrical Engineering,
         Purdue University, Lafayette, Ind., April 1972.

[10.127] Saridis, G. N., and Stephanou, H. E., "A Hierarchically
         Intelligent Control for a Bionic Arm," Proc. 1975
         Conference on Decision and Control, Houston, Texas, Dec.
         1975.

[10.128] Sebestyen, G., *Decision Making Processes in Pattern
         Recognition,* Macmillan, New York, 1962.

[10.129] Shanmugam, K., and Breipahl, A. M., "An Error Correcting
         Procedure for Learning with an Imperfect Teacher," *IEEE
         Trans. Systems Man Cybern.,* SMC-1, 223-229 (1971).

[10.130] Sklansky, J., "Threshold Training of Two-Mode Signal
         Detection," *IEEE Trans. Inform. Theory,* IT-11, 353-362
         (1965).

[10.131] Sklansky, J., "Learning Systems for Automatic Control,"
         IEEE Trans. Automatic Control, (1) 6-20 (1966).

[10.132] Slagle, J. R., *Artificial Intelligence; The Heuristic
         Programming Approach,* McGraw-Hill, New York, 1971.

[10.133] Smith, F. B., Jr., et al., "Trainable Flight Control System
         Investigation," FDL-TDR-64-89, Wright-Patterson Air Force
         Base, Ohio, 1964.

[10.134] Smith, F. W., "Contact Control by Adaptive Pattern-
         Recognition Techniques," Tech. Rept. No. 6762-2, Stanford
         Electronic Lab., Stanford University, Calif., April 1964.

[10.135] Smith, F. W., "Pattern Classifier Design by Linear
         Programming," *IEEE Trans. Computers,* C-17, (3) 367-372 (1968).

[10.136] Specht, D. F., "Generation of Polynomical Discriminant
         Functions for Pattern Recognition," *IEEE Trans. Electron.
         Computers,* EC-16, (3) 308-319 (1967).

[10.137] Stein G., Private communication, Honeywell Inc.,
         Minneapolis, Minn., March 1975.

[10.138] Tamura, S., Higuchi, S., and Tanaka, K., "On the Recognition
         of Time-Varying Patterns Using Learning Procedures," *IEEE
         Trans. Inform., Theory,* IT-17, (4) 443-452 (1971).

[10.139] Tilley, E. A., and Barron, R. L., "On Barron's Self-
         Organizing Control," *IEEE Trans. Systems Man Cybern.*,
         SMC-1, (1) 84-85 (1971).

[10.140] Tou J., and Hill, J. D., "Step Towards Learning Control,"
         Preprints JACC, 1966, p. 12.

[10.141] Tsypkin, Ya. A., *Foundations of the Theory of Learning
         Systems,* translated from Russian by Z. J. Nikolic, Academic
         Press, New York, 1973.

[10.142] Tsypkin, Ya. Z., "Adaptation, Training and Self-Organization
         in Automatic Systems," Automation Remote Control, 27, (1),
         14-51 (1966).

[10.143] Tsypkin, Ya. Z., "Self-Learning - What is it?," *IEEE Trans.
         Automatic Control,* AC-13, (6), 608-612 (1968).

[10.144] Tsypkin, Ya. Z., "Generalized Learning Algorithms,"
         Automation Remote Control, (1), 86-92 (1970).

[10.145] Tsypkin, Ya. Z., "Automatic Training Systems," *Automation
         Remote Control,* (4), 560-575 (1970).

[10.146] Van Ryzin, J., "A Stochastic Posteriori Updating Algorithm
         for Pattern Recognition," Tech. Rept. 121, Dept. of
         Statistics, Stanford University, Stanford, Calif., Oct.1966.

[10.147] Vaysbord, E. M., and Yudin, D. B., "Multi-extremal Stochastic
         Approximation," *Engineering Cybern.,* (5), 1-10, Sept.-Oct.
         1968.

[10.148] Wagner, T. J., "The Rate of Convergence of an Algorithm for
         Recovering Functions from Noisy Measurements Taken at
         Randomly Selected Points," *IEEE Trans. Systems Sci. Cybern.*,
         SSC-4, (2), 151-155 (1968).

[10.149] Waltz, M. D., and Fu, K. S., "A Heuristic Approach to
         Reinforcement Learning Control Systems," *IEEE Trans.
         Automatic Control,* AC-10, (4), 390-394 (1965).

[10.150] Watanabe, S. (ed.), *Frontiers of Pattern Recognition,*
         Academic Press, New York, 1972.

[10.151] Watanabe, S. (ed.), *Methodologies of Pattern Recognition,*
         Academic Press, New York, 1969.

[10.152] Wee, W. G., and Fu, K. S., "A Formulation of Fuzzy Automata
         and Its Application as a Model of Learning Systems," *IEEE
         Trans. Systems Sci. Cybern.,* SSC-5, (3) 215-223 (1969).

[10.153] Wee, W. G., and Fu, K. S., "An Extension of the Generalized
         Inverse Algorithm to Multiclass Pattern Classification,"
         IEEE Trans. System Sci. Cybern., SSC-4, (2) 192-193 (1968).

[10.154] Whitney, D. E., "Resolved Motion Control of Manipulators and
         Human Prosthesis," *IEEE Trans. Man-Machine Systems*, MMS-10,
         (2), (1969).

[10.155] Whitney, D. E., "State Space Models of Remote Manipulation
         Tasks," Proc. 1969 International Joint Conference on
         Artificial Intelligence, Washington D.C., 1969.

[10.156] Widrow, B., "Generalization and Information Storage in
         Networks of Adaline 'Neurons'," in *Self-Organizing Systems*
         (M. C. Yovits, G. T. Jacobi, and C. D. Goldstein, eds.),
         Spartan Books, Washington, D.C., 1962.

[10.157] Widrow, B., Mantey, P. E., Griffiths, L. J., and Goode, B.
         B., "Adaptive Antenna Systems," *Proc. IEEE*, 55, (12), 2143-
         2159 (1967).

[10.158] Widrow, B., and Smith, F. W., "Pattern Recognizing Control
         Systems," *Computer and Information Sciences*, (J. T. Tou, and
         R. H. Wilcox, eds.), Spartan Books, New York, 1964.

[10.159] Wiener, N., *Cybernetics*, Wiley, New York, 1960.

[10.160] Winograd, T., "Procedures as a Representation for Data in
         a Computer Program for Understanding Natural Language,"
         Tech. Rept. Project MAC TR-84, MIT, Cambridge, Mass., Feb.
         1971.

[10.161] Winston, P. E., "Learning Structural Descriptions from
         Examples," Tech. Rept. Project MAC TR-76, MIT, Cambridge,
         Mass., 1970.

[10.162] Witten, I. H., "Finite-Time Performance of Some Two-Armed
         Bandit Controllers," *IEEE Trans. Systems Man Cybernetics*,
         SMC-3, (2), 194-197 (1973).

[10.163] Wolverton, C. T., and Wagner, T. J., "Asymptotically
         Optimal Discriminant Function for Pattern Classification,"
         IEEE Trans. Inform. Theory, IT-15, (2), (1969).

[10.164] Yacubovich, V. A., "Finitely Convergent Algorithms for the
         Solution of Countable Systems of Inequalities and Their
         Application in Problems of the Synthesis of Adaptive
         Systems," *Soviet Physics-Dokl.*, 14, (11), 1051-1054 (1970).

[10.165] Yacubovich, V. A., "On a Method of Adaptive Control Under Conditions of Great Uncertainty," Preprints Fifth World Congress of IFAC, Paris, July 12-21, 1972.

[10.166] Yacubovich, V. A., "On Adaptive (Self-Learning) Systems of Some Class," Preprints, Fourth World Congress of IFAC, Warsaw, Poland, June 16-21, 1969.

[10.167] Yau, S. S., and Schumpert, J. M., "Design of Pattern Classi Classifiers with the Updating Property Using Stochastic Approximation Techniques," *IEEE Trans. Computers*, $\underline{C-17}$, (4), 861-872 (1968).

[10.168] Zadeh, L. A., "Fuzzy Sets," *Inform. Control*, $\underline{8}$, (3), 338-353 (1965).

[10.169] Zadeh, L. A., "Similarity Relations and Fuzzy Ordering," *Inform. Sci.*, $\underline{3}$, (2), 177-200 (1971).

[10.170] Zadeh, L. A., "Outline of a New Approach to the Analysis of Complex Systems and Decision Processes," *IEEE Trans. Systems Man Cybern.*, $\underline{SMC-3}$, (1), 28-44 (1973).

[10.171] Zadeh, L. A., Tanaka, K., Fu, K. S., and Shimura, M. (eds.), *Fuzzy Sets and Their Applications*, Academic Press, New York, 1975.

SUBJECT INDEX

A

Active Feedback, 139, 264, 297, 300
Actively Adaptive S.O.C., 297-303, 393, 395, 396, 399
Adaptation, 17, 395
Adaptive Control, 14, 18, 24, 257
Adaptive Control Definitions, 16-17
Analog Search Identification, 35
Artificial Intelligence, 23, 319, 446, 450
Automatic Control, 10
Automatic Control Definition, 10
Automaton, Adaptive, 350-356, 394
Automaton, Expediency of, 330
Automaton, Fuzzy, 447, 449, 460, 462-463
Automaton, Growing, 332-335, 398, 405-418, 423
Automaton, Learning, 312
Automaton, Optimality of, 330
Automaton, Stochastic, 311, 327-330, 344, 394, 449
Automaton, Variable Structure, 330-332, 398

B

Bayesian Estimation, 84, 86, 93, 449
Bayes Rule, 84, 176
Behavioral Methods, 18, 23, 309, 312, 319, 327, 398, 447
Bioengineering, 24, 233, 309
Biological Systems, 24
Bionic Systems, 23, 24, 446, 447, 450-453, 459, 464
Boundary Control, 371

C

Central Limit Theorem, 233
Certainty Equivalence Principle, 144
Closed-loop Control, 138, 140, 141, 287
Closed-loop Identification, 199, 284, 287
Control Action, 321, 332, 360
Control Coordinator, 455, 462
Control Intervals, 39
Controllability Subindices, 287, 289
Cost function, Per-Interval, 213, 214
Cost-to-go, 248, 252, 273, 298
Crosscorrelation function, 191, 193, 196
Crosscorrelation Methods, 188, 191, 192, 195, 218, 223, 315, 398
Cybernetics, 250

D

Decision Probabilities, 167, 169
Deterministic Control, 10, 144, 282
Differential Generator, 143, 145, 147, 436
Dirac Delta Function, 171, 194, 260
Dirichelet Condition, 370
Discrete Interval Binary Noise (DIBN), 192, 194, 195, 197, 198, 217
Dual Control, 14, 139, 140, 159, 160, 163, 164, 165, 171, 174, 175, 185, 233, 234, 263, 303, 467